Case Studies in Ethics and HIV Research

T0206040

Sana Loue

Earl C. Pike

Case Studies in
Ethics and HIV Research

 Springer

Sana Loue
Case Western Reserve University,
Cleveland, OH

Earl C. Pike
AIDS Taskforce of Greater Cleveland,
Cleveland, OH

ISBN-13: 978-1-4419-4393-4   e-ISBN-13: 978-0-387-71362-5

Printed on acid-free paper.

9 8 7 6 5 4 3 2 1

springer.com

Preface

This book represents the compilation of efforts by researchers across the country, each of whom is dedicated not only to the prevention and elimination of HIV infection, but also to the conduct of research according to the highest ethical principles. The authors of the case studies have graciously agreed to share their experiences in conducting research, which raised questions for them and will motivate us to further inquiry and examination.

The views that are presented in this text are diverse and readers may or may not agree with the analyses of the editor-authors or the authors of the case studies. We do not aim for agreement among readers, but rather, the studied analysis of the ethical issues raised in the conduct of HIV research. We clearly emphasize the protection of the individuals participating in such research, as well as their communities, and view research not as an enterprise undertaken by researchers, but rather as a negotiated exchange between researchers, participants, and communities that also involves interplay with funding sources, ancillary partners, and governments.

We clearly do not address all ethical quandaries that may arise during the course of an HIV-related investigation, and it would be impossible to do so within one text. We have focused in the first section on issues that underlie not only HIV research, but all research involving human participants. The second portion of the text examines ethical issues related to the design and methodology of research studies. Each chapter is followed by a case study authored by a prominent HIV/AIDS researcher. These case studies explore such topics as the selection of study participants, the standard of care to be provided to individuals in the control group of a study, the personal safety of the researcher in the field and its relation to the safety of participants, the retention of study participants, and the conduct of research at multiple study sites. The authors of each case study describe the context of the issues that arise, the ethical issues that they confronted, the process that they utilized to arrive at a solution, and the ethical principles that guided their decision.

Parts III and IV of the text is similarly structured, with each chapter followed by a case study. Part III focuses on the conduct of research with specific populations, including children, cognitively impaired individuals, minority-identified individuals, communities outside of the United States, and individuals who are often identified in relation to their activities, such as sex work and injection drug use. Part IV discusses various issues that arise in the context of training both researchers and community members, including the ownership of data and confidentiality, among others.

As researchers and educators, we must constantly challenge ourselves to examine not only the science of what we do, but also the processes that we utilize, and their impact on our research participants, their relations, and their communities. We owe no less.

Sana Loue

Earl C. Pike

Acknowledgments

We gratefully acknowledge the substantial and significant contributions of the researchers who have shared their experiences with us and our readers through the development of their case studies. The studies from which these case studies were derived were often made possible through funding from various institutes of NIH and/or foundation grants; these are mentioned specifically by the authors of the case studies following their contributions.

Contents

Case Studies Contributed By

Andrea Boyd, Healthy Youths Program, Institute for Juvenile Research, University of Illinois at Chicago, Chicago, IL

David D. Celentano, Department of Epidemiology, John Hopkins School of Public Health, Baltimore, MD

Geri R. Donenberg, Healthy Youths Program, Institute for Juvenile Research, University of Illinois at Chicago, Chicago, IL

Andrew Fullem, Family Health International, Durham, NC

Cheryl Gore-Felton, Department of Psychiatry and Behavioral Sciences, Stanford University School of Medicine, Stanford, CA

Oscar Grusky, Department of Sociology, University of California, Los Angeles, CA

Kristi Y. Jordan, Department of Psychiatry, University of Illinois at Chicago, Chicago, IL

Moses Kamya, Department of Medicine, Makerere University, Kampala, Uganda.

Linda S. Lloyd, Alliance Healthcare Foundation, San Diego, CA

Sana Loue, Department of Epidemiology and Biostatistics, Case Western Reserve University, School of Medicine, Cleveland, OH

David W. Lounsbury, Department of Psychiatry & Behavioral Sciences, Memorial Sloan-Kettering Cancer Center, New York, NY

Patricia Marshall, Department of Bioethics, Department of Medicine, Case Western Reserve University, School of Medicine, Cleveland, OH

Janet W. McGrath, Department of Anthropology, Case Western Reserve University, Cleveland, OH

Nancy Méndez, Center for Minority Public Health, Case Western Reserve University, Cleveland, OH

Paulette Murphy, Department of Psychiatry & Behavioral Sciences, Memorial Sloan-Kettering Cancer Center, New York, NY

Daniel P. O'Shea, HIV, STD, and Hepatitis Branch, Public Health Services, County of San Diego Health and Human Services Agency, San Diego, CA

Bruce Rapkin, Department of Psychiatry and Behavioral Sciences, Memorial Sloan-Kettering Cancer Center, New York, NY

Janice Robinson, Department of Pediatrics, Albert Einstein College of Medicine of Yeshiva University, Bronx, NY

B. R. Simon Rosser, Division of Epidemiology & Community Health, University of Minnesota School of Public Health, Minneapolis, MN

Martha Sajatovic, University Hospitals Case Medical Center & Department of Psychiatry, Case Western Reserve University, School of Medicine, Cleveland, OH

Ingrid Vargas, Center for Minority Public Health, Case Western Reserve University, Cleveland, OH

Christopher Whalen, Department of Epidemiology and Biostatistics, Case Western Reserve University, School of Medicine, Cleveland, OH

Chapter 1
Human Rights, International Guidelines, and HIV Research

Introduction

When the first cases of what is now known as HIV (or AIDS) were reported by the Centers for Disease Control in 1981, the sexual orientation of those who were ill was noted as a factor that might be of epidemiological significance. Subsequent reports indicated, sometimes erroneously, that the new illness might be clustered among Haitians, injection drug users, commercial sex workers, African Americans and Latinos, and other marginalized communities. That the illness was also noted among newborns and infants, transfusion recipients, and people with clotting factor illnesses only confused the picture, but very quickly in the life of the epidemic, public associations about disease and risk were cemented: whatever it was that was causing such symptoms and death was somehow linked with marginalized identities and communities. Because the disease itself evoked in the general public a wide array of restrictive and discriminatory responses and policies, and because the disease from the outset was so strongly associated with "otherness," the human rights of people with HIV/AIDS—or of individuals or communities who were often assumed to be at elevated risk—has been, and will likely continue to be, a topic of intense intrapersonal, interpersonal, and societal conflict.

Very recently Gruskin and Dickens (2006) addressed the differences and similarities between human rights and ethics. Historically, they asserted, "human rights are meant to guide the actions of governments, whereas ethics in health care much more broadly encompass concern for the specific actions, aspirations, and individual health care workers, researchers, and organizations." They proceed to point to the foundational documents of each, noting that the 1964 Helsinki Delcaration, which concerned itself primarily with research on human subjects, serves as the intellectual basis for modern bioethics, and that the 1948 UN Universal Declaration of Human Rights sets out a broad range of rights for which transnational consensus—in theory, if not in practice—can be reached.

The distinction between ethics and human rights is blurred by the reality that health care practice is shaped by governments and other public institutions, and by the reality that health care workers and researchers, not to mention research participants, are actors in political processes designed to influence the policies, and specifically the health care and public health policies, of public institutions and governments. It is further blurred by the reality that the various actors involved in medical research and care, including care recipients and providers and funding institutions, are awarded various degrees of authority and power in the provision of care and the conduct of research: an equality, among those involved, cannot be assumed.

The resolution of these tensions and clarification of these issues is certainly more than an academic enterprise: the stakes are vast beyond measure: "If no effective vaccine or cure is found within the next 20 years, areas of the world that are now witnessing explosive epidemics or are in their second or third wave of HIV infection may well find themselves harder hit—and more deeply transformed—than Europe was by the Black Death" (Garrett, 2005: 63). For many in the field, the depth and width of possible consequences, on the one hand, and the resource inequities within which those consequences are played out, on the other hand, have argued for a human rights perspective on HIV/AIDS from the beginning. To Jonathan Mann, for example—a pioneer in the reframing of equations about rights and health—the epidemic was always been about rights: "People come first. Treatment in developing countries is about equity and human rights" (Mann, 1998: 6). This chapter argues for an approach in which ethics and human rights are interdependent, even when the sphere of action may be, in one case, interpersonal, and in another case, governmental. In other words, a more integrated approach would contend that practitioner and research ethics are dependent on fully elaborated and universal understandings of human rights, and that broad definitions of human rights are only meaningful if guided by fully elaborated and universal understandings of how they should be realized in specific and variable clinical applications.

HIV/AIDS and Human Rights

It is worth noting that one of the first "declarations" outlining what might be viewed as the human rights of people with HIV/AIDS was first articulated by people with HIV/AIDS themselves in an historic meeting in Colorado in 1983. What is now referred to as "The Denver Principles" still echoes throughout the discourse on HIV/AIDS and human rights today, even among those not directly familiar with the document (National Association of People with AIDS, 1983).

The Denver Principles were not based on theoretical or academically-developed notions of human rights; they rather emanated from the lived experience of those assembled in an informal gathering of mostly gay and bisexual men early on in the epidemic, at a point in time during which no meaningful treatment was available and public fear and hostility were high. As such, the document reflects an aware-ness of the possibility of rapid death for many who were diagnosed, and the need to legitimize the individual's right to access alternative and experimental therapies.

But the document is especially notable for its articulation of both rights and responsibilities, and for the "call" it issues to physicians and researchers. The following provides the most germane excerpts from the document.

We recommend that health care professionals:

• Who are gay, come out, especially to their patients who have AIDS.
• Always clearly identify and discuss the theory they favor as to the cause of AIDS, since this bias affects the treatment and advice they give.
• Get in touch with their feelings (fears, anxieties, hopes, etc.) about AIDS, and not simply deal with AIDS intellectually.
• Take a thorough personal inventory and identify and examine their own agendas around AIDS.
• Treat people with AIDS as whole people and address psychosocial issues as well as biophysical ones.
• Address the question of sexuality in people with AIDS specifically, sensitively, and with information about gay male sexuality in general and the sexuality of people with AIDS in particular.

We recommend that all people:

• Support us in our struggles against those who would fire us from our jobs, evict us from our homes, refuse to touch us, separate us from our loved ones, our community, or our peers, since there is no evidence that AIDS can be spread by casual social contact.
• Do not scapegoat people with AIDS, blame us for the epidemic, or generalize about our lifestyles.

We recommend that people with AIDS:

• Be involved at every level of AIDS decision-making and specifically serve on the boards of directors of provider organizations.
• Be included in all AIDS forums with equal credibility as other participants, to share their own experiences and knowledge.
• Substitute low risk sexual behaviors for those that could endanger themselves or their partners, and we feel that people with AIDS have an ethical responsibility to inform their potential sexual partners of their health status.

People with AIDS have the right:

• To as full and satisfying sexual and emotional lives as anyone else.
• To quality medical treatment and quality social service provision, without discrimination of any form, including sexual orientation, gender, diagnosis, economic status, age, or race.
• To full explanations of all medical procedures and risks, to choose or refuse their treatment modalities, to refuse to participate in research without jeopardizing their treatment, and to make informed decisions about their lives.
• To privacy, to confidentiality of medical records, to human respect, and to choose who their significant others are.
• To die and to live in dignity.

That concepts of privacy and confidentiality, personal freedom, dignity, autonomy, and personal agency were inscribed into an HIV/AIDS foundational document may appear surprising, given the tenor of the times, but clearly the Denver Principles reassert earlier principles, established in earlier documents, but specifically reframed to address HIV/AIDS. If the UN Declaration on Human Rights functions as the foundational document upon which subsequent advocacy for human rights rests, then the Denver Principles can be said to serve as the foundational document that profoundly influenced subsequent under- standings of the rights of people with HIV/AIDS and the advocacy that has arisen from that understanding—and that first articulated what would become an evolving discourse about the appropriate relation between those who provide care and treatment or conduct research, and those who are treated or serve as research "subjects." The new ethic, not previously articulated in medical care and research, was partnership.

Much more recently the UNAIDS International Guidelines on AIDS and Human Rights, in the 2006 Consolidated Version, established 12 broad guidelines that together addressed the need for national planning and frameworks within countries; the importance of community consultation at all levels; review of pub- lic health law and policies; review and reform of criminal law as needed; adoption of antidiscrimination measures; the provision of legal services to people with HIV/AIDS; the need to create enabling environments for women, children, and other vulnerable groups; and the need for monitoring and enforcement of human rights standards. Among the specific human rights principles the Guidelines enumerates as relevant to HIV/AIDS—which it bases on a wide number of inter- national and regional instruments—are:

• The right to non-discrimination, equal protection, and equality before the law;
• The right to life;
• The right to the highest attainable standard of physical and mental health;
• The right to liberty and security of person;
• The right to freedom of movement;
• The right to seek and enjoy asylum;
• The right to privacy;
• The right to freedom of opinion and expression and the right to freely receive and impart information;
• The right to freedom of association;
• The right to work;
• The right to marry and found a family;
• The right to equal access to education;
• The right to an adequate standard of living;
• The right to social security, assistance, and welfare;
• The right to share in scientific advancement and its benefits;
• The right to participate in public and cultural life;
• The right to be free from torture and cruel, inhuman or degrading treatment or punishment.

A number of antecedents and antecedents have helped fill in the foundational dossier of HIV/AIDS and human rights, including the National Bioethics Advisory Commission, the Council for International Organizations of Medical Sciences, the Helsinki Declaration, the UNAIDS' *Ethical Considerations in HIV Preventive Vaccine Research,* and the Nuffield Council on Bioethics. These, integrated and summarized, point to key themes in the discussion of HIV/AIDS and human rights:

1. respect for persons, decision-making autonomy, and self-determination;
2. beneficence and nonmaleficence; and,
3. distributive justice, or distribution of participant burdens and benefits in research.

But the relevant dossier of HIV/AIDS and human rights—and the ethical guidelines which inform more universal declarations—cannot be restricted to the common list of international agreements and protocols. Hellsten (2005), as an example, outlines the various levels and domains informing medical ethics in Tanzania:

First, there are international regulations. Second, there are national laws (such as the law forbidding abortion and the sexual offence act that, e.g., makes female genital mutilation, rape, etc. illegal) and national policies (such as the National HIV/AIDS policy that guides national strategies to combat the epidemic). Third, there are the local belief and values systems and their related social ethics . . . Often these different norms are inconsistent with each other, or their value base may appear to clash. With the simultaneous political and economic transition from a collectivist socialist economy to a more individualist market economy, the values and practices in health, healthcare, and human well-being in general become confused: Cultural and traditional claims are tangled together, and the requirements of a market economy are set against the collectivist promotions of public health as well as the protection of individual rights (Hellsten, 2005: 258).

Thus, not only are international, Western standards in potential conflict with traditional, local values, but what is defined as "traditional" is itself in flux, being continually shaped and redefined by larger political and economic forces that themselves have underlying values. International declarations and documents are therefore necessary, but not sufficient; they are abstractions that require moral interpretation within the relevant context (Benatar, 2002).

At its base the most vexing application of "human rights" to HIV/AIDS has to do with the meaning of distributive justice in an HIV/AIDS context, for it is in relation to this application that we find the most intricate interplay between individual rights and collective responsibility, and varying moral philosophies and political-economic ideologies. That is, while there is widespread agreement (if occasional passionate disagreement) about such broadly-defined rights as life, liberty, and the pursuit of happiness, or even the right to health, there is far less agreement about individual rights that incur a social obligation:

By confining human rights to a set of rights that support freedoms associated with free market economics, and redefining socioeconomic rights as 'aspirations,' the current global order seeks to establish a set of values that legitimates particular kinds of social behavior.

Importantly, the most powerful actors associated with globalization seek to free themselves of costs and duties seen as too burdensome and as an unnecessary barrier to the prosecution of their interests (Evans, 2004: 23).

Foremost among the direct human rights issues associated with HIV/AIDS in this regard is the "right" to Western standards of care in the treatment and management of HIV infection, and Western tools and resources for prevention. For while most will concede that we "should" collaborate to make sure that all the world's people living with HIV/AIDS have access to such care, and that all the world's citizens should have access to the full toolkit of available prevention technologies, that aspiration is not embedded in social contracts that guarantee those rights. In this regard, a brief review of Rawls' fundamental concept of justice as fairness is warranted.

The principle of justice, which is closely associated with the concept of human rights, is concerned in Rawls' moral philosophy, with fairness. Rawls' formula for ensuring justice is familiar to most:

All social values—liberty and opportunity, income and wealth, and the social bases of self-respect—are to be distributed equally unless an unequal distribution of any, or all, of these values is to everyone's advantage (Rawls, 1999: 54).

Within the Rawlsian framework there is the proposition—and this could very well apply to researchers and scientists—that talent and knowledge are not necessarily private possessions:

The difference principle represents, in effect, an agreement to regard the distribution of natural talents as in some respects a common asset and to share in the greater social and economic benefits made possible by the complementarities of this distribution. Those who have been favored by nature, whoever they are, may gain from their good fortune only on terms that improve the situation of those who have lost out." (Rawls, 1999: 87).

Among the "social and economic benefits" to which Rawls refers we can fairly add the benefit of health.

This particular requirement places considerable burdens on resource-rich countries and institutions, since there is no doubt that the application of the Rawlsian principle to HIV/AIDS resource distribution would result in something close to a consensus that everyone should 1) have access to treatment and prevention technology, and 2) that resource-rich countries and institutions have the capacity to underwrite those costs without a drastic change in the quality of life in the developed world. This particular requirement also places, admittedly, burdens on medical researchers, because it insists that the research study itself— whatever question it might be addressing—and the selection of research subjects should depend entirely on scientific benefit to individuals and communities, and not on the economic goals of pharmaceutical companies, or the easy availability of research participants. It also suggests that funding should follow the person-centered or community-centered science, rather than the other way around; in practice—as we have seen recently with microbicide research and investigation of new HIV prevention-technologies—if a product or process does not show promise of profit, it is often difficult to secure needed funding for research. Kok-Chor Tan (1997: 62) offers an interdependency perspective, asserting that the "intricate economic, social, and political interdependencies of the global community draw nearly everyone, some more deeply than others, into social arrangements with each other." For Deborah Zion (2004), this implies that individuals and groups that have benefited from HIV/AIDS research have an obligation to redress the imbalances that may have resulted.

But even on the simple matter of distributive justice for HIV/AIDS-related commodities, not to mention issues such as intellectual property, there is bound to be wide disagreement, because there is wide disagreement on the core principles. On the one hand, we have the perspective of Peter Singer (1996), who asserts that we must morally prevent something bad from happening if we have the power to do so without sacrificing anything of comparable moral value. But John Arthur (1996), among others, has criticized what he views as the excessive demands or burdens of Singer's "obligation to assist;" Singer's proposition, according to Arthur, would require individuals to donate a single kidney or a single eye, given that, in some meaningful sense, two of each are not needed. And Slote (1996) suggests that the "obligation" has limits—individuals, for example, should not have to sacrifice major life plans.

And as the debate about the human rights principles informing commodity distribution continues, John Harrington (2002: 1425) argues that individual human rights—the principle of autonomy or self-determination as a central ethic in the societal response to HIV/AIDS—may be weakening as "the normalization and re-medicalisation of HIV/AIDS, as well as the changing profile of the at-risk population have opened up new possibilities for coercion." Viewing the evolution of HIV/AIDS ethics and policy responses from a specifically European perspective, he notes that only two jurisdictions—Bavaria and Sweden—elected to take a coercive approach to the epidemic in the early days, by adopting measures such as compulsory testing for sex workers and non-European immigrants, shutting down gay clubs and saunas, and implementing extensive contact tracing. But the opposite approach was utilized in the remaining countries of Western Europe, where gay activists, in particular, played a pivotal role in the development of social policy that was profoundly autonomy-focused (Harrington, 2002).

With the advent of antiretrovirals and others treatments, however, HIV/AIDS has become a treatable chronic condition, although it remains an incurable one. As a result, observers have identified a "re-medicalisation" or "normalization" of AIDS. Normalization can here be understood in two senses. First, responses to the disease are increasingly conditioned by a series of medical standards or norms; through this extension of medical knowledge the profession has reasserted control over what had been an area of heightened lay involvement. Second, as a chronic disease AIDS is now a routine, unexceptional part of the social, as well as the medical, landscape in Western countries." These shifts, Harrington (2002) notes, have led to parallel policy shifts; for example, expanded calls for and implementation of routine testing; the establishment of legal frameworks that facilitate compulsory treatment; and the criminalization of a range of risk behaviors. How these evolving understandings of the interaction between medical and state power will influence the contours of research ethics is unclear, though it is worth speculating that a greater willingness to accept coercive treatment under certain conditions could also lead to diminished respect for autonomy and self-determination as guiding ethics in HIV/AIDS research.

For Slack and colleagues (2000), there is the enduring challenge of translation. They have argued that researchers often approach the ethical principles of autonomy, beneficence and justice that are embedded in research practices such as informed consent and the protection of confidentiality as "add-ons" to scientific procedures rather than intrinsic components.

At the intersection of plague and urgency, the continuing debate about human rights and HIV/AIDS rages on, sometimes oblivious to the fact that the research participants in a village in say, Kenya, or for that matter, the research participants in a clinical trial in Washington, DC, may have no idea what the experts are arguing about. Still, there is no consensus on what issues or problems constitute a "human rights" dimension of HIV disease, in part because widely varying concepts of "social justice" and "human rights" are often blurred. The concept of social justice, however, often addresses inequities in the distribution of resources or commodities, and strategies for equitable distribution. The concept of human rights generally addresses the problem of intrinsic eligibility and the question of which resources or commodities are to be equitably distributed, and may or may not address means for doing so. This discussion utilizes the narrower perspective of human rights – while a broader approach, that embraces the strategic and policy dimensions of resource distribution, is a critical question in HIV/AIDS, it is beyond the scope of the present volume. But we must acknowledge, as Fortin indicates, that "Ethical discourse must continually articulate anew the moral basis for medical action, since the presumptive pursuit of science is in itself not enough," (Fortin, 1991). To do so, we must attempt to begin a catalog of what "rights," or rather, what domains of rights, ought to be included in the discussion of HIV/AIDS and human rights in both an international, and a developed-country, international analysis.

Eligibility

The notion of eligibility to rights has been firmly encoded in a variety of documents, from the Nuremberg Code to more recent United Nations declarations. Nevertheless, it deserves to be restated, especially in light of continuing debates that some individuals—current drug addicts, for example—may not enjoy categorical eligibility for certain HIV/AIDS-related resources, such as antiretroviral therapy.

Some exclusion proposals are based on the proposition, now discredited, that some individuals may not have the capacity to effectively utilize some forms of treatment or participate in research. Antiretroviral therapy, indeed, can be complicated, and demand a degree of personal volition and organization. Others have suggested that where there are variances in treatment utilization, the underlying issue is not so much personal, or personally categorical differences, but differences in the degree and scope of social supports available to the individuals involved, that will enable them to successfully manage therapy.

Both of these positions occlude a more fundamental, ethical position: that generally, rights are not behaviorally-earned, but existentially endowed. There are exceptions, to be sure, made in various ways in societies around the world: a convicted murderer, for example, may lose her or his liberty, the right to vote, and so on. But as a general rule, such expectations have not extended to health resources, or since they are viewed as (1) more intrinsic, and (2) part of a bilateral ethical relationship between the individual and society. That is, the highest ethical principles of many societies recognize that deprivation of health resources to individuals exacts a cost not merely to the individuals involved, but to society as a whole, since there is a recognition that it debases the core humanity of all people to stand by idly when health resources can be made available, but they are deliberately withheld, to the individual.

When it comes to the question of "eligibility" for HIV/AIDS-related health resources, therefore, it is difficult to name individuals or classes of individuals who would not enjoy categorical eligibility. Exclusions, therefore, must be viewed as intrinsically unethical and immoral.

While not an ethical considerations per se, it deserves to be said that exclusions from eligibility would also be counterproductive from a public health perspective, since, in essence, universal, stand-of-care treatment for all persons with HIV/AIDS would substantially reduce global infectivity: treatment is, indeed, prevention, since it lowers the amount of the virus that can be transmitted to others.

It also deserves to be said that there is likely to be widespread disagreement, from ideological perspectives, about eligibility. Societies differ in their treatment of, for example, prison inmates or gay men, and social policies within different societies may justify exclusionary treatment. But it is equally important to underscore the fact that exclusionary ideological justifications, where they exist, are not founded upon exclusionary bioethics. Medical experimentation on prisoners, or post-death harvesting, without consent, of the organs of prisoners who are executed by the State, can find political apologists, but not bioethics apologists. Again, there is no moral or ethical basis for exclusion of eligibility for HIV/AIDS health-related reasons for any of the world's citizens living with the disease.

Resources

What a commonly-accepted agreement has been developed that articulates who is eligible for human rights, we must articulate what the resources for which they are eligible. Here, resource restrictions will confound the application of rights, but the absolute availability of resources, and their allocation across populations of eligible persons, may be a question of strategy or social justice than the more fundamental question of rights. It will suffice here to identify those key resources that ought to be included in an inventory of rights, with regard to the question of mechanisms for their distribution.

Commodities

Certain commodities comprise the basic list of "resources" associated with HIV/AIDS. Those commodities certainly include a reasonably full list of the medications or treatments that slow down HIV replication—antiretroviral medications—as well as medications and treatments used to prevent or treat opportunistic infections associated with HIV infection. There is an extensive and growing list of such medications, and universal access has been constrained by cost, but it is important to acknowledge that cost alone is not a prohibitive barrier: it is within the capacity of national and international financing institutions, working collaboratively with pharmaceutical companies, to absorb those costs without undue burden on the global community. What is lacking is the political will, particularly among governments in donor states: "For donor states the best option is to bite the bullet and spend heavily not only on HIV/AIDS prevention, care, and treatment, but also on development aimed at bringing the poor world into the global economy, so that it might eventually derive sufficient wealth to pay for the great expenses involved in coping with HIV/AIDS" (Garrett, 2005: 63).

The list of HIV/AIDS-related commodities to which eligible persons have rights also includes an expanding list of prevention tools and technologies, most specifically, latex condoms and barriers, but also more generally, family planning tools. This list is likely to expand soon as effective microbicides, and even vaccines, are developed and deployed. The right to access effective prevention tools, however, has been subject to wide variation in application; local ideology, which can itself change over time, has created an uneven global patchwork of rights applications, so that in some communities or regions condoms are readily accessible to nearly all, while in others, they are actively discouraged and access to highly restricted. These same variations are notable in availability of and access to clean syringes and methadone maintenance therapy for opiate users.

While these constitute core rights-related commodities in relation to HIV/AIDS, it must be addressed as well whether assurance of an individual's right to those commodities is sufficient by itself to ensure their adequate utilization. As has been oft-noted, being able to secure a needed antiretroviral medication that requires refrigeration doesn't mean much unless the individual has a refrigerator. It is here that the emerging and sometimes contentious debate about what constitutes HIV/AIDS rights-related commodities is now most vigorous: while there is very broad consensus that all people should enjoy access to adequate medical treatment, and some consensus that prevention tools and technologies need to be available, there is very little agreement about whether all eligible persons have shared rights to the commodities that can make utilization of core commodities meaningful— such as clean, decent housing; reliable transportation; and nutritionally adequate food. Such disagreement is even seen in US domestic policy: while nearly everyone agrees on the need to subsidize medications so that universal access can be achieved, there is far less concordance about whether there is a public responsibility to subsidize HIV/AIDS-related housing, food, and transportation services that will make availability of medications truly achievable. We must be careful not to conflate the range of imperatives, in the fight against AIDS, into a single strategic goal: providing drugs. As Barnett and Whiteside (2002: 364) have concluded, "Treatment is not really the starting point of the problem; it is the end-state. To always think about treatment is to remain distanced from the social and economic origins of illness and well-being."

Personal Freedoms

The second set of resources related to HIV/AIDS human rights includes an extensive list of personal freedoms. This includes freedom of personal expression; the freedom to form and maintain intimate relationships; freedom of association and community affiliation; the freedom to make informed decisions about treatment and prevention options (including the freedom to refuse treatment); freedom of speech and protest; and the freedom to engage in consensual sexual behaviors. These personal freedoms are, or should be, constrained by collective freedoms only if those collective freedoms are founded on rational, scientifically-derived principles. One does not have the freedom to work in a setting where there are no people with AIDS, for example, because there is no scientific basis for fears of infection through casual contact. But a person with HIV/AIDS might be legitimately denied, through legal prohibitions, from having unprotected sexual intercourse with others who do now know their status, because there may a legitimate basis for concern about possible transmission, and a legitimate claim by the second party to be able to make informed choices about personal risk.

Foremost among the personal freedoms associated with HIV/AIDS have been freedom of personal expression—especially the freedom to express oneself in ways that do not conform with dominant or traditional normative expectations for male and female gender—and the freedom to engage in intimate, sexual relationships of the individual's choosing. Indeed, in many parts of the world, the exercise of such freedoms places the individual at real and immediate risk of not merely violence perpetrated by other individuals, but state-sanctioned violence and murder as well. The curtailment of such freedoms has and will continue to have a profound effect on our ability to mount an effective response to the global HIV/AIDS epidemic.

Protections

The HIV/AIDS-related resources to which individuals have a valid claim to access also includes a range of state- and community-enforced protections. This is a slightly different resource than personal freedoms; one may enjoy the freedom of personal expression as a result of local community norms, but still be subject to violations of that freedom because of the lack of structural mechanisms that will prohibit and sanction violations. Of particular significance in HIV/AIDS is the presence or absence of local and national structural mechanisms that will protect the rights of women and girls, and of lesbian, gay, bisexual and transgender persons. In terms of gender, laws about property rights, spousal and relationship violence and rape, employment, education, mobility, medical and family planning decision-making, and divorce will all have a profound affect on opportunities to reduce HIV case rates among women and girls, and opportunities to treat women and girls who are already infected. As Moodie (2000) has recently suggested, the most important risk factor for women in Sub-Saharan Africa is being married. If there is any truth at all to the assertion—and there almost certainly is—then we will be severely hampered in our ability to fight the epidemic unless women and girls are guaranteed greater freedom and autonomy in social relations, and in intimate relationships.

Contracts

The successful implementation of ethical and effective research requires a series of relationships: between researchers and research participants; with local communities; between developed-country and developing-country IRBs, and more. Whether cemented by formal agreements—in the way informed consent agreements might exemplify—or understood as a matter of informal agreement or understanding, the availability of such contracts, defining and constraining the limits and benefits of relationships, can be viewed as one of the resources to which those who are eligible for HIV/AIDS-related resources (that is, everyone who is infected) can be said to have a legitimate "right." The WHO/CIOMS guidelines, for example, address not only the urgency for research within communities in which disease is endemic but also the need to solicit personal consent *and* the active engagement of community leadership; the need to work in collaboration with local public health authorities; the need to educate the community fully about the aims of the research and potential hazards or inconveniences; the need for multi-disciplinary review and auditing that involves community participants; and the need to fully address the ethical dimensions of externally-sponsored research (Council of International Organizations for Medical Sciences, 1993). While "protections" offer, as a resource, legal frameworks in which the expectation of ensured human rights can be realized; "contracts" refers to transparently established duties and aspirations that can guide the series of realizations within which research unfolds. The right is two-fold: first, that such duties and aspirations are articulated in the first place; and two, that their terms are clearly available to all parties in the relationships mediating research.

We can and should wish, if eligibility for HIV/AIDS resources is universally conceded, and if the commodities, personal freedoms, protections, and contracts of which those resources consist can be reasonably guaranteed, that new, rights-based relationships between unequal partners will begin to alter the fundamental inequalities that now exist. The gaps—between donor nations and impoverished nations; between powerful hegemonic Western medical and political discourse and local knowledge; between the status of Western researchers and developing-world research participants; between standards of care; and many others—are profound, and continuing to work "around" them, in the short-term, in order to carry out short-term research objectives, will only continue to confound our efforts to carry out research that is ethical in every respect, to the core. Baldwin (2005: 289) anticipates as much when he asks, "Being optimistic, may we hope for new relations between First and Third Worlds, partly prompted by AIDS? Certainly the increasing recognition that the epidemic has international security implications, that it threatens to destabilize a large part of the globe, encourages the industrialized West not to ignore its developing cousins. The movement to override market principles in the pricing of new medicines in the Third World bespeaks an unprecedented—however limited—sense of international solidarity. To be sure, public health issues, broadly speaking, are being recognized as unsolvable within national borders."

References

Arthur, J. (1996). Rights and the duty to bring aid. In W. Aiken & H. LaFollette (Eds.), *World Hunger and Morality*, 2nd ed. (pp. 39–50). Englewood Cliffs, New Jersey: Prentice Hall.

Baldwin, P. (2005). *Disease and Democracy: The Industrialized World Faces AIDS*. Berkeley, CA: University of California Press.

Barnett, T. & Whiteside, A. (2002). *AIDS in the 21st Century: Disease and Globalization*. New York, New York: Palgrave MacMillan.

Benatar, S.R. (2002). Reflections and recommendations on research ethics in developing countries. *Social Science and Medicine, 54*, 1131–1141.

Council of International Organizations for Medical Sciences. (1993). *International Guidelines for Biomedical Research Involving Human Subjects*. Geneva, Switzerland: Author.

Evans, T. (2004). A human right to health? In N.K. Poku, & A. Whiteside (Eds.) *Global Health and Governance: HIV/AIDS* (pp. 7–25). New York, NY: Palgrave MacMillan.

Fortin, A.J. (1991). Ethics, culture, and medical power: AIDS research in the Third World. *AIDS & Public Policy Journal, 6(1)*, 15–24.

Garrett, L. (2005). The lessons of HIV/AIDS. *Foreign Affairs, 84(4)*, 51–64.

Gruskin, S. & Dickens, B. Human rights and ethics in public health. *American Journal of Public Health, 96(11)*, 1903–1905.

Harrington, J.A. (2002). The instrumental uses of autonomy: A review of AIDS policy and law in Europe. *Social Science and Medicine, 55*, 1425–1434.

Hellsten, S.K. (2005). Bioethics in Tanzania: Legal and ethical concerns in medical care and research in relation to the HIV/AIDS epidemic. *Cambridge Quarterly of Healthcare Ethics, 14*, 256–267.

Mann, J. (1998). HIV/AIDS, micro-ethics, and macro-ethics. *AIDS Care*, 10(1), 5–6.

Moodie, R. (2000). Should HIV be on the development agenda? *Development Bulletin, 52*, 6–8.

National Association of People with AIDS. (1983). The Denver Principles. Last accessed January 18, 2007; Available at http://www.napwa.org/Denver.html

National Bioethics Advisory Committee. (2001). *Ethical and Policy Issues in International Research: Clinical Trials in Developing Countries*. Bethesda, Maryland: Author.

Nuffield Council on Bioethics. (2002). *The Ethics of Research Related to Healthcare in Developing Countries*. London: Author.

Rawls, J. (1999). *A Theory of Justice*, rev. ed. Cambridge, Massachusetts: Belknap Press.

Singer, P. (1996). Famine, affluence, and morality. In W. Aiken & H. LaFollette (Eds.). *World Hunger and Morality*, 2nd ed. (pp. 26–38). Englewood Cliffs, New Jersey: Prentice Hall.

Slack, C., Lindegger, G., Vardas, E., Richter, L., Strode, A., & Wassenaar, D. (2000). Ethical issues in vaccine trials in South Africa. *South African Journal of Science, 96(6),* 291–295.

Slote, M. The morality of wealth. In W. Aiken & H. LaFollette (Eds.). *World Hunger and Morality*, 2nd ed. (pp. 125–127). Englewood Cliffs, New Jersey: Prentice Hall.

Tan, K.C. (1997). Kantian ethics and global justice. *Social Theory and Practice, 23(1),* 53–73.

UNAIDS. (2000). *Ethical Considerations in HIV Preventive Vaccine Research.* Geneva, Switzerland: Author.

UNAIDS. (2006). *International Guidelines on HIV/AIDS and Human Rights: 2006 Consolidated Version.* Geneva, Switzerland: Author.

World Medical Association. (2000). Declaration of Helsinki. 52nd World Medical Association General Assembly. Edinburgh, Scotland: Author.

Zion, D. (2004). HIV/AIDS clinical research, and the claims of beneficence, justice, and integrity. *Cambridge Quarterly of Healthcare Ethics, 13,* 404–413.

Chapter 2
U.S. Regulations
and HIV-Related Research

The Development of U.S. Regulations Governing Research

Even from the earliest years of the 20th century, the medical literature has been cognizant of the need for protections of humans involved in research. Osler in 1907 cautioned that humans were not to be used in experiments until after the safety of a new drug or procedure was established in animals; that the full consent of patient prerequisite to application of a new therapy; that patients entrusted to the care of physician were not to be recruited for experimentation unless the new therapy had the potential to result in direct benefit to patient; and that the participation of healthy volunteers was permissible, subject to the requirements of full knowledge of the circumstances and their agreement to participate (Bynum, 1988).

Despite these cautionary notes, American history is replete with examples of research that is unethical, not only by today's standards, but even applying the standards iterated by Osler. These include:

- Experimentation with female slaves to cause and repair vesico-vaginal fistulas (Sims, 1894)
- Injection of sterilized gelatin into two young boys and feeble-minded girls (Abt, 1903)
- Injection of tuberculin solution into more than 164 children less than 8 years old, most in orphanage (Belais, 1910; Hammill, Carpenter, and Cope, 1908)
- Tuskegee Syphilis Study, 1932–1972
- Vanderbilt Nutrition Study (Hagstrom et al., 1969)
- Fernald State School Experiments, 1946–1953 (Welsome, 1999)
- University of Cincinnati Radiation Experiments (Welsome, 1999)
- Holmesburg Prison Experiments, 1956–1969 (Hornblum, 1998)
- Atlanta Malaria Experiments, 1946 (George, 1946)
- Diethylstilbesterol experiments, 1940s (National Institutes of Health, 1991)
- Willowbrook Hepatitis Experiments, 1950–60s (Beecher, 1970)
- Tearoom Trade (Humphreys, 1970)

TABLE 2.1. Provisions of the Nuremberg Code

Voluntary consent is essential.

The experiment must yield fruitful results for the good of society.

The experiment should be based on the results of animal experimentation and a knowledge of the natural history of the disease under study to justify performance of the study.

The experiment should be conducted to avoid all unnecessary physical and mental suffering and injury.

In general, no experiment should be conducted where there is a priori reason to believe that death or disabling injury will occur

Proper precautions must be taken to provide adequate facilities to protect the participant against the risk of injury, disability, or death.

The experiment may be conducted by only scientifically qualified persons.

The participant may end the experiment.

The researcher must be prepared to end the experiment at any time.

Many of these studies were initiated or continued even after the promulgation of the Nuremberg Code, formulated in response to the experiments conducted by the Nazi physicians (Table 2.1). It was not until after the public exposure of the Tuskegee Syphilis Study in 1972 and subsequent congressional hearings in 1973 that U.S. regulations on the protection of human participants in research were completely revamped (Thomas and Quinn, 1991), resulting in greater explicit protections for participants. These basic precepts have been modified somewhat, as scientific research has become increasingly complex and as our understanding of diverse cultures has become more sophisticated. As an example, the Helsinki Declarations allow consent for participation in research to be provided by a surrogate in circumstances in which the individual lacks capacity to do so him-or herself. This is critical, for instance, in research relating to childhood disorders, mental illnesses, and diseases such as Alzheimer's, for if we needed the consent of the individual him- or herself, we would be unable to conduct research into the cause, treatment, and prevention of such disorders.

Ethical Principles Governing Research

Embodied within the Nuremberg Code are three basic principles: respect for persons, beneficence, and justice. To these, some scholars have added a fourth, non-maleficence, which is the corollary to the principle of beneficence. Their purpose is ultimately to protect individuals from harm and to respect individuals' autonomy.

Our concept of informed consent to participate in research derives from the first principle of respect for persons. Valid informed consent requires that the individual be provided with information, that he or she understand the information provided, that the individual have capacity to provide consent, and that the consent given be voluntary. In general, capacity requires that the individual have the ability to evidence a choice, the ability to understand relevant information, the ability to appreciate situation and its consequences, and the ability to manipulate information rationally. Voluntariness requires that the decision to participate be free from force, coercion, duress, fraud, or misrepresentation.

The Integration of Ethical Principles into Regulation

Federal regulations have attempted to integrate into law the ethical principles that were enumerated in the Nuremberg Code and subsequent international documents. The following is a partial listing of the regulatory and policy provisions that reflect these principles.

- Federal Policy for Protection of Human Subjects: "Common Rule," adopted by 17 agencies, Subpart A, Part 46, Title 45 of Code of Federal Regulations
- Food and Drug Administration, 21 CFR 314.126 (1985)
- FDA, Guideline for the Study and Evaluation of Gender Differences in the Clinical Evaluation of Drugs (1993)
- FDA, Accelerated Approval of New Drugs, 21 CFR 314.500 et seq.
- FDA, Exception from Informed Consent Requirements for Emergency Research, 21 CFR 50.24
- NIH, Research Involving Impaired Human Subjects: Clinical Center Policy for the Consent Process (1986)
- NIH, Report of the Human Embryo Research Panel (1994)

The "Common Rule," mentioned above, requires that the following information be provided to prospective research participants as part of the informed consent process:

- A statement that it is research
- An explanation of purposes of study
- The expected duration of participation
- A description of procedures
- The identification of any procedures that are experimental
- A description of reasonably foreseeable risks or discomforts
- A description of reasonably expected benefits
- The disclosure of appropriate alternative treatments
- The extent to which the confidentiality of records will be maintained
- For research involving more than minimal risk, an explanation of compensation, medical treatments
- An explanation of whom to contact with any questions relating to the study
- A statement that participation is voluntary and that the participant may discontinue his or her participation at any time without penalty or loss of benefits to which otherwise entitled

Additional elements of information may be provided, including statements indicating that treatment may involve unforeseeable risks, the circumstances under which participation may be involuntarily terminated, any additional costs to participant, the consequences of a participant's decision to withdraw, a statement that significant new findings will be provided to participant, and/or the number of participants in study.

In addition to enumerating the requirements for informed consent, the regulations also establish the requirement for review of research protocols by institutional review boards, their composition, and requirements for documenting their procedures and decisions. Importantly, the regulations specify the primary mission of the institutional review boards: to ensure that risks of the research outweighed by the anticipated benefits, to ensure that the rights and welfare of research participants are protected, and to ensure that informed consent will be obtained by adequate and appropriate means.

Special protections are also enunciated for pregnant women, children, and prisoners. These protections range from the inclusion on the reviewing institutional review board of individuals with specified expertise, such as in the case of research with prisoners, to increasing requirements for valid consent as the level of risk involved increases in the case of research involving children.

Ensuring Compliance

Federal regulations also set forth various provisions designed to ensure compliance with the regulations and the ethical principles that they embody. The misuse of human participants in research falls within the jurisdiction of the Office of Human Research Protections (http://www.hhs.gov/ohrp/policy/). (Regulations also address scientific misconduct, which refers to the "fabrication, falsification, plagiarism, or other practices that seriously deviate from those that are commonly accepted within the scientific community for proposing, conducting, or reporting research." A discussion of scientific misconduct is beyond the scope of this text. The issue of scientific misconduct is overseen by the Office of Research Integrity; interested readers are urged to consult the website at http://ori.dhhs.gov/misconduct/index.shtml for information.)

U.S. Regulations and HIV Research

Enhanced Protections

The protections provided by the regulations may be particularly important in the context of HIV-related research. Having AIDS or being HIV-positive is often a stigmatized condition because it is "an enduring characteristic that relegates an infected individual to a socially recognized, negatively evaluated category" (Herek, 2002: 595). Those with HIV are stigmatized because HIV infection is often perceived to be the result of the individual's own behavior, such as in the case of male-male transmission or injection drug use; because the condition is fatal and incurable; and because the condition is believed to pose a risk to others (Herek, 2002). Those who are infected may attempt to "pass" in an effort to avoid the potential social and legal consequences associated with being infected (Goffman, 1963).

It is not surprising, then, that various policy statements, specific to the conduct of HIV-related research, have been issued by the federal government to supplement the regulations. Many of the cautionary notes contained in these policy statements underscore the importance of confidentiality in the conduct of HIV-related research. The former Office for Protection from Research Risks issued in 1984 policy guidance for institutional review boards reviewing HIV-related protocols (United States Department of Health and Human Services, 1984). This document urged that HIV-related protocols include provisions to

- Inform research participants of positive antibody tests
- Inform research participants of significant new findings developed during the course of the research which may relate to subjects' willingness to continue to participate
- Advise participants of the limits of confidentiality
- Separate identifiers from the data collected for the study, with linkage restored only when necessary to conduct the research
- Provide that no lists be maintained of individuals declining to participate in the research
- Minimize the recording of data from the studies in individuals' medical records
- Advise study participants of any local laws that may require the disclosure of AIDS status prior to their volunteering for the study
- Establish procedures for information disclosure in emergency situations involving the health either of research subjects or others
- Establish procedures for responding to requests by third parties who have authorizations for disclosure of information signed by subjects (United State Department of Health and Human Services, 1984).

In addition, the guidance provided to IRBs stressed in particular, three ethical issues to be addressed: (1) fairness in the distribution of the risks and benefits of research, such "that age, competence, experience, education, position, lifestyle, etc., are not used to determine eligibility for entrance into a study unless these factors are necessary for the research design;" (2) the maximization of benefits and the minimization of harm, requiring that special care be taken to prevent accidental or careless disclosure of information that could result in harm to the participant; and (3) that the rights of potential research participants to make informed judgment be respected through appropriate provisions of the informed consent procedure.

Subsequent policy statements issued by the Public Health Service (PHS) provide that, in research conducted or supported by the PHS, individuals must be informed of their HIV serostatus and be provided with appropriate counseling unless "there are compelling and immediate reasons that justify not informing a particular individual that he or she is seropositive," in which case the IRB must be informed; it is required by the protocol; or a variation is required for research conducted at foreign sites (United States Department of Health and Human Services, 1988).

Exceptions based on the protocol design require that the investigator demonstrate to the IRB that

(a) research subjects will be informed of their risk of infection; (b) research subjects will receive risk reduction counseling whether or not they receive their test results; (c) there is good reason to believe that a requirement for test notification counseling whether or not they receive their test results; (c) there is good reason to believe that a requirement for test result notification would significantly impair collection of study information that could not be obtained by other means; an (d) the risk/benefit ratio to individuals, their partners, and society will be periodically reevaluated by the IRB so that the study might be revised or terminated if it is determined that it is no longer justifiable to allow subjects to continue to participate without receiving their HIV test results (United States Department of Health and Human Services, 1988).

Exceptions to the policy based on protocol design or specific circumstances at foreign sites require review and approval by the head of the appropriate agency, following receipt of IRB approval.

A further amendment to these policies was promulgated on 1990 to provide for sex- and needle-sharing partners of HIV-infected infected persons of their exposure to HIV (United States Department of Health and Human Services, 1990). The policy applies only to clinical activities conducted at PHS facilities that are conducted by PHS personnel, where there is a physician-patient relationship or other provision of health care. These facilities include the NIH Clinical Center, hospitals and clinics of the Indian Health Service, and other PHS facilities that conduct similar clinical activities. The policy further suggests that this be carried out whenever possible in conjunction with local public health authorities.

Where is Justice?

The regulations as currently formulated provide significant guidance with respect to the requirement that individuals be afforded adequate information on which to base a decision regarding participation in research. Special protections are provided for some classes of persons who may be especially vulnerable due to a lack of capacity or coercion. Institutional review boards are charged with the responsibility of assuring that the risks of participation are outweighed by the potential benefits of the research. What the regulations and policy statements fail to do is (1) provide guidance on how to operationalize the principle of justice, apart from advising institutional review boards "that age, competence, experience, education, position, lifestyle, etc., are not used to determine eligibility for entrance into a study unless these factors are necessary for the research design" (United States Department of Health and Human Services, 1984) and (2) provide guidance on the relative weight to be afforded to each of the underlying ethical principles in a given situation in which the maximization of these principles appears to be in conflict.

The *Belmont Report* (National Commission for the Protection of Human Subjects of Biomedical and Behavioral Research, 1979: 7–8) noted

Justice is relevant to the selection of subjects of research at two levels: the social and the individual. Individual justice in the selection of subjects would require that researchers . . . not offer potentially beneficial research only to some patients who are in their favor.

Injustice may appear in the selection of subjects, even if individual subjects are selected fairly by investigators and treated fairly in the course of research. This injustice arises from social, racial, sexual, and cultural biases institutionalized in society. . . .

Although individual institutions or investigators may not be able to resolve a problem that is pervasive in their social setting, they can consider distributive justice in selecting research subjects.

One must question, though, whether this operationalization of justice is sufficient in view of the lack of universal health care coverage and the likelihood that a portion of study participants will be unable to access the standard of care provided to them during the course of a study due to a lack of any or adequate health care coverage.

The second question relating to justice is best explained by way of a real-life example. It was reported that a sex worker in a research study suffered complications during the labor and delivery of her infant. The fetus dies in utero and, lacking funds and relying on the advice of the birth attendant, the sex worker delayed seeking medical care. Her peers found her several days later near death. They collected sufficient funds to pay for her emergency caesarean section and one day's worth of antibiotics, after which the funds were exhausted. They approached the expatriate investigator from the U.S. for additional funds to pay for more antibiotics; he/she contributed the equivalent of US$10. The sex worker dies before the antibiotic treatment was resumed. The investigator was reprimanded by the institutional review board for having provided the funds for antibiotics, believing that this was coercive (Fitzgerald and Behets, 2003).

This situation illustrates the tension that may exist in attempting to simultaneously maximize two ethical principles: that of voluntariness, to ensure that individuals do not make a decision to participate that they would not otherwise because the potential incentives are too good to refuse, and that of beneficence, to maximize good and minimize harm. Fitzgerald and Behets (2003: 68) reject the legitimacy of this result, arguing that our

system of ethics regulation has truly arrived at an Orwellian state when an ethics committee reprimands a health-care professional for providing care to a dying sex worker, when fear of coercion takes precedence over the imperative to save a human life. . . .

Rather, they argue, HIV prevention investigators enter into a relationship with the participants who they enroll and, consequently, also enter into their world of suffering and social injustice. Investigators must, consequently, define and prioritize their responsibilities within the social context of their research participants and their research (Fitzgerald and Behets, 2003).

References

Abt, I. (1903). Spontaneous hemorrhages in newborn children. *Journal of the American Medical Association, 40,* 284–293.

Beecher, H. K. (1970). *Research and the Individual: Human Studies.* Boston: Little, Brown, and Company.

Belais, D. (1910). Vivisection animal and human. *Cosmopolitan, 50,* 267–273.

Bynum, W. (1988). Reflections on the history of human experimentation. In S.F. Spicker, I. Alon, A. de Vries, & H.T. Englehardt, Jr., eds., *The Use of Human Beings in Research: With Special Reference to Clinical Trials* (pp. 29–46). Dordrecht, Netherlands: Kluwer Academic.

Fitzgerald, D.W. & Behets, F. M-T. (2003). Women's health and human rights in HIV prevention research. *Lancet, 361,* 68–69.

George, J. (1946). Atlanta's malaria project. *Atlantian, 6,* 14–17, 43.

Goffman, E. (1963). *Stigma: Notes on the Management of Spoiled Identity.* Englewood Cliffs, New Jersey: Prentice-Hall.

Hagstrom, R.M., Glasser, S.R., Brill, A.B., & Heyssel, R.M. (1969). Long term effects of radioactive iron administered during human pregnancy. *American Journal of Epidemiology, 90,* 1–10.

Hammill, S.M., Carpenter, H.C., & Cope, T.A. (1908). A comparison of the von Pirquet, Calmette and Moro tuberculin tests and their diagnostic value. *Archives of Internal Medicine, 2,* 405–447.

Herek, G. M. (2002). Thinking about AIDS and stigma: A psychologist's perspective. *Journal of Law, Medicine, & Ethics, 30(4),* 594–607.

Hornblum, A.M. (1998). *Acres of Skin: Human Experiments at Holmesburg Prison.* New York: Routledge.

Humphreys, L. (1970). *Tearoom Trade: Impersonal Sex in Public Places.* Chicago: Aldine.

National Commission for the Protection of Human Subjects of Biomedical and Behavioral Research. (1979). *The Belmont Report: Ethical Principles and Guidelines for the Protection of Human Subjects of Research.* Washington, D.C.: United States Department of Health, Education, and Welfare [DHEW Pub. No. OS 78–0012].

National Institutes of Health. (1991). NIH Workshop: Long-Term Effects of Exposure to Diethylstilbesterol (DES), Falls Church, Virginia, April 22–24. Sponsored by the Office of Research on Women's Health, National Cancer Institute, National Institute of Child Health and Development and the National Institute of Environmental Sciences. Cited in Institute of Medicine. (1994). *Women and Health Research: Ethical and Legal Issues of Including Women in Clinical Studies,* vol. 1. Washington, D.C.: National Academy Press.

Thomas, S.B. & Quinn, S.C. (1991). The Tuskegee Syphilis Study, 1932 to 1972: Implications for HIV education and AIDS risk education programs in the black community. *American Journal of Public Health, 81(11),* 1498–1504.

United States Department of Health and Human Services. (1984). OPRR Reports: Guidance for Institutional Review Boards for AIDS Studies. Last accessed December 8, 2006; Available at http://www.hhs.gov/ohrp/humansubjects/ guidance/hsdc84dec.htm.

United States Department of Health and Human Services. (1990). PHS Policy on Partner Notification. Last accessed December 8, 2006; Available at http://www.hhs.gov/ohrp/humansubjects/guidance/hsdc90may.htm.

United States Department of Health and Human Services. (1988). PHS Policy on Informing Those Tested About HIV Serostatus. Last accessed December 8, 2006; Available at http://www.hhs.gov/ohrp/humansubjects/guidance/hsdc88jun.htm.

Chapter 3
The Informed Consent Process

Introduction

On the morning of June 10, 1964, the United States Senate began—after a series of impassioned speeches—the roll call for cloture on a filibuster against the proposed Civil Rights Act. When the clerk reached the name of Senator Clair Engle of California, and called out his name, there was silence in the chamber. The Senator was mortally ill with a brain tumor, and was unable to speak. But he was present, and wanted to cast his vote. Raising a crippled arm, he pointed to his eye—to signify "aye." The filibuster was ended by a four-vote margin, and nine days later the Senate approved the Civil Rights Act itself—one of the most significant legislative actions of the 20th century.

This is not an anecdote about informed consent per se, but it is a story of the fluidity and contextuality of consent, assent, affirmation, or approval generally—and one with enormous potential consequences. In the U.S. Congress, there are no rules or protocols governing how Member decisions, under uncertain circumstances, are validated or authorized. Within the vacuum created by the absence of rules, the body has, over the years, made numerous adaptations, based on informally shared understandings of what is acceptable. It may well have been that Sen. Engle was complaining of eye or head discomfort and so signifying by pointing to the discomfort; it may have been that Sen. Engle was trying to say, "I'm watching this vote with great interest"; it may have been an involuntary movement of his arm. All of these interpretations, and many more, are possible. What is important, however, is that his signal, by universal agreement, was interpreted as a "yes" vote to end the current filibuster and move the Civil Rights Act forward.

Such an interpretation, though enormously consequential, could hardly meet the more specific tests that should be met in securing informed consent for medical research; nor could such an interpretation adequately address the underlying principles upon which current understandings of informed consent for medical research rests. But it does highlight the multiplicity of meanings that confound the effort to secure consent or agreement, even under the application of rigid rules for doing so. Consent, as an ideal, is easily understood; as a condition that has or has not been met, it is or can be maddeningly unreliable.

The classical principles of research ethics—autonomy, beneficence, justice, universality, and rationality—remain highly instructive. "Autonomy"—from the Greek, meaning, self-rule—is dependent on the notion of personhood, which in the present-day American context means individual rights, self-determination, and privacy (De Craemer, 1983). Beneficence obliges the researcher to a relationship that will or has the potential to benefit individuals or classes of individuals or, at the very least, does not cause harm. Justice is predicated on the researcher's duty to help allocate either resources or burdens, among all people, in a way that is fair. Universality – or reciprocity – limits the researcher to acts that he or she would be willing to have performed on him/herself under similar circumstances. Rationality limits the researcher to acts that can be logically reasoned and justified and numerous codes, guidelines, protocols, laws, and practice requirements help elucidate—and at the same time, demand further clarification.

The Nuremberg Code requires, as a precondition for consent, that the person must "1) be so situated as to be able to exercise free power of choice; 2) have the legal capacity to give consent; 3) have sufficient . . .comprehension to make an enlightened decision; and 4) have sufficient knowledge on which to decide" (In Lebacqz and Levine, 1982: 757). But arguably, in many realms of HIV/AIDS research, these preconditions may not be fully met, and informed consent does not exist or has not existed: the precondition of "free power of choice," alone, would, on the face of it, have the capacity to potentially exclude all women in all gender-stratified societies (and there are many) where women are not free to make autonomous choices about their own bodies; or would, on the face of it, have the potential to exclude as research participants all those individuals who elect to join a trial because of the free health care offered for participation in that trial—an inducement that could be said punish non-involvement in clinical trials by withholding health care. Much more demands clarification and definition, including a workable understanding of "autonomy" that serves as a precondition for the Nuremberg requirements.

The Helsinki Declaration provided further guidance on ethical considerations related to biomedical research, as did the 1974 National Research Act, and the Belmont Report. Because of the rise of several forms of knowledge in the past several decades—anthropological understandings of cross-cultural communication; feminist understandings of power and coercion in male/female relationships, or within male-dominated societies; post-modern critiques of the power of medical authority in relation to the patient; cultural criticism that explored inequality and identity in terms of race, sexual orientation, and age; and analyses by political economists who scrutinized power relationships between peoples and national bodies, and the degree to which representative individuals reflected macrocosmic relations in their microcosmic interactions—all these, and more, have, at the same time, offered powerful and needed interrogations of biomedical ethics, and thoroughly muddled the application of bioethical principles in field practice.

The Bioethical Principles

Autonomy

The first challenge is the notion of autonomy itself. Some languages, such as Bantu, do not include a term that adequately corresponds to the English word "person," which seems, at least at first blush, to invalidate, at least in Western eyes, the notion of "the self" upon which personhood, and therefore autonomy, depend (De Craemer, 1983). Thus, "the notion of who gives informed consent— heads of households, elders, individuals persons, the tribe, the ministry of health, or the government—needs to be asked with cultural sensitivity," and, one might add, with a willingness to adopt flexible protocols that will ensure that fundamental principles of consent, and cultural styles and norms, are both equally respected (Barry, 1988).

Faden and Beauchamp (1986: 238) contend that "X acts autonomously only if X acts 1) intentionally, 2) with understanding, and 3) without controlling influences." But "without controlling influences" is a vast and subjective territory; for many of the world's girls and women, "controlling influences" are extensive, and reside everywhere from the intimate chambers of domestic life, to the local or national chambers of certified law and authority.

Respect for autonomy implies that individuals are different in a variety of ways, including how, and on what schedule, they make decisions, and communicate those decisions to others. Respect for autonomy also requires a profound appreciation for individuals with categorical vulnerabilities—children, the frail elderly, and so on—and with individual vulnerabilities: cognitive and emotional differences, and so on. Key to informed consent while conducting research in international settings is researcher respect for the notion that "autonomy" in many contexts may include an "enlarged self," a concept that is historically at odds with Western individualism, but not at all at odds with the core principle of respect for autonomy, since such respect also implies respect for how different individuals may variably define their own sense of autonomy.

Tindana, Kass, and Akweongo (2006) explored the role of community leaders in the consent process, the issue of trust, perceptions about the consent process, and participant beliefs about the benefits of research. In particular, research participants acknowledged the role of community/tribal leaders in "authorizing" consent for a community. One traditional chief explained that role not merely as one of authority, but responsibility as well: "It is because I own the land and I look after the people. I am the chief so if [researchers] come and do bad work and there is a quarrel, it will not be good . . ." (Tindana, Kass, and Akweongo, 2006: 3). The authorizing role of community leaders, however, does not allow the chief to compel participation. As one male participant said: "When the chief informs us and we don't want to participate, we can refuse. If you agree with the chief then you can participate, that is when you agree but if you don't want, he won't force you." Remaining unexplored in this study, however, are the various forms of coercion that lie on a continuum; between compulsion and prohibition, there are a wide number of mechanisms community leaders can use to influence individual decisions to participate. Also unexplored is how "authorization" combines with gender, age, and other differences; presumably, and quite possibly, the word of community leaders could carry more or less weight, depending on who the leader is "authorizing."

All of the female participants in the study reported consulting with their husbands before participation. By itself, that tells us little, unless compared to the degree with which male participants consult with wives before participating in the study. More significant is the varying accounts women gave about those consultations—the degree to which husbands directed the decision. As one woman said, "I will go and tell my husband that the [researchers] have come to ask me to join their study and so I want to let him know about it and if he says I should go then I will go, but if he refuses, I won't go." Another woman, however, reported that "I make that decision and then my husband will also agree and then he will ask me to participate. If I don't want to participate, my husband cannot force me to participate."

For Moodley, cultural differences are significant: "The important concepts of informed consent, risk/benefit ratio and fair treatment of trial participants are interpreted differently in traditional, rural African communities, where a moderate form of communitarianism referred to as 'Ubuntu' or 'communalism' is still prevalent" (Moodley, 2002: 197). As a practical strategy—since consent is still vital, but may be mediated through a community context—the provision of waiting periods might address both universalist requirements and local conditions: after explanation of a consent form, a potential research participant might be given a period of time to think it over, and presumably, consult with others, before signing.

Respect for autonomy also requires the researcher to make allowance for cognitive, emotional, and sensory differences—for interpretive and communication variables. Presentation of information in a variety of formats may be required, as well as adequate, pre-consent education.

Beneficence

Concern for enhancing benefits and nullifying harm has taken on heightened significance in the case of HIV/AIDS, on both a small scale (the individual) and a global scale (development and deployment of new medications, and diffusion of the highest standards of care). The urgency of the epidemic itself, combined with widespread political activism to reduce scientific and governmental "red-tape" or barriers to care access, have accelerated research and the development and deployment of new treatments: there is enormous pressure, with so many of the world's people with HIV/AIDS still not receiving treatment, to move efficacious medications out of trials and into bodies as soon as possible, and to make available, to all, the medications already proven to be effective. But on the global level, there is a lingering awareness, that has ebbed and flowed over the course of the last 25 years, that pricing, marketing, and legal actions (especially as they relate to patents and intellectual property in general) on the part of a relative handful of Western-based pharmaceutical companies can have a significant, trickle-down affect on benefit/harm ratios for people with HIV/AIDS "out there," especially those living in the most resource-poor parts of the world. The nature of HIV/AIDS as a truly global epidemic, and the powerful forces of globalization itself, have forced us to think about the complex relationships associated with the principle of beneficence as it relates to HIV/AIDS.

For example, research participants from both developed and developing countries, as well as researchers from developing countries, may not be cognizant of the financial arrangements and market implications of the research they are involved in (Thomas, 1998). And research participants are generally unaware of their role in a process that might constitute a kind of product endorsement. While pharmaceutical companies may invest financial capital (and take financial risk) in research, that could result in significant profit for the companies, research participants invest significant human capital (and take personal risks), but do not participate in an equitable allocation of those profits. If financial investors are not bound by a requirement that dictates that their motives must be purely altruistic, human capital investors should not be expected to be required to be purely altruistic, either.

Controversies related to "harm" have been addressed elsewhere in this volume, especially in the use of placebo-controlled studies. Opinion about the ethics of placebo control ACTG 076 studies have varied, and not just among developed country researchers. Salim S. Abdool Karim (1998: 565), a South African researcher, concluded

Although the ACTG 076 regimen of therapy is the standard of care in some countries, it is not the international standard, such as set by the World Health Organization. Providing high-quality care to the control arm without providing the ATGC 076 regimen of zidovudine cannot then be construed as causing undue risk or harm to the study participants. No therapy that they may otherwise receive is being withheld from study participants . . . It is, therefore, my opinion that the placebo control arm is ethically justifiable.

The gold standard in clinical trial design remains the randomized controlled trial (RCT), in which research participants are randomized to two or more treatment arms and followed until a predetermined outcome or clinical endpoint has been achieved. To further prevent bias or contamination, some trials are conducted as double-blinded trials, in which neither the researcher nor the research participant knows which treatment, or arm, the participant is participating in. In an RCT, a new or experimental treatment is compared with a control treatment (the current standard for any particular illness). In the absence of a "standard treatment," the new or experimental treatment can be compared to either observation or placebo. Many believe that placebo controls are unethical for life-threatening conditions, or if there is an effective, available treatment that can prolong or improve life or health. Placebo-control trials, however, can be ethically justified for *non*-threatening conditions, as long as the research participant has been fully informed (1) about alternative treatment, (2) that the proposed treatments will not increase the risk of mortality, and (3) there will be no irreversible harm to the participant (Pitler, 2002).

The challenge, again, is communicating what may be viewed as complicated and culturally "foreign" concepts about benefits and harms to individuals in internationalized settings, and doing so in a manner that also respects the need for full disclosure of other, macrocosmic "benefits" and "risks" to which the individual, however remotely, may be a participant—full disclosure, in other words, of the investments all actors are making, and the possible profits (better personal health, enhancing market share and earnings) they seek to realize. Some commentators have offered general suggestions. Kilmarx and colleagues (2001), for example, have made the following recommendations for ensuring informed consent:

1. conducting a comprehension assessment before conducting clinical trials,
2. training of staff on communication and counseling, and the absence of language and cultural barriers,
3. easily readable and understandable language in consent forms,
4. utilization of special media, such as video, to help inform participants,
5. ongoing notification of participants that they are free to leave the study at any time, and
6. data collection to monitor participants' understanding.

These same recommendations are relevant to the conduct of research and the informed consent process in U.S.-based HIV/AIDS research.

To these, more narrow guidelines concerning *methodology*, one could add the more specific requirement to communicate key *content*, which would, at minimum, include (1) the nature of the research being conducted and its anticipated benefits and risks for the individual, and, (2) the planned investments and anticipated benefits for other parties involved, including industry, government, international bodies, and others. It is troubling to recognize that there is still no significant discussion of the possible right a research participant might have to know that others can stand to amass benefits or suffer losses as the result of the research participants' investment of his or her body.

Justice

Justice, as a generalized obligation to assist in the fair allocation of resources and burdens, is a much more difficult concept to apply, in no small measure because the definitions of distributive justice, and how it is to be instituted, vary considerably. The Rawlsian conception of the difference principle—

The difference principle represents, in effect, an agreement to regard the distribution of natural talents as in some respects a common asset and to share in the greater social and economic benefits made possible by the complementarities of this distribution. Those who have been favored by nature, whoever they are, may gain from their good fortune only on terms that improve the situation of those who have lost out (Rawls, 1999: 87).

—intrinsically supports an equality of benefits that is not widely assured in current HIV/AIDS research, for reasons stated previously: profits are not shared equally in a macrocosmic analysis. But it may very be that profits secured as a result of financial investments, professional progress secured as a result of intellectual investments on the part of researchers, and health benefits secured as a result of physical investments by research participants (bodies in trials) all achieve a relative equality, if—and it is a substantial pre-condition—the essential Rawlsian requirement for justice has been met:

All social values—liberty and opportunity, income and wealth, and the social bases of self-respect—are to be distributed equally *unless an unequal distribution of any, or all, of these values is to everyone's advantage* [emphasis added.] (Rawls, 1999: 54).

What the application of a principle of justice argues for most clearly in the case of informed consent is transparency and negotiated risks and benefits; that is, "justice" necessitates that each party to the research for which informed consent is sought understand, and consent to, the multiplicity of risks and benefits, investments and profits, of which the research is constructed.

There is a reasonable argument to be made for this position; operationalizing such transparency and negotiated terms, however, would no doubt be met with considerable resistance. Let us suppose, for example, that the hypothetical pharmaceutical company PharmaCom is providing funding to a researcher, Dr. Smith, to conduct clinical trials on the clinical effectiveness of a new antiretroviral medication, and that the research is to be conducted in Tanzania. Let us also suppose that all other ethical conditions and reviews have been satisfactorily met, that the terms under which informed consent is to be sought and secured, for Tanzanian research participants, have been developed with the highest regard for traditional ethical questions involving consent. But let us also suppose that 1) PharmaCom is the target of current international product boycott, because of pricing and marketing practices that have made AIDS medications more difficult to access, and/or that 2) Dr. Smith, herself, came under particularly strong criticism by US and international bioethicists for her role in conducting a placebo-controlled study of interventions designed to prevent perinatal transmission of HIV – a study that some bioethicists compared to Tuskegee. Do the research participants in the planned Tanzanian study have the right to know such information, and if so, do researchers have a positive obligation to ensure that such information is known, before informed consent is satisfied? Would the answer be the same, or different, if the potential research participant were residing in the United States?

Such questions have not been adequately considered, and deserve further discussion. One can reasonably predict that most Americans would agree that they have a right to know certain information about the practice history of their physician, and certain information about the financial or management practices of their health care institutions—these anticipated claims of rights, then, ought to be extended to research settings, and globally. But they are not.

The principle of justice, more narrowly, obligates the researcher to provide full information, and ensure comprehension of that information, about the specific research study in which the participant is being asked to enroll. Potential benefits must be reasonably explained, but without guarantees. Potential harms, on all levels, must be elucidated. Participants must be fully informed of their right to leave the study, and the benefits and risks that will maintain at any point of study participation, including the risks and benefits that will likely exist if the participants elect an early withdrawal. And particularly in the case of HIV/AIDS, when study participants often receive free medical care and monitoring while involved in the study, participants must be fully informed of what access to care, if any, they will receive after the study is concluded.

Universality/Reciprocity

The key problem in the application of the principle of universality to HIV/AIDS is "under similar conditions." Similar, as in "living in poverty in Tanzania," or similar as in "living with HIV infection"? The higher standard will of course be met with the latter similarity, and it is, generally, the universality that should be applied. There is no reason to conclude that an ill person, anywhere in the world, desires any less passionately the fullest possible health that she or he can achieve. But the more specific question of whether or not the principle of universality has been met when an HIV + Tanzanian considers enrolling in a US-sponsored clinical trial is much more difficult to ascertain, for the reality of power relationships (both symbolic power and resource power) is always present in that

[m]edicine is not the simple administration of culturally or morally neutral procedures and technologies. Western medicine reproduces a cultural order, and in its dealing with the needs of the nonwhite, the nonmale, and the non-Western, medicine is beset with problems generated by its own hubris and perceived universalism (Fortin, 1991: 18).

Here, as elsewhere, the debate between relativist and fundamentalist perspectives is again renewed. Lisa Newton (1990). contends that the assumption that the informed consent requirement is itself a universal ethical standard constitutes ethical imperialism at its worst. Ruth Faden and Carol Ijsselmuiden argue the opposite: "Appeals to cultural sensitivity . . . are no substitute for careful moral analysis. We see no convincing arguments for a general policy of dispensing with, or substantially modifying, the researcher's obligation to obtain first-person consent in biomedical research in Africa." They go on to suggest that relativist perspectives may "have relied on limited and dated anthropologic literature that does not reflect the rapid cultural changes brought by colonialism and independence, warfare, and urbanization," and, it could be added the powerful capacity of globalization to hegemonize Western cultural values" (Faden and Ijsselmuiden, 1992: 833.) We might add, in light of the earlier discussion about justice, that there does not seem to be any convincing argument for a policy that attenuates the content of consent outside Western territories; that is, whatever information (about personal risks and benefits, about institutional risks and benefits, and so on) deemed necessarily disclosable to research participants in US and European settings must also be deemed necessarily disclosable everywhere else in the world.

At the least, scientists must take care not to allow the provision of "under similar conditions" to serve as an ethical caveat that permits the impermissible. "Desperation," after all, is a condition, and one that would induce the average person to engage in, or concede to, sometimes horrific acts in order to find some escape. And the combination of brutalizing poverty and mortal illness would, for many of the world's people so circumstanced, certainly qualify as desperation. Actions the individual concedes to, ought of a relative position of powerlessness and perceived or real choicelessness, must but subject to strict ethical interrogation.

Rationality

The assurance that a proposed study is predicated on sound and logical scientific thinking and understanding will generally be located in the research proposal itself, and review processes—whether initiated by governments-as-funders, or industry-as-funder—will be particularly keen to scrutinize the logic, since they are also to be asked to make resource investments based on that logic. But the conduct of research in international settings must also recognize forms of knowledge or rationality, and consider the impact that divergent forms will have on the proposed of informed consent.

It is a reasonable expectation, that as part of the process of informing for consent, that researchers would and should attempt to impart the basic theoretical and mechanical/physiologic bases upon which a proposed study is founded, and that the research participants, in consenting, can attest to a meaningful personally meaningful understanding of those bases. But there is no doubt that the transfer of conceptual knowledge as one of the responsibilities inherent in informing for consent can be a challenge, especially if there is no simple correspondence between conceptual knowledge sets in cross-cultural settings. For example, it may be extremely difficult in some contexts to convey an understanding of the concept of randomization, particularly in contexts in which the understanding of disease differs from Western scientific theories. According to Christakis (1988), in such circumstances it may be impossible to obtain truly informed consent. Difficult; sometimes yes. "Impossible" has yet to be proven.

Much the preceding would seem to argue for something that, in the end, looks less like "informed consent" as a static, achieved *condition*; and more like "education for co-participation in research" that looks more like an evolving, bilateral, sequenced, culturally competent *process* that rests on a fundamental equality of partners, and a consequent equality of a full range of rights in the implementation of the research itself—a process that, to be sure, includes necessary verifications of ethically-derived "consent" for the narrow purpose of proceeding with a proposed study, but that does not truncate that process, or the ethical interrogations attached to the process, in the interest of "getting the job done." To a number of observers, the evolving research environment has had the effect of "reducing consent" to "a participant's signature on a form obtained before study enrollment" (Kass, Sugarman, and Schoch-Spana, 1996; Sugarman, Popkin, Fortey, and Rivera, 2001).

But what if the signature on a page, or the gesture toward an eye to signify agreement, were but one platform in a more elaborate, multi-layered architecture of the informed consent process, one that, to bring forth John Rawls again, viewed not merely material resources, but knowledge itself as common property? Would not we seek, under such conditions, to carry out much more "informing," more nuanced informing, more identity-sensitive informing, and more informing that exposed the multiple bodies of knowledge, institutional relationships, resource allocations, reciprocities and exceptions, that invariably underlie, directly or indirectly, the proposed research in question? And for whom, other than those who would seek to ration and apportion such knowledge, and thereby apportion the resources and power that may and often does extend from such rationing, would object?

Practice

Woodsong and Karim (2005) have outlined a useful conceptual framework, based on the experience of the HIV Prevention Trials Network funded by the National Institutes of Health, that can serve as a useful template for developing and managing ethical issues related to informed consent. It is based on a "principled approach to working in partnership with communities . . . to facilitate the 2-way communication required to achieve mutual understanding of research endeavors" (Woodsong and Karim, 2005: 412).

The model addresses a pre-enrollment phase, the enrollment phase, and post-enrollment. The pre-enrollment phase includes "activities conducted to determine how to convey research protocol concepts, recognize community concerns about HIV, and respect community norms and expectations for individual versus group decisionmaking" (Woodsong and Karim, 2005: 412), including involving community leaders and stakeholders in informational meetings that introduce the proposed research; involvement of community representatives in pilot activities designed to test knowledge and understanding assumptions; and the solicitation of core concerns or clusters of concerns originating from the community. Concrete tasks include the development of draft consent forms and supporting documents, and testing of consent verification.

With the increasing and increasingly valid concern about verification of true understanding and unencumbered consent, there is a parallel concern about the development of consent forms and supporting documents that have linguistic and literacy integrity—materials that are not merely appropriate, but effective, for the intended audience of prospective research participants. Pre-enrollment testing of materials and protocols allows researchers to identify gaps between intended understanding and actual comprehension. While the growing length of documents has been a concern, this should not be permitted to trump the more fundamental goal of comprehension (Institute of Medicine, 2003).

Providing that adequate educational activities have been outlined in the pre-enrollment phase, and that consent forms and supporting documents and strategies have been developed as the result of an interactive, inclusive community-researcher process, the enrollment phase should address strategies for explaining and securing consent. The use of a variety of explanatory formats, combined with waiting periods after explanation and other approaches, can help ensure cultural and gender-based sensitivities. While researchers may be limited by the fact that many research sponsors now insist on the use of standardized consent forms, they can be viewed as a floor rather than a ceiling; the signature, in most cases, is not insufficient, even if it meets legal obligations. Normal rules for verifying cultural and cognitive sensitivity—such as back translation of local-language translation, gender-specific focus group reviews, and the like—should apply as well.

In the process of authorizing specific forms as valid or invalid verifiers of true consent, we should also scrutinize the role that the form itself, or the interactions that precede the form itself, may or may not play in the lives of individuals. Tindana's, Kass', and Akweongo's (2006) study, for example, asked research participants about their perceptions of informed consent, and at least one aspect of their findings deserves more exploration: the way in which some research participants found meaningful value in a copy of the signed consent form itself. A few said the consent forms "would always remind them that they had taken part in the study, suggesting that they valued the forms, even if they did not understand them." Additionally, for some participants, "the consent form not only symbolized participation in the research but also was viewed as a 'ticket' for future research benefits" (Tindana, Kass, and Akweongo, 2006). These are powerful, extra-legal meanings, and the symbolic power of the form itself needs to be more carefully examined, through an ethical lens, as an incentive for research participation.

But if, as indicated before, informed consent is not viewed as a static, achieved condition, but rather as an evolving, ongoing process with clearly recognizable benchmarks (such as the necessary signature on the necessary form), then post-enrollment is critical, and implies that the researcher is still obliged to attempt to deepen participant understanding of the multiplicity of issues related to, and preceding, the study in question. As Woodsong and Karim (2005) note, it is critical that comprehension of informed consent be maintained throughout the study period but, even further, the researcher's obligation to engage in a bilateral process of sharing multiple knowledges, with participants and participants communities, should not be limited to the chronology of the study:

We argue that attention to both individual and community contexts, over time, is necessary to achieve the *spirit* [emphasis ours] of informed consent—a reflection of respect for individuals and autonomy. Working with community members can aid in the creation of a consent form that is comprehensible to potential participants and an informed consent process that remains active throughout the study. In addition, it can contribute to a process that addresses potential issues such as undue inducement, psychological risks, lack of awareness of the Western scientific model, and differences in concepts of autonomy and rational decisionmaking. Finally, it can strengthen the research effort through recruitment and retention of participants who better understand their roles and responsibilities in the study and can thus better adhere to the study protocol. Thus, improved consent will benefit all who are involved in public health. (Woodsong and Karim, 2005: 417).

To "better understand their roles and responsibilities," we must add, "better understand their rights and their potential agency in relation to those rights," in light of our earlier discussion about the potential moral necessity to ensure that research participants understand risks and benefits as a part of a larger, more complex series of power and resource exchanges between individuals, governments, and industry—exchanges that, microcosmically and macrocosmically, are further conditioned as a result of gender essentialism, the imperial power of culture, and the postmodern understanding of medicine as a sphere for allocating and regulating power. The principle of universality, understood to mean, at minimum, that everyone everywhere has the right to the same body of knowledge that could, if fully understand, sway the decision whether or not to participate in the trial, must be preserved. If potential developed-world research participants have the implied right and the culturally-embedded opportunity to access a wealth of information about governments, researchers, and industry, so should every prospective research participant in the developing world. The burdens, for the research community, have therefore increased, and will likely increase even further. But the ethical imperative to maximize the values of autonomy, beneficence, justice, universality, and rationality far outweighs the trouble involved, and in the end, can go a long way to restoring trust between natural allies—researchers, governments, industry, individuals, and communities—who wish to work together to solve the overwhelming challenges of research that will change the course of the modern catastrophe of HIV/AIDS.

References

Barry, M. (1988). Ethical considerations of human investigation in developing countries: The AIDS dilemma. *New England Journal of Medicine, 319(16)*, 1083–1086.

Christakis, N.A. (1988). The ethical design of an AIDS vaccine trial in Africa. *Hastings Center Report*, 18, 31–37.

De Craemer, W. (1983). A cross-cultural perspective on personhood. *Milbank Memorial Fund Quarterly, 61*, 19–34.

Ehrenreich, J. (1978). Introduction. In J. Ehrenreich (Ed.). *The Cultural Crisis of Modern Medicine* (pp. 1–35). New York: Monthly Review Press.

Faden, R.R. & Beauchamp, T.L. (1986). *A History of Informed Consent*. Oxford: Oxford University Press.

Faden, R.R. & Ijsselmuiden, C.B. (1992). Images in cultural medicine. *New England Journal of Medicine, 326(12)*, 830–833.

Fortin, A.J. (1991). Ethics, culture, and medical power: AIDS research in the Third World. *AIDS and Public Policy Journal, 6(1)*, 15–24.

Institute of Medicine. (2003). *Responsible Research: A Systems Approach to Protecting Research Participants*. Washington, D.C.: National Academy Press.

Karim, S.S.A. (1998). Placebo controls in HIV perinatal transmission trials: A South African's viewpoint. *American Journal of Public Health, 88(4)*, 564–566.

Kass, N., Sugarman, J., & Schoch-Spana, M. (1996). Trust, the fragile foundation of contemporary biomedical research. *Hastings Center Reports, 26*, 25–29.

Kilmarx, P.H., Ramjee, G., Kitayaporn, D., & Kunasol, P. (2001). Protection of human subjects' rights in HIV-preventive clinical trials in Africa and Asia: Experiences and recommendations. *AIDS, 15 (suppl 5)*, S73-S79.

Lebacqz, K. & Levine, R.J. (1982). Informed consent in human research: Ethical and legal aspects. In W.T. Reich (Ed.), *Encyclopedia of Bieothics* (p. 757). New York: Free Press.

Moodley, K. (2002). HIV vaccine trial participation in South Africa – An ethical assessment. Journal of Medicine and Philosophy, 27(2), 197–215.

Newton, L. (1990). Ethical imperialism and informed consent. *IRB: A Review of Human Subjects Research, 12(3)*, 11.

Nuremberg Code. (1946). In K. Lebacqz & R.J. Levine. (1982). Informed consent in human research: ethical and legal aspects. In W.T. Reich (Ed.). *Encyclopedia of Bioethics* (p. 757). New York: The Free Press.

Pitler, L.R. (2002). Ethics of AIDS clinical trials in developing countries: A review. *Food & Drug Law Journal, 57*, 133–153.

Rawls, J. (1999). *A Theory of Justice*, rev. ed. Cambridge, Massachusetts: The Belknap Press.

Sugarman, J., Popkin, B., Fortney, J., & Rivera, R. (2001). International perspectives on protecting human research subjects. In National Bioethics Advisory Commission. *Ethical and Policy Issues in International Research: Clinical Trials in Developing Countries*, Vol. 2. (pp. E1-E11). Bethesda, Maryland: National Bioethics Advisory Commission.

Thomas, J. (1998). Ethical challenges of HIV trials in developing countries. *Bioethics, 12(4)*, 320–327.

Tindana, P.O., Kass, N., & Akweongo, P. (2006). The informed consent process in a rural African Setting: A case study of the Kassena-Nankana District of Northern Ghana. *IRB: Ethics & Human Research, 28(3)*, 1–6.

Woodsong, C. & Karim, Q.A. (2005). A model designed to enhance informed consent: Experiences from the HIV Prevention Trials Network. *American Journal of Public Health, 95(3)*, 412–419.

World Medical Association Declaration of Helsinki. (2000). Ethical Principles for Medical Research Involving Human Subjects, 52nd WMA General Assembly. Edinburgh, Scotland.

Chapter 4
Working with Boards and Committees: ECs, DSMBs, CABs

This chapter focuses on the roles of the various boards with which investigators, communities, and/or research participants may interact during the course of HIV-related research. These include ethics review committees (ECs, known as institutional review boards, or IRBs, in the United States), data safety and monitoring boards (DSMBs), and community advisory boards (CABs). Depending upon the nature of a particular study, any one or all of these boards may be required.

Ethics Review Committees

Purpose, Composition, and Functioning

Purpose. The World Health Organization has suggested in its *Operational Guidelines for Ethics Committees That Review Biomedical Research* (2000: 1) *(WHO Guidelines)* that the primary purpose of ethics review committees should be "to contribute to safeguarding the dignity, rights, safety, and wellbeing of all actual or potential research participants." These *WHO Guidelines* stress, in particular, the need for such review committees to consider the principle of justice, which requires the equitable distribution of the burdens and benefits of research among all groups and classes in society.

The *WHO Guidelines* recommend the establishment at institutional, local, and national levels of ECs that are independent, multidisciplinary, multisectorial, and pluralistic, and that they be provided with adequate administrative and financial support to enable them to fulfill their purpose. As an example, the Thailand Ministry of Public Health formed an ethics review committee more than 10 years ago. This committee is charged with the responsibility of reviewing all proposals to conduct clinical trials that are generated by staff in the Ministry and any hospital or institute that is within the Ministry's jurisdiction (Chokevivat, 1998). Medical schools and research centers that are not within the Ministry's jurisdiction generally maintain their own institutional ethics review committee.

Composition. The WHO Guidelines recommend that the review committee reflect diversity in expertise, age, gender, and community concerns. It has also been suggested that ethics review committees reviewing HIV-related protocols include HIV health consumers among its members (National Health and Medical Research Council, 2005). Members of the review committee are to be free from bias and conflicts of interest. In this context, the *WHO Guidelines* define "conflict of interest" as arising

When a member (or members) of the EC holds interests with respect to specific applications for review that may jeopardize his/her (their) ability to provide a free and independent evaluation of the research focused on the protection of the research participants. Conflicts of interest may arise when an EC member has financial, material, institutional, or social ties to the research (World Health Organization, 2000: 21).

Functioning. The WHO Guidelines indicate that the EC should establish and publish its procedures for the submission of applications for review; that meetings be conducted at regularly scheduled times; that decisions be made in accordance with pre-specified procedures and criteria at duly constituted meetings at which a quorum is present; that decisions be communicated on a timely basis to the investigator whose proposal has been reviewed; and that appropriate provisions are made for the archiving of the documents pertaining to the EC functioning and decisionmaking, including agendas, minutes, and copies of all materials submitted by the investigators.

The scope of the EC's review is quite broad and encompasses the scientific design and conduct of the study, the recruitment process, the provisions developed to ensure the protection of the research participants, the procedures to protect the confidentiality of the data, the informed consent process and community considerations. Table 4.1, below, provides additional detail with respect to the elements constituting each of these domains.

Ethics Review Committees in the United States: Institutional Review Boards

In the United States, a federal requirement mandating the establishment of IRBs was promulgated in 1974, following the revelation of numerous instances of unethically conducted scientific research studies. By 1991, almost all federal agencies that funded research involving human participants had adopted these or similar regulations (Gordon, Sugarman, and Kass, 1998).

Under the regulations of the federal Department of Health and Human Services (HHS), before any human subjects research can be conducted, the institution must provide the department or agency a written Assurance that it will comply with the requirements of the federal policy relating to the conduct of research involving human subjects; the Assurance must be approved by the department or agency; and the institution must certify to the department or agency head that the research has been reviewed and approved by a properly constituted IRB (Code of Federal Regulations, 2006). The Food and Drug Administration (FDA), whose regulations governing research involving human participants differ somewhat from those of HHS, does not require the filing of an Assurance. The discussion

Table 4.1. Elements of study to be reviewed by ethics review committees

Scientific design and conduct of the study	Recruitment	Protection of participants	Protection of confidentiality	Informed consent process	Community considerations
Appropriateness of the study design in relation to the study objectives	Characteristics of the proposed study population	Suitability of investigator's qualifications to conduct study	Description of persons having access to data	Description of the process	The impact on and relevance to the community of the research
		Plans to withhold/withdraw standard therapies for research and justification to do so		Identification of individuals responsible for obtaining consent	
Potential to reach sound conclusions with smallest number of participants	Means/mechanism for initial contact and recruitment	Provisions for medical care for participants during and following research study	Measures to be taken to ensure confidentiality and security of information	Adequacy, completeness, and understandability of the written document	
Justification of risks versus benefits for the individual participants and the community		Adequacy of medical supervision and psychosocial support for participants			
Justification for the use of a control arm	Means/mechanism for providing complete information to prospective participants	Steps to be taken if participants voluntarily withdraw from study		Justification for intent to include as participants individuals who cannot provide their own consent	
		Criteria for extended use, emergency use, compassionate use of study products			
Criteria for premature withdrawal of participants		Arrangements and procedures for communication of study-related information to participant's physician			
Criteria for suspending or terminating the study				Assurances that individuals will receive information during the course of participation that is relevant to their participation	
	Inclusion criteria	Description of plans to make research product available to participants following close of study			
Adequacy of provisions for monitoring the research, including the formation of a data safety monitoring board	Exclusion criteria	Description of financial costs to participants			
		Rewards and compensation to be provided to participants		Provisions for responding to participant questions and complaints during the course of the study	
Adequacy of the study site and staff		Provisions for compensation and/or treatment for participants in the event of injury, disability, or death attributable to participation in research			
Mechanisms/plan for disseminating study findings					

that follows focuses on the requirements of HHS; readers are urged to consult the federal regulations specifically for further details relating to the requirements of research that falls within the jurisdiction of the FDA.

IRB Composition. Each IRB is required by federal regulation to have at least five members (Code of Federal Regulations, 2006). In addition, the regulations require that the board be comprised of individuals with varying backgrounds in order "to promote complete and adequate review of research activities commonly conducted by the institution." This diversity must reflect variations in education, discipline, race/ethnicity, and sex. All members must have adequate training and expertise; as a body, the board should reflect expertise in the scientific discipline of the research to be reviewed, as well as relevant regulations, law, and standards of conduct. At least one member of the IRB must be an individual from outside of the institution who is not an immediate family member of someone affiliated with the institution in which the IRB functions. Additional membership requirements may apply depending upon the nature of the research and/or the population to be recruited into the study.

IRB Functions. The IRB review process ideally reflects the qualities of independence, transparency, and competency (Cho, 2003). The IRB is required to operate independently from the investigators, the institution, the professional community, and any other undue influences. Independent review ensures, to the extent possible, that the researcher will not take advantage of or abuse the research participants.

Federal regulations require that IRBs maintain copies of all of the research proposals that have been reviewed, any scientific evaluations that accompany the proposals, approved sample informed consent documents, progress reports about the research that have been submitted by the investigators, and any reports of injuries to the research participants (Code of Federal Regulations, 2006). In addition, IRBs must keep minutes of their meetings that detail the attendance at the meeting, the actions taken by the IRB, the number of votes for and against and in abstention with respect to each action, the basis for approval or disapproval of a research protocol, and a summary of any disputed issues and their resolution. Documentation must be kept of all continuing review activities, all correspondence between the IRB and investigators, the names and specified characteristics of the IRB members, the written procedures to be followed by the IRB, and statements of significant new findings that are provided to research participants.

In view of these federally mandated functions, it is not surprising that IRBs face a staggering workload. A total of 491 IRBs were found to be responsible for the annual review of 284,000 reviews, of which 105,000 were initial reviews, 116,000 annual reviews, and 63,000, amendments to the original protocols (Office of Extramural Research, 1998). The annual full-board meeting time ranged from 9 to 50 hours, with an average per-protocol discussion time of 21 minutes for low-volume IRBs and 3 minutes for high-volume IRBs.

IRB Functioning. IRBs have often been criticized for a lack of efficiency. To some extent, this results from their overwhelming workload. It has also been attributed to the evolution of research from federally funded single center research studies to large, multicenter trials of complex treatments and interventions (Randal, 2001; United States Department of Health and Human Services, 1998) and the requirement by the local IRB of each institution that the protocol undergo local review at the respective site. These multiple reviews are expensive in terms of both time and money, often yield conflicting IRB decisions necessitating multiple rounds of revisions and resubmissions, and frequently resulting in increased length and reduced readability of the informed consent forms (Burman et al., 2003; Dziak, Anderson, Sevick, Weisman, Levine, and Scholle, 2005; Vick, Finan, Kiefe, Neumayer, and Hawn, 2005). One research team involved in the conduct of an eight-site observational substance abuse treatment study estimated that the cost of each supplemental IRB action, defined as every IRB review that was required following approval by the "home" IRB, was $56,191 in 2001 dollars, consuming 16.8% of the total research grant budget for all activities during the same time period (Humphreys, Trafton, and Wagner, 2003). Centralized institutional board review for multicenter trials has been suggested as a possible remedy to this inefficiency and the lack of consistency across local IRBs reviewing the same protocol (Christian et al., 2002; Vick, Finan, Kiefe, Neumayer, and Hawn, 2005).

Concerns have also been expressed regarding the scope of IRB review, which has expanded from its original focus on risky medical and behavioral research to include studies involving interviews, journalism, and the secondary use of publicly available data, all of which are characterized by minimal or no risk to the research participants (Gunsalus et al., 2006). Some scholars have argued that, by regulation, IRBs' authority does not extend to specific types of activities, such as protocols intended to effectuate improvements in local health care processes (Nerenz, Stoltz, and Jordan, 2003). Some have asserted that the review board is limited to a consideration of safety and ethical issues, while still other scholars have contended that the scientific merit and trial design of an investigation are appropriate for IRB review because these issues are directly related to an assessment of the risk-benefit ratio (Christian et al., 2002). It has been hypothesized that IRBs based at medical schools are more likely to review the scientific merits of a protocol than are IRBs based in academic departments in liberal arts colleges, based on a belief that there is greater inherent risk associated with participation in clinical trials and studies using invasive therapeutic techniques, as compared with behavioral investigations (Rosnow, Rotheram-Borus, Ceci, Blanck, and Koocher, 1993).

Even within the same review committee, perspectives may differ depending upon the expertise and biases of the individual committee members. In a study of one IRB's review of 124 submitted research protocols during a five-year period, it was found that the IRB was more likely to request revisions of the informed consent documents for pharmacological studies as compared to nonpharmacological studies, but had significantly more concerns relating to the protocol itself in nonpharmacological studies (Sansone, McDonald, Hanley, Sellborn, and Gaither, 2004). An earlier study of psychologist-respondents' opinions about the costs and benefits of participation in hypothetical studies found that those who emphasized the benefits were more likely to be male and to have been employed in research-oriented contexts, while those focusing more on the risks of participation were more likely to be female, employed for shorter periods of time, and employed in service-oriented contexts (Kimmel, 1991).

IRB Decisions and Their Impact on Participants and Communities. The controversies regarding IRB functioning that are noted above have direct implications for research participants. Decisions of IRBs may affect whether individuals at risk of HIV or those who are infected may participate in studies, the extent to which they understand the procedures involved in a study in which they enroll, individuals' rights and remedies as participants, and participants' access to information about the study following its conclusion.

Study Participation. As an example, consider the findings from a study of IRBs and their policies regarding research consent by adolescent minors (Mammel and Kaplan, 1995). This survey of 600 IRB chairs in the U.S. yielded 183 fully scorable responses. Almost three-quarters (70%) of the responding IRBs required parental consent for all research on minors; 52% required parental consent for even a simple satisfaction survey, and only 29% would waive parental consent for an anonymous HIV seroprevalence study. What this means practically is that almost three-quarters of the responding IRBs would have denied adolescents the ability to participate in an anonymous HIV seroprevalence survey without their parents' consent. Because HIV is most frequently transmitted through unprotected intercourse and shared contaminated injection equipment, it is likely that many youth would not wish to involve their parents in this process, regardless of their level of sexual and drug-using activity. The data that would have resulted from such a study may have been critical to health planners and HIV educators in order to allocate resources and develop appropriate, targeted intervention efforts.

Understanding the Informed Consent Form. The Tuberculosis Trials Consortium (TTC) was funded by the Centers for Disease Control (CDC) and Prevention to conduct research on the treatment of tuberculosis (Burman et al., 2003). The sites included academic medical centers, Veterans' Administration Medical Centers, and public health departments. The CDC required that the protocols and informed consent forms of all participating institutions be approved first by the CDC, then by the respective local IRB, and again by the CDC. Additional changes in the protocol and/or informed consent forms required approval at both the local and centralized (CDC) level. Local review required a median of 104.5 days and resulted in a median of 46.5 changes per consent form, increased length, and an inappropriately high reading grade level in 41% of the forms. The authors of the study concluded that their findings supported the Institute of Medicine's observation that the variability in the research protocol and informed consent forms resulting from duplicative review by participating institutions in a multicenter trial may actually detract from participant protections.

Many HIV-related studies may include participants with lower levels of education and/or reading ability (Woodsong and Karim, 2005). If they are unable to understand the informed consent form because of its readability level, their consent will not be truly "informed." It can be argued that informed consent is a continuing process and the research team has the obligation to explain aspects of the protocol that are unclear in the consent form. However, it is also possible that the level of the prospective participant's understanding is such that he or she cannot formulate appropriate questions.

Participant Remedies. In yet another study, researchers extracted from the websites of 123 medical schools the language used in their informed consent templates that relates to research-related injury (Paasche-Orlow and Brancati, 2005). Of these 123 websites, 106 contained adequate information to permit analysis. They found that more than one-third of the schools (39%) did not offer coverage for medical bills when the research was industry-sponsored, such as a pharmaceutical company, and more than three-quarters (78%) provided no coverage when the research was funded by an entity other than industry. One-half of the 22 schools providing medical coverage for research-related injuries occurring in the context of non-industry-funded research provided coverage only for emergency bills. The authors of the study concluded that federally funded research at the majority of U.S. medical schools "fails to protect subjects from the financial burden of research-related injury" (Paasche-Orlow and Brancati, 2005: 175).

This has direct implications for participants in HIV-related research, some of whom may suffer research-related injuries. For instance, an individual participating in a clinical trial of a new HIV treatment might experience an unforeseen allergic reaction or, worse yet, suffer an anaphylactic response requiring immediate medical attention. Individuals without any health insurance coverage, with coverage that requires the payment of deductibles and co-pays prior to coverage, and/or with coverage that has a annual or lifetime cap on payments, may suffer considerable expense.

Dissemination of Research Findings. Current international guidelines for the ethical conduct of research specify that research findings should be made known to the professional and participant communities following conclusions of the investigations (Council of Organizations for Medical Sciences, 1991, 2002). It appears, however, that this does not occur in many cases and that research review committees rarely address this issue. In a study of IRB-approved consent forms relating to acute lymphoblastic leukemia studies, it was found that only 2 of 202 consent forms offered participants the option of receiving study results, 5 (2.8%) indicated that participants had a right to receive a summary of the research results, and 10 (5.5%) included unambiguous language indicating that new information would be provided to participants after the close of the study (Fernandez, Kodish, Taweel, Shurin, and Weijer, 2003).

In another study of the determinations of one ethics review committee in Spain, it was found that the committee approved 158 of the 166 protocols it reviewed for clinical trials (Pich, Carné, Arnaiz, Gómez, Trilla, and Rodés, 2003). The recruitment rate was lower than anticipated in 45% of the trials, and only 64% of the trials were completed in accordance with the original protocol. The results of only 31% of the studies were published in peer-reviewed journals; the findings of 27% of the studies were presented at scientific conferences. The authors of the study admonished that

public dissemination of clinical-research results is an important ethical requirement, and . . . RECs [research ethics committees] are in a privileged position, along with institutions, research funders, editors, and consumers, to ensure it (Pich, Carné, Arnaiz, Gómez, Trilla, and Rodés, 2003: 1016).

New Directions in Working with IRBs.

1) *Law and Policy.* Investigators, IRBs, and research participants alike are increasingly confronted with ethical dilemmas resulting from changes in law and policy that impact the conduct of HIV research. As an example, during the early years of the HIV/AIDS epidemic and HIV-related research, research participants could choose to be told of their HIV serostatus. Later, the NIH adopted a policy mandating the disclosure of HIV test results to research participants. As a result of this shift, individuals enrolled in a study could have declined notification of their test results, only to find that their decision was later superseded by an NIH policy directive. IRBs and investigators were confronted with a situation in which the voluntary informed consent of participants declining disclosure was open to question, with a consequent shift in the risks and benefits of study participation.

Similar situations have arisen due to changes in states' criminal laws mandating the disclosure of one's HIV serostatus to sexual and needle-sharing partners. As an example, an individual may have decided to participate in HIV-related research believing that the benefits of study participation, such as early knowledge of HIV status in order to obtain appropriate medical treatment, may have outweighed the attendant risks of participation, such as stigmatization and a possible loss of confidentiality due to state health department reporting requirements. However, the balance of risks and benefits may have shifted significantly for the individual during the course of the study as a result of a change in state law criminalizing the failure to disclose one's serostatus to specified partners.

In recent years, there has been a noticeable increase in the initiation of lawsuits against researchers, their institutions, and the IRBs that have reviewed the protocols that are the focus of the legal actions (Mello, Studdert, and Brennan, 2003). It has been asserted that this increase in litigation is associated with both the increase in private industry funding of research and apparent investigator conflicts of interest, which may be tied, in part, to the industry funding (Icenogle and Dudek, 2003; Kiskaddon, 2005; Lo, Wolf, and Berkeley, 2000). However, lawsuits have also been initiated against IRBs for failure to comply with the governing federal regulations relating to review procedures (Robertson v. McGee, 2001). It has been suggested that, depending upon the specific situation, lawsuits could be initiated against investigators, their institutions, and/or the ethics review committee on the basis of strict product liability, fraud and misrepresentation, intentional and negligent infliction of emotional distress, battery, lack of informed consent, violations of civil rights, violations of state consumer protection statutes, and violations of the Health Insurance Portability and Accountability Act (HIPAA) (Icenogle and Dudek, 2003;

Price and Lemons, 2002). IRBs' fear of litigation may underlie the frequently-voiced complaint of investigators that the committees often act as a police force rather than a protector of the rights of research participants (Christakis, 1988).

The case of *Grimes v. Kennedy Krieger Institute, Inc.* (2001), although not HIV-specific, is relevant here. The court in that case described the research as follows:

Johns Hopkins University . . . created a nontherapeutic research program whereby it required certain classes of homes to have only partial lead paint abatement modifications performed, and in at least some instances . . . arranged for the landlords to receive public funding by way of grants or loans to aid in the modifications. The research institute then encouraged, and in at least one of the cases . . . required, the landlords to rent the premises to families with young children . . . [T]he children and their parents . . . were from a lower economic strata and were, at least in one case, minorities.

The purpose of the research was to determine how effective varying degrees of lead abatement procedures were. . . . [I]t was anticipated that the children, who were the human subjects in the program, would, or at least might, accumulate lead in their blood from the dust, thus helping the researchers to determine the extent to which the various partial abatement methods worked (*Grimes v. Kennedy Krieger Institute, Inc.*, 2001: 811–813)

In reviewing the case, the court specifically found that the IRB had abdicated its responsibility to protect the participants and had, instead, assisted the researchers in avoiding their obligations under federal regulations. The court also found that (1) the informed consent of the parents to have their children participate in the study was invalid because the researchers had not provided the parents with full information; (2) the signed consent form included express representations of both the research institute and the parents and, consequently, created a bilateral contract whose terms the researchers were bound by; and (3) there may exist a special relationship between researchers and their research participants, giving rise to special duties, a breach of which may constitute the basis for negligence. Importantly, the court implicitly reprimanded the IRB involved in this research in stating:

When it comes to children involved in nontherapeutic research, with the potential for health risks to the subject children in Maryland, we will not defer to science to be the sole determinant of the ethicality or legality of such experiments (*Grimes v. Kennedy Krieger Institute, Inc.*, 2001: 855).

This finding seems to suggest that courts may be increasingly willing to review not only the elements of the consent to determine its validity, but also the context in which that consent is sought and obtained. Researchers and their institutions may be required to justify in a legal context their selection of research participants and the methods that they employ to recruit and retain them.

2) *Diverse Cultures*. IRBs are increasingly being asked to review research protocols pertaining to studies that are to be conducted in groups that are outside of their members' expertise and experience. As an example, HIV vaccine trials to be conducted in sub-Saharan Africa or in Southeast Asia involve populations and community norms with which the IRB members may have little familiarity. However, they are charged with the responsibility of reviewing the proposed research in sufficient detail and with sufficient knowledge to assess the adequacy

of the informed consent process, the risks and benefits to potential participants, and the acceptability of the mechanisms that are proposed for the protection of the research participants.

It has become increasingly common in such circumstances for IRBs to rely on outside experts as consultants. This is particularly true in situations involving review of non-English informed consent documents, in which a translator may be necessary to ensure that the non-English and English documents reflect the same meaning.

3) *Access to Study Participation*. IRBs have been challenged for their denial of access to the participation in research by individuals who may not obtain a direct benefit for themselves (Kiskaddon, 2005). The denial of access in such circumstances may stem from the need to protect vulnerable persons in the context of research and IRBs' fear of litigation. However, the

principle of justice, as well as the principle of respect for persons, require promoting high-priority research for communities of people, regardless of the prospect of direct, individual benefit. Being "left out" of the progress toward the development of safe and effective treatments does not respect that class of persons, nor does it serve the principle of justice (Kiskaddon, 2005: 931).

This would suggest that in balancing protection and access, IRBs must consider both the possibility of direct benefit to the individual research participant to the class of persons as a whole. Because IRBs appear to be more focused on an assessment of individual risks and benefits, particularly in the context of clinical research (cf. Kiskaddon, 2005; Sansone, McDonald, Hanley, Sellborn, and Gaither, 2004), both researchers and communities will be challenged to focus IRBs' attention, as well, on the benefit to be derived by relevant communities.

Data Safety Monitoring Boards (DSMBs)

Ethics review boards clearly serve an important role research oversight. However, the increasing complexity of research design, such as multicenter randomized clinical trials, and the ethical issues that accompany their conduct, including, for instance, issues relating to the existence of clinical equipoise, risk-benefit analyses, the inclusion of vulnerable participants, and the possibility that one or more arms of a trial may experience greater benefit or risk during the course of the trial, have necessitated the development and establishment of an alternative, more consistent mechanism to address such issues. Accordingly, DSMBs were developed in order to monitor on an ongoing basis the data collected during the course of a study (Gordon, Sugarman, and Kass, 1998).

In order to provide adequate monitoring, the DSMB must include experts in all of the disciplines needed to ensure participant safety including, if relevant to the study, clinical trials experts, biostatisticians, bioethicists, and clinicians who are knowledgeable about the disease and the treatment or intervention that are the focus of the study. The DSMB generally meets in an open session with the investigators and subsequently in a closed session during which the members review the emerging data. The members must

- Evaluate the progress of the trial, including data quality, recruitment, accrual and retention, participant risks and benefits, performance at the various trial sites, and scientific or therapeutic developments that could affect the participants' safety or the study's ethicality;
- Make recommendations to the investigators, the IRB, and/or the institution regarding the need to continue with or terminate one or more arms of the study, or the entire study; and
- Protect the confidentiality of the data and the results of the monitoring.

Examples of the situations that a DSMB might encounter include the following, adapted from those provided by NIH in its 1998 policy memorandum.

Phase I: A phase I trial of a new drug or agent often involves relatively high risk to a small number of participants. The investigator and occasionally others may have the only relevant knowledge regarding the treatment because these are the first human uses. The study investigator may be required to perform continuous monitoring of participant safety and report frequently to a designated individual or entity having oversight responsibility.

Phase II: Phase II trials follow Phase I trials. As a result, there is often more information regarding risks, benefits and monitoring procedures. However, more participants are involved and the toxicity and outcomes are confounded by disease process. The level of monitoring that will be required may be similar to that of a Phase I trial or the Phase I level of monitoring may be supplemented with individuals with expertise relevant to the study who might assist in interpreting the data in order to ensure patient safety.

Phase III: A Phase III trial is often designed to compare a new treatment to a standard treatment or to no treatment (placebo). Participants may be randomized to a particular arm (experimental treatment, standard treatment, or placebo) and the data may be masked. These studies usually involve a large number of participants who are followed for longer periods of treatment exposure. There may be long term effects resulting from longer exposure to the study agent or there may be significant safety or efficacy differences between the control and study groups for a masked study. A DSMB may perform monitoring functions to regularly assess the trial and offer recommendations concerning its continuation.

DSMBs have the authority and, indeed, the responsibility, to stop a trial or recommend the cessation of a study if it finds that one group is either benefiting significantly more than the other group(s) receiving alternate treatments or experiencing adverse effects at a significantly greater rate than the other groups. As an example, consider the study known as Syntex 1654, which was designed to evaluate the efficacy of oral ganciclovir in preventing HIV-related cytomegalovirus compared to placebo. This investigation was terminated following the finding of the DSMB that individuals receiving the oral ganciclovir had a significant clinical advantage compared to those being administered placebo (Hillman and Louis, 2003). In contrast, however, a similar study comparing oral ganciclovir and placebo that was being conducted concurrently, known as CPCRA CMV, was not terminated early because the DSM monitoring this study did not find that there was a significant benefit to those receiving the drug compared to those receiving placebo. Instead, study participants were presented with three alternative courses of action. They could (1) continue taking their assigned blinded study drug (oral ganciclovir or placebo); (2) they could stop taking their assigned blinded study drug (oral ganciclovir or placebo) and receive open-label oral ganciclovir; or (3) they could stop taking their assigned blinded study drug (oral ganciclovir or placebo) and not take open-label ganciclovir. In all options, participants were advised that they could continue with their study visits in the CPCRA CMV study if they wished to do so.

A review of randomized clinical trials that were stopped early because of results favoring the intervention found that 17, or 12% of the 143 such trials reported in the scientific literature between 1975 and 2004, were trials related to the treatment of HIV/AIDS (Montori et al., 2005). In almost three-quarters of the 143 trials, the decision to terminate the trial was made by the executive committee of the study, based on a recommendation from the relevant DSMB. In another 6%, the DSMB itself made the decision to stop the trial (Montori et al., 2005).

Studies may be stopped for other reasons, as well. On March 11, 2005, Family Health International (FHI) stopped the Nigerian arm of the tenovir PREP trial, in consultation with its DSMB (Singh and Mills, 2005). FHI found that the study team at that site was not able to comply with the required operational and laboratory procedures at the level necessary to conduct the study, which was critical to ensure the participants' safety and data quality.

DSMBs in the United States

In June 1998, the National Institutes of Health (NIH) issued a policy on data and safety monitoring requiring the oversight and monitoring of all clinical trials (including Phase I physiologic, toxicity, and dose-finding studies; Phase II efficacy studies; and Phase III comparative trials) in order to ensure the safety of participants and the validity and integrity of the data. The policy specified that the monitoring was to be commensurate with both the risks and the size and complexity of the trials. This policy, however, is in addition to, not instead of, the requirements for data safety monitoring that may be required by institutional review boards, the Food and Drug Administration (FDA), and any special NIH guidelines, such as the Guidelines for Research Involving Recombinant DNA Molecules.

Additional NIH guidelines issued in 2000 require that investigators seeking to implement Phase I or Phase II clinical trials submit a general description of the data and safety monitoring plan as part of the research application, subject to review as part of the NIH review process. In addition, these guidelines require the inclusion of a detailed monitoring plan as part of the protocol, the submission of the plan to the local IRB, and review and approval of the plan by the funding Institute and Center (IC) before the trial begins. The 2000 policy further requires that all monitoring plans include a description of the reporting mechanisms of adverse events to the IRB, the FDA and the NIH.

The policy statement recommended that, for multisite Phase I and II trials, investigators organize a central reporting entity responsible for the preparation of summary reports of adverse events for distribution among sites and the IRBs. It was further suggested that grantee institutions with a large number of clinical trials develop standard monitoring plans for Phase I and II trials, which investigators could then include in their submissions to the NIH, with the added caveat that these plans must be tailored to be appropriate to the specific investigation (National Institutes of Health, 2000).

Community Advisory Boards

The Structure and Function of CABs

Community advisory boards have often been employed in the context of community-based participatory research, also known as community-centered research (Cox et al., 1998; Israel et al., 1998). This approach assumes that research is to be conducted as a partnership between the researchers and the community, with active community input and engagement in all aspects of the research process (Melton et al., 1988).

On a federal level, recognition of the importance of community advisory boards came about as the result of significant conflicts between AIDS researchers and community groups during the early years of the AIDS epidemic (Shilts, 1987). These conflicts culminated in a demand by activists for the right to be included in the discussions relating to the development and conduct of clinical trials (Phillips, 1995), and the issuance by the National Institute of Allergy and Infectious Diseases of a policy recommending that all grantees establish a CAB in connection with HIV-related clinical trials (Spiers, 1991a, b, c). By 1996, local CABs were required at each site of the NIH AIDS Clinical Trials Group (ACTG), the multicenter research network established in 1987 for the development and evaluation of HIV/AIDS treatments (Siskind, 2004).

Two different models for HIV research-related CABs have been identified. The "broad community" model, which is frequently observed in Thailand and Zimbabwe, takes a long-term view of its role and, consequently, focuses its attention on such issues as sustainability, independence in funding and accountability, and the promotion of community-initiated research (Morin, Maiorana, Koester, Sheon, and Richards, 2003). Membership consists of individuals from a broad spectrum of the community, including religious and political leaders, educators, and representatives from nongovernmental organizations (nonprofit organizations). In contrast, the "population-specific" model tends to be concerned with a specific research protocol and with the needs of specific groups at increased risk of HIV infection in the context of that protocol, such as injection drug users. Representatives on such CABs are often drawn from the population/community participating in the research (Morin, Maiorana, Koester, Sheon, and Richards, 2003). CABs of both models exist at the local and national levels.

CABs have been developed in conjunction with both clinical trials for the treatment of HIV in infected persons and with HIV prevention trials. CABs developed for clinical trials have frequently included HIV-infected persons, while those focusing on the prevention of transmission to uninfected individuals often reflect the interests of various communities that may be at increased risk for HIV infection (Morin, Maiorana, Koester, Sheon, and Richards, 2003). CABs are frequently relied upon to provide guidance in the development of the informed consent process, the design of research protocols, and the design of recruitment and enrollment procedures; and to serve as a bridge between research participants and the research team (Loue and Méndez, 2005; Morin, Maiorana, Koester, Sheon, and Richards, 2003; Strauss et al., 2001). CAB members often confront and discuss ethical issues in the context of these roles, including the provision of care and treatment to those screened for trial participation, the protection of vulnerable populations, and the value of the proposed research to the host community (Morin, Maiorana, Koester, Sheon, and Richards, 2003).

The recruitment and retention of CAB members can be challenging due to differences in language, educational and literacy levels, and experience in the larger community, as well as difficulties in identifying representatives from the community (Siskind, 2004). Additional barriers to member participation and retention include the use of technical terms by researchers; limitations of time and funds for child care and travel (Silver et al., 1996); and increasing severity of illness of HIV-seropositive members.

The existence of a mission statement has been associated with better attendance and more active participation of CAB members (Chovnick, 2005; Cox et al., 1998). Individual motivations for participation on a CAB vary, but include a commitment to fighting HIV/AIDS; a sense of legitimacy; the opportunity to contribute something meaningful to the community; and material rewards such as reimbursements, lunches, and stipends (Morin, Maiorana, Koester, Sheon, and Richards, 2003). These material rewards have, however, been somewhat controversial. Researchers have reported that the provision of such benefits to CAB members in a rural area of western Kenya prompted concerns regarding the objectivity of the CAB members because the rewards were provided in the context of a high local level of unemployment and poverty (Odhiambo et al., 2004).

Ethical Issues in the Formation of CABs

Numerous ethical issues may arise during the course of forming and maintaining CABs. These include (1) the selection of community representative: who has the right to speak for a "community" that does not exist; (2) conflicts resulting from the multiple roles of providers as advisory board members, research interviewees, and providers to study participants; and (3) conflicting priorities of community members and researchers (Loue and Méndez, 2005).

Selecting Community Representatives. The identification of individual(s) who are to be empowered as members of the CAB to speak for the "community" raises significant ethical issues. These issues are best demonstrated by way of an example.

Suppose that an investigator wishes to develop an HIV prevention intervention for members of a specific ethnic community. The selection of an individual who is out of touch with that portion of the community that is the focus of the research may not be attuned to the sensitivities of prospective participants so that, inadvertently, prospective participants could be harmed by a lack of attention to their concerns. Conversely, overemphasis on the community's sensitivities could result in the termination of the research, thereby depriving individuals of not only the burdens of the research, but its benefits as well. In both instances, the principle of respect for persons is violated because individuals are either not adequately protected from harm, or are not afforded an opportunity to decide for themselves whether or not to participate. In the latter instance, the principle of justice, which seeks an equitable distribution of the benefits and burdens of research, may also be violated.

How community is defined in the context of specific research is therefore critical, particularly because diverse groups may define "community" quite differently. In a study designed to assess whether community was defined in similar ways by diverse groups, the investigators asked African Americans in Durham, North Carolina, gay men in San Francisco, California, injection drug users in Philadelphia, Pennsylvania, and HIV vaccine researchers across the United States what the word "community" meant to them (MacQueen et al., 2001). In identifying common themes across the responses, the authors found a common definition of community: a group of people who share diverse characteristics and who are linked by social ties, share common perspectives, and engage in joint action in geographical settings or locations. However, they also noted that in some ways these groups experienced community differently. For instance, for the gay men in San Francisco, a shared sense of history and perspective constituted the dominant theme, followed by a sense of identity with the location. However, for the African Americans in Durham and the injection drug users in Philadelphia, locus was the principal element of community, followed by joint action and social ties (MacQueen et al., 2001).

Multiple Roles of CAB Members. CABs often include representatives from community-based organizations, social service organizations, and consumer groups who are selected for membership on CABs specifically because of their connections to and relationships with the communities that are the focus of the HIV research. However, on a practical level, this means that some of the CAB members may have professional relationships with individuals who are participants in particular studies that are the focus of the CAB's attention. Ethical issues may arise where the CAB member wishes to access for their use in a clinical context data collected from a particular participant in the research context. Although the intentions of the CAB member may be to benefit the participant, a disclosure of research data could violate participant trust in the investigation and the investigative team; damage relationships between members of the CAB, the study team, the participants, and the larger community; and violate the terms of the protocol and the human subjects protections that have been established.

Various strategies can be utilized to reduce the likelihood of such requests and/or to address such requests when they are made. All CAB members can be advised at the commencement of their participation that no disclosures of research data will be made, other than in the aggregate. Participants can be given the option to have specific information conveyed by the research team to named providers with a signed, written request/release of information from the participant.

Conflicting Community-Researcher Priorities. Members of a CAB may be particularly sensitive to the needs of their community members and, because of this perspective, may have goals that differ from those of the research team. For instance, researchers may feel that additional data are needed to understand how to best design and test an HIV prevention intervention. Community members may see that many individuals are ill from HIV, that access to care is limited, and that a growing number of individuals in their community are becoming infected; in other words, the need is *now.* There may be a consequent impetus to implement a program without the benefit of conducting the formative research necessary to culturally and linguistically tailor its elements to the specific community that is to receive the program. The resolution of such disagreements often requires negotiation between the research team and the CAB and other members of the community of interest in order to address everyone's concerns and fashion an approach that is acceptable to the majority.

References

Burman, W., Breese, P., Weis, S., Bock, N., Bernardo, J., Vernon, A., & Tuberculosis Trials Consortium. (2003). The effects of local review on informed consent documents from a multicenter clinical trials consortium. *Controlled Clinical Trials, 24,* 245–255.

Cho, S-Y. (2003). Stages of institutional review board activities. *Journal of Korean Medical Science, 18,* 1–2.

Chokevivat, V. (1998). The current status of clinical trials in Thailand. *Drug Information Journal, 32,* 1235S-1241S.

Chovnick, G. (2005). A plan for maintaining a successful community advisory board. Presented at the National HIV Prevention Conference, June 12–15, Atlanta, Georgia [abstract no. TP-104].

Christakis, N.A. (1988). Should IRBs monitor researcher more strictly? *IRB: A Review of Human Subjects Research, 10(2),* 8–10.

Christian, M.C., Goldberg, J.L., Killen, J., Abrams, J.S., McCabe, M.S., Mauer, J.K., & Wittes, R.E. (2002). A central institutional review board for multi-institutional trials. *New England Journal of Medicine, 346(18),* 1405–1408.

Code of Federal Regulations. (2006). Title 45, part 46.

Council of Organizations for Medical Sciences. (2002). *International Ethical Guidelines for the Conduct of Biomedical Research Involving Human Beings.* Geneva, Switzerland: CIOMS.

Council of Organizations for Medical Sciences. (1991). *International Guidelines for Ethical Review of Epidemiological Studies.* Geneva, Switzerland: CIOMS.

Cox, L.E., Rouff, J.R., Svendsen, K.H., Markowitz, M., & Abrams, D.I. (1998). Community advisory boards: Their role in AIDS clinical trials. *Health & Social Work, 23,* 290–297.

Dziak, K., Anderson, R., Sevick, M.A., Weisman, C.A., Levine, D.W., & Scholle, S.H. (2005). Variations among institutional review board reviews in a multisite health services research study. *Health Services Research, 40(1),* 279–290.

Fernandez, C.V., Kodish, E., Taweel, S., Shurin, S., & Weijer, C. (2003). Disclosure of the right of research participants to receive research results: An analysis of consent forms in the Children's Oncology Group, *Cancer, 97(11),* 2904–2909.

Gordon, V.M., Sugarman, J., & Kass, M. (1998). Toward a more comprehensive approach to protecting human subjects. *IRB: A Review of Human Subjects Research, 20(1),* 1–5.

Grimes v. Kennedy Krieger Institute, Inc. (2001). 782 A.2d 807.

Gunsalus, C.K., Bruner, E.M., Burbules, N.C., Dash, L., Finkin, M., Goldberg, J.P., Greenough, W.T., Miller, G.A., & Pratt, M.G. (2006). Mission creep in the IRB world. *Science, 312,* 1441.

Hillman, D.W. & Louis, T.A. (2003). DSMB case study: Decision making when a similar clinical trial is stopped early. *Controlled Clinical Trials, 24,* 85–91.

Humphreys, K., Trafton, J., & Wagner, T.H. (2003). The cost of institutional review board procedures in multicenter observational research. *Annals of Internal Medicine, 139(1),* 77.

Icenogle, D.L. & Dudek, W.H. (2003). IRBs, conflict and liability: Will we see IRBs in court? Or is it when? *Clinical Medicine & Research, 1(1),* 63–68.

Israel, B.A., Schulz, A.J., Parker, E.A., & Becker, A.B.. (1998). Review of community-based research: Assessing partnership approaches to improve public health. *Annual Review of Public Health, 19,* 173–202.

Kimmel, A.J. (1991). Predictable biases in the ethical decision making of American psychologists. *American Psychologist, 46,* 786–788.

Kiskaddon, S.H. (2005). Balancing access to participation in research and protection from risks: Applying the principle of justice. *Journal of Nutrition, 135,* 929–932.

Lo, B., Wolf, L.E., & Berkeley, A. (2000). Conflict of interest policies for investigators in clinical trials. *New England Journal of Medicine, 343,* 1616–1620.

Loue, S. & Méndez, N. (2005). A community-based participatory approach to the development of HIV prevention for severely mentally ill Latinas. Conference of the International Society for Urban Health, Toronto, Canada, October 26–28.

MacQueen, K.M., McLellan, E., Metzger, D.S., Kegeles, S., Strauss, R.P., Svotti, R., Blanchard, L., & Trotter II, R.T. (2001). What is community? An evidence-based definition for participatory public health. *American Journal of Public Health, 91,* 1929–1938.

Mammel, K.A., & Kaplan, D.W. (1995). Research consent by adolescent minors and institutional review boards. *Journal of Adolescent Health, 17,* 323–330.

Mello, M.M., Studdert, D.M., & Brennan, T.A. (2003). The rise of litigation in human subjects research. *Annals of Internal Medicine, 139,* 40–45.

Melton, G.B., Levine, R.J., Koocher, G.P., Rosenthal, R., & Thompson, W.C. (1988). Community consultation in socially sensitive research: Lessons from clinical trials of treatments for AIDS. *American Psychologist, 43,* 573–581.

Montori, V.M., Devereaux, P.J., Adhikari, N.K.J. et al. (2005). Randomized trials stopped early for benefit. *Journal of the American Medical Association, 294(17),* 2203–2209.

Morin, S.F., Maiorana, A., Koester, K.A., Sheon, N.M., & Richards, T.A. (2003). Community consultation in HIV prevention research: A study of community advisory boards at 6 research sites. *Journal of Acquired Immune Deficiency Syndromes, 33,* 513–520.

National Health and Medical Research Council. (2005). Research Involving HIV/AIDS. NHRC Human Research Ethics Handbook. Australia: Author. Available at http://www7.health.gov.au/nhmrc/publications/hrecbook/02_ethics/18.htm. Last accessed January 27, 2006.

National Institutes of Health. (2000). Further Guidance on a Data and Safety Monitoring for Phase I and Phase II Trials (Notice OD-00–038), June 5. Bethesda, Maryland: Author.

National Institutes of Health. (1998). NOH Policy for Data and Safety Monitoring. Bethesda, Maryland: Author.

Nerenz, D.R., Stoltz, P.K., & Jordan, J. (2003). Quality improvement and the need for IRB review. *Quality Management in Health Care, 12(3),* 159–170.

Odhiambo, F.O., Amornkul, P.N., Vandenhoudt, H. et al. (2004). Experiences of a rural community advisory board with HIV research in western Kenya. Presented at the International Conference on AIDS, July 11–16 [abstract no. ThPeC7578].

Office of Extramural Research, National Institutes of Health. (1998). *Evaluation of NIH Implementation of Section 491 of the Public Health Service Act, Mandating a Program of Protection for Research Subjects.* Bethesda, Maryland: Author.

Paasche-Orlow, M.K., & Brancati, F.L. (2005). Assessment of medical school institutional review board policies regarding compensation of subjects for research-related injury. *American Journal of Medicine, 118,* 175–180.

Phillips, L. (1995). A seat at the table. *Trustee, 48,* 10–13.

Pich, J., Carné, X., Arnaiz, J-A., Gómez, B., Trilla, A., & Rodés, J. (2003). Role of a research ethics committee in follow-up and publication of results. *Lancet, 361,* 1015–1016.

Price, E. & Lemons, A. (2003). Clinical trials: Protecting the subject, avoiding liability, and managing risk. *Health Law Digest, 30(1).*

Randal, J. (2001). Examining IRBs: Are review boards fulfilling their duties? *Journal of the National Cancer Institute, 93,* 1440–1441.

Robertson v. McGee. (2001). No. 4:01CV60 (N.D. Okla., filed Jan. 29).

Sansone, R.A., McDonald, S., Hanley, P., Selbom, M., & Gaither, G.A. (2004). The stipulations of one institutional review board: A five year review. *Journal of Medical Ethics, 30,* 308–310.

Shilts, R. (1987). *And the Band Played On: Politics, People, and the AIDS Epidemic.* New York: St. Martin's Press.

Silver, S., Brodeur, S., Cunningham, J. et al. (1996). Evolution of a participant-directed Women's Interagency HIV Study community advisory board. Presented at the International Conference on AIDS, July 7–12. [abstract no. Tu.D.2763]. Available at http://gateway.nlm.nih.gov/robot_pages/MeetingAbstracts/102218827.html. Last accessed July 22, 2006.

Singh, J.A. & Mills, E.J. (2005). The abandoned trials of pre-exposure prophylaxis for HIV: What went wrong? *PLoS Medicine, 2(9),* e234.

Siskind, R. (2004). Models for community input [slides]. Bethesda, Maryland: Division of AIDS, National Institute of Allergy and Infectious Diseases.

Spiers, H.R. (1991a). Community consultation and AIDS clinical trials: Part I. *IRB: A Review of Human Subjects Research, 13,* 7–10.

Spiers, H.R. (1991b). Community consultation and AIDS clinical trials: Part II. *IRB: A Review of Human Subjects Research, 13,* 1–6.

Spiers, H.R. (1991c). Community consultation and AIDS clinical trials: Part III. *IRB: A Review of Human Subjects Research, 13,* 3–7.

Strauss, R.P., Sengupta, S., Quinn, S.C., Goeppinger, J., Spaulding, C., Kegeles, S.M., & Millett, G. (2001). The role of community advisory boards: Involving communities in the informed consent process. *American Journal of Public Health, 91,* 1938–1943.

United States Department of Health and Human Services, Office of the Inspector General. (1998). *Institutional Review Boards: A Time for Reform.* Washington, D.C.: Author.

Vick, C.C., Finan, K.R., Kiefe, C., Neumayer, L., & Hawn, M.T. (2005). Variation in institutional review processes for a multisite observational study. *American Journal of Surgery, 190,* 805–809.References

Woodsong, C. & Karim, Q.A. (2005). A model designed to enhance informed consent: Experiences from the HIV Prevention Trials Network. *American Journal of Public Health, 95(3),* 412–419.

World Health Organization. (2000). *Operational Guidelines for Ethics Committees That Review Biomedical Research.* Geneva, Switzerland: Author.

Chapter 5
Researcher-Participant Relations

Introduction

When researchers at the Centers for Disease Control and the National Institutes of Health first realized, after receiving initial reports of illness among gay and bisexual men in large U.S. urban areas, that they were dealing with identities and practices with which they were entirely unfamiliar, the challenges of surveillance and research were immediately and radically multiplied. Very few people in the surveillance and research communities knew much about gay male sexuality and behavior, except for those handful of physicians and scientists who were gay themselves. They were forced to learn very quickly—to develop models and styles of researcher-participant relationships or partnerships that would help them understand HIV pathogenesis, and how to manage it.

Three competing forces have been at work in HIV/AIDS-related researcher-participant relations since the beginning of the epidemic, and though those forces have evolved over time, they remain at work today.

The first is what might be called affiliation or alliance: LGBT and LGBT-friendly physicians and researchers were confronted not merely with a professional challenge, but a disease that was, for some, also simultaneously and rapidly killing friends and loved ones. In the early days of the epidemic in large U.S. LGBT communities, gay and bisexual men especially – whether they were scientists or factory workers – were caring for, and burying, friends in rapid succession, and that led to an overwhelming sense of dread and urgency.

The second force at work might be called "disinfection," and typified responses at higher administrative or bureaucratic levels of public and private research bodies. Those responses often effaced the particular identities, communities, and practices disproportionately associated with the unfolding epidemic by refusing to acknowledge gay sexuality and injection drug use as objective realities, and by demonstrating reluctance to conduct research on those objective realities. Practices such as anal intercourse or "booting" drugs while injecting might engender uneasy discussion and debate, but they did not engender, nearly often enough, scientific and culturally competent research that could investigate more fully their connection to HIV transmission.

And the third force, often noted as one of the most powerful and enduring features of the HIV/AIDS epidemic affecting nearly every aspect of personal and societal responses—a force that might be called self-enfranchisement—was the early and vigorous assertion of rights and identities by people with HIV/AIDS themselves.

The recognition of people with HIV/AIDS as simultaneous "self" and "other," and the parallel recognition that practices believed contained at the margins (for example, anal intercourse or illicit drug use) were closer to the center than anyone willingly admitted, have made the proximity between various actors in the epidemic—the sick, the caregivers, the social commentators, and the public at large—more fluid, and often closer, than most wanted to affirm, much less acknowledge. But these very proximities have also influenced the relationship between researchers and participants, sometimes blurring the boundaries, sometimes adding a dose of close-knit passion that has helped invigorate and accelerate research.

At least, that is, in the West. It has not been entirely different in the case of research conducted in international settings, for HIV/AIDS researchers, as a general rule, remain a passionate and personally-engaged bunch, for all kinds of reasons, and those personal investments have tended to carry over when research began moving out of strictly US and European laboratories, and into a wider range of field settings, especially in African states. But it is also true that the "proximate passion" has been no doubt tempered by the conduct of transnational research in which the researcher, and the participant, are increasingly more likely to be "other" than "we."

It is these two sets of realities—a history of close relationships between various actors, on the one hand, and the reality of more distant relationships in international research, which must be more carefully monitored for the possibility of ethical shortcomings, on the other hand—that are addressed in this chapter.

US Context and Research Shift

The early relationship between care and research providers and care and research consumers in the United States is best summarized and illustrated in an extended passage from a Keynote Address given by Larry Kramer to the annual meeting of the Association of Nurses in AIDS Care in 2003. Kramer, one of the co-founders of the AIDS Coalition to Unleash Power (ACT-UP), reflects on the early days of the epidemic, and the relationships between patients, doctors, nurses, and researchers. What emerges is an image of a tightly-knit family, in which conflict and disagreement is so vocally expressed precisely because of such a profound sense of togetherness against an external threat:

You remember the timetable of these past years, I'm sure. How government and science and medicine—but particularly government, particularly Ronald Reagan nationally and Ed Koch locally—did not out of compassion or vision or even the most minimum of humanitarian instincts lead the fight. Patients were forced to innovate and revolutionize in desperate attempts to stay alive, much less change the system. Then came next when it dawned on us very harshly that we were going to have to take matters into our own hands because no one else would help us. We had to teach ourselves everything we could about the hateful system that was totally ignoring us. We had to learn everything we could about the shitty government and the bureaucracies it spawns—the Food and Drug Administration (FDA), the NIH, the controlled clinical trials—how research was and more specifically was not done, and we had to learn how to threaten drug companies to stop dawdling and get to work (they did not believe there was any money in AIDS drugs) and to cough up some drugs or else. New drugs were literally yanked out of them: Scientists were literally backed up against the wall and, in a couple of cases, blackmailed into releasing what we knew they had or we would take the pills we had smuggled out of their labs and have them duplicated elsewhere ourselves. Drove them nuts. We got ddI. And we demanded, for the first time in research history, that the placebo control be eliminated on all trials on us. It is inhumane and stupid to expect dying patients to take a drug that might be a placebo.

"Make no mistake: All the drugs that are out there saving lives today are out there because of ACT UP and its supporters, out there because we got angry, and we tied up traffic, and had sit-ins in executives' offices, and chained ourselves together so that drug company trucks delivering their products couldn't leave their factories and warehouses. We threw fake blood; and we carried caskets with actual dead bodies in them and threw them on to the White House lawn; and we had a huge condom manufactured and completely covered Jesse Helms' house in Arlington with it; and we locked ourselves into the offices of another one of our hated enemies, what was then called Burroughs Wellcome, and we wouldn't come out, they had to blast us out, which cost them tens of thousands of dollars; and with forged fake IDs, we invaded and stopped trading on the floor of the New York Stock Exchange (which had never been done before) when Wellcome raised the price of AZT; and we invaded the offices of the FDA, and the NIH—oh did we not give holy shit and hell to the NIH more than once? We really shaped up the NIH, and Dr. Faucci, whom I early on labeled in print a murderer more than once and who is now one of my dearest, most loving friends, gives us credit for doing just that, changing the entire way science and medicine are now practiced (Kramer, 2003).

These are the reflections of an advocate who clearly claims full autonomy and equality when faced with the personal and community challenge of the epidemic, who expects to be treated as a co-equal participant in the common effort to solve the problems at hand, and who owns both a sense of personal responsibility, and the necessary intellectual and interactive skills, to take action. Kramer was and remains an actor not by virtue of rights assigned or recognized by others, but by virtue of an intense awareness of endogenous rights.

While there remain many skilled, articulate, self-asserting community advocates living with HIV/AIDS in the United States today, the horrible truth is that many have, since the 1980s, died, and others have exhausted their personal capacity for advocacy and sustained self-assertion. And as the locus of attention for the conduct of research has shifted from the North to the South, a new set of relationships has begun to emerge, one that has added additional challenges in the conduct of ethical research. As Anne-Christine D'Adesky (2004: 10) puts it, with discernable (but perhaps not inappropriate) cynicism, "With the new money pouring into the international AIDS arena, it seems like everyone is beginning to chase research and treatment dollars—be they scientists, public health officials, agencies, physician groups, non-profit organizations, community and activist groups. What matters most is how new programs will directly benefit those living with HIV and AIDS, and those very ill or dying in hospitals or bedrooms, desperately waiting for medicine."

The personal dimensions of provider-consumer-researcher relationships have not dissolved as a result of that shift:

I imagine . . . that nurses at Casino, Bestlands, and clinics like them in Johannesburg, Kampala, Lilongwe, Abidjan, and Lusaka continue to struggle first and foremost with the problematic of gender, actively and alternately interpreting, reinforcing, and challenging the definitions of heterosexuality, femininity, and masculinity that they receive from ministers of health, researchers, donors, and patients. If research is to be useful and to contribute to a productive reframing of knowledge about HIV, one strategy might be to assist health care workers in evaluating the validity of, and then building upon, their own perceptions of women's opportunities to change heterosexuality (Booth, 2004: 144).

But clearly those personal relationships have *changed* in profound ways, in part because the conduct of HIV/AIDS research in international settings has tended to place considerable wealth and privilege, and abject poverty and powerlessness, side by side—not only because of the glaring inequalities between citizens of the North and citizens of the South, but because of stark inequalities and injustices within developing countries as well.

The historical legacies of exploitation cannot be forgotten; they are alive yet today, and will influence the construction of responsible ethics, especially since the urgency of the epidemic, its human toll, continues to put pressure on research to develop rapid solutions to seemingly intractable virological challenges:

Exploitation by industrialized countries of the human and natural resources of the developing world has a long and tragic history. It has never been difficult for economically wealthy countries to justify their acts by citing, for example, the supposed genetic or moral inferiority of those exploited. Substituting economic inferiority," [or intellectual or cultural inferiority] "in these olds arguments makes the enterprise no less offensive (Kim, 1998: 838).

And the postmodern critique of medical power, in which medicine itself has been too often deployed as simply another manifestation of colonial or imperial power in relations between countries or peoples, is particularly relevant, and particularly reenergized by still-corporeal memories and historical associations, in the case of research sponsored by the North and conducted in the South (Baker, 1998).

Empowerment

Over-employed, at times to the point of irrelevance, the concept of empowerment is nevertheless an instructive starting point in an attempt to construct a meaningful ethics of researcher-participant relations. For it is the first obligation: that the researcher not engage in processes that are ultimately disempowering to the participants; by extension, and to redress some of the conditional inequalities that are more than likely to characterize that relationship from the outset, those processes should work to expand and increase the individual's sense of power, agency, and autonomy as well. Baylies (2004: 179) defines the empowering process as

one whereby an individual or group gains greater control over the uncertainties of their environment (be they social, psychological, physical, economic, or whatever) and is able to act on it to greater advantage. But as the environment changes and the actors, resources, and relations change within it as well, the extent of control may change, as may evaluation of what constitutes greater advantage. Thus empowerment is perhaps best understood as a process of applying in respect of a specific context, as contingent upon time, place, and sphere of action or thought and as relational in respect of the roles, capacities, and resources which may be brought to bear in any particular instance. It is typically partial and, to some degree, ephemeral by virtue of its contingent nature, not necessarily expanding or increasing over time but both variable and sporadic. People do not become empowered once and for all.

If research participants truly feel, and actually are, empowered as a result of their enrollment in research, then several realities are likely to prevail. They will certainly feel respected, as undue and complex individuals, not merely for their unidimensional role as research subjects. Their input and critical insights, in the design and implementation of research, and about the social, economic, political, and cultural realities in which HIV/AIDS and research takes place, will be viewed as co-equivalent in value with those of the researcher. They will feel valued as a source of knowledge, even though that knowledge may be founded on differently-evolved discourses than those of the researchers. They will feel positively appreciated for their role and contributions as participant-ethicists; that is, as individuals who are not merely static placeholders in the creation of ethical formulae, but as individuals who help devise, negotiate, and critique the ethical rules and protocols by which the research is governed. And, of course, they will be accorded, as a natal prerogative, the full range of human rights available to (at least some) in the North—including the right of personal objection, and personal freedom of association with others to mobilize community activism, even if that criticism is directed toward research sponsors or research institutions themselves. In other words, they will understand their liberty, as does Larry Kramer, to receive medical care, to participate in medical research, and then, and even perhaps on the very same day or in the next sentence, to vigorously protest any feature of the care or research he finds objectionable—without risking continuing participation in care or research.

Indeed, what is unusual about HIV/AIDS is the extraordinary degree to which activists and the ill and dying have informed the contours of evolving bioethical understandings; there is now much more agreement than there has been in a long time that researchers cannot rely solely on self-policing to correct ethical lapses or violations, and recognition that research necessarily takes place in, and is best served ethically by taking place within, a public sphere characterized, often, by a wide array of diverse and contentious voices, including those of research participants.

Edejer (1999: 440) puts it more succinctly, framing it more specifically to the research study context:

Increasingly, there is an awareness that the success of North-South research collaboration should be judged not solely on the results of scientific research activities. This awareness must be coupled with a learning approach to create a sustainable, mutually beneficial working relationship, that aside from advancing science must address inequity and put local proprieties first, develop capacity with a long term perspective and preserve the dignity of local people by ensuring that the benefits of research will truly uplift their stasis.

General Researcher Obligations

It is easy enough to suggest that research participants ought to feel empowered, and that the relationship between the researcher and the research participant ought to address multiple levels of interaction and the associated realities that inform interactions, but what, then, is the researcher obliged to do? First, she or he must recognize that the professional relationship is also a social relationship, overlain and undergirded by multiple meanings. For Kok-Chor Tan (1997), the economic, social, and political interdependencies of the global community draw nearly everyone into social arrangements with each other. For Deborah Zion (2004), this implies that individuals who have benefited from HIV/AIDS research have a duty to redress the resulting imbalances created by this involvement.

The researcher may wish, in order to maintain an artificial construct of self and other, to view himself or herself as operating from a coolly analytical position that is "professional" in its most dispassionate meaning, but no ethical or research objective is served by such dehumanization, and the research participant is likely, regardless, to view the researcher more multi-dimensionally—at least, and perhaps somewhat abstractly, as someone who occupies a particular position within symbolic and real hierarchies of power and privilege. The relationship is, intrinsically, social and multi-dimensional (and quite probably fluid as well), and must be recognized as such.

Neither is the researcher free, in his or her relationships with participants, to accept or reject specific cultural elements based on whether they will help or hinder the research, or, for that matter, the way in which they do or do not correspond with the researcher's own values about a wide range of issues, including gender, sexuality, and individual autonomy versus group decision-making or expectations (Nairn, 1993).

And finally, as a general duty—and to elaborate on Zion's point—researchers have a positive obligation to comprehend, as fully as possible, the structural conditions and historic events that have helped determine the lives of the research participants with whom they work, who are often literally sitting in front of them across a desk or table, and to accept (1) that the researcher or researcher's society may have in fact benefited from the structural conditions or historic events that have caused harm to the research participant's community or person, and (2) to take responsibility for actively rectifying those conditions in order to correct the injustice or disparity. This point may strike to how research itself is conceived, and how the research question is developed; an acknowledgement of the structural or ecological dimensions of HIV/AIDS would argue forcefully for research that takes local and non-local ecologies into account (Parker, Easton, and Klein, 2000).

A fascinating study by Decosas (1996) provides powerful evidence of the need to recognize the role of larger historical, structural, and ecological factors in the conduct of HIV/AIDS research, and by implication, the degree to which those factors have shaped and conditioned the lives of individual research subjects. In an analysis of how the HIV/AIDS epidemic in Krobo, Ghana was facilitated by historic events, he was able to demonstrate how the Akosombo dam—a large project, constructed in the 1960s with the support of extensive international aid—led to new patterns of migratory work, agricultural disruption, and dismembered families that then contributed to a substantial increase in commercial sex work, and therefore growing HIV infection, decades later. Nothing happens, we are reminded, outside an evolving ecology of interactions and consequences, and the person in front of us, ready to sign a consent form, embodies and circulates much of that ecology.

Specific Obligations

Beyond general recommendations, researchers are obligated to a number of more specific actions with regard to research participants. Some of these are enumerated below.

1. The researcher has an ethical obligation to provide the participant with the highest possible standard of medical care (Wolf and Lo, 1999). There is yet considerable disagreement about which standard of care—that provided in the Western countries, or that more commonly expected in developing countries—should be provided, and still yet more disagreement about whether the standard of care should be presumed to include ancillary services that enhance medical outcomes, such as are available in the United States and elsewhere (Angell, 1997; Lurie and Wolfe, 1997; Varmus and Satcher, 1997).

   One way to consider this dilemma is to reflect on what might be termed a relative parity of investments. If the research process in one in which a number of actors invest—governments and industry with financial backing, researchers with time and intellectual capacity, participants with their bodies—then we should be able to expect a relative parity of investments, and a relative parity of projected benefits. It is the general hope of governments that research outcomes will maximize health for populations, and it is the general hope of industry that research will maximize profits (without treating it as a potentially distasteful word). Government and industry do not qualify those expectations based on the location of research; that is, they do not say, "we will settle for a diminished degree of maximized community health, or a diminished level of expected profit, if the research is conducted in a developing country." Should, then, the research participant be expected to concede to a diminished level of care—that is, a standard of care that is less than one would expect to find in the West—because the research is conducted in a developing country? This is a fundamental question that must be resolved at the very outset, for it will determine the level of resource investment required to guarantee responsible research.

2. The researcher is, of course, required to protect the confidentiality of the research participant, but more broadly, the researcher is required to support the participant in the personal process of disclosure and privacy that will serve the participant's best interests within the community and cultural contexts in which the participant lives. There is no question that some individuals may suffer serious harm in some settings, even death, if their HIV status is known to others; there is also no question that in many settings implied disclosure can result from the smallest inadvertent breach—say, using a particular door to enter a clinic, or possessing infant formula with no community-known way of being able to pay for it. Every possibility should be thoroughly considered by the researcher, who then has an obligation not merely to develop appropriate protective strategies, but to fully disclose possible breaches, and strategies designed to prevent them, with research participants.

3. The researcher should work to involve individual participants, to the highest degree possible, in all phases of research design, evaluation, implementation, and dissemination, and to create and utilize the educational and translational materials and processes that will make such involvement possible and meaningful. If, indeed, the conduct of research is a partnership based on a rough correspondence of investments, than the research participant has a right to co-participate, to the highest degree possible and desirable, in the design, implementation, and evaluation of research that will directly affect the participant; further, the participant has the right to disseminate, to other members of his or her community, research findings that could directly benefit the lives of those about who the research participant cares.

In the specific area of vaccine trials, a number of truly excellent tools have been developed to help inform members of the community about vaccines, vaccine trials, and ethical decisionmaking in relation to those trials; they help "give the power back" to potential research participants in terms of knowledge and access to the "language of the academy." (AIDS Vaccine Advocacy Coalition, 2005; Godwin and Csele, 2005; International Council of AIDS Service Organizations, 2006). Organizations such as the International Council of AIDS Service Organizations (ICASO) have developed useful materials, specifically on vaccine research, that explain the research process, potential points of influence and decisionmaking for community advocates, and the kinds of questions that advocates or watchdogs can pose about local trials. Unfortunately, the available material—and there is too little of it—would not be generally accessible to many learners. Such materials, in a variety of formats, local languages, and reading levels, need to be more widely available, in order to equip civil society groups and individual advocates with tools to understand and influence the management of local research (International Council of AIDS Service Organizations, 2002).

4. The research should attempt to involve participants in all phases of the design and monitoring of ethical protocols, and to develop and utilize the educational and translational materials and processes that will make such involvement possible and meaningful. The participant has the right to review and critique the foundational and operational ethics of the research, and the right to make substantive contributions to the trial's research ethics that fully address the participant's ethical, moral, and philosophical concerns.

5. The participant has an intrinsic right to full disclosure and transparency of relationships and interests, to the degree allowable by law and reasonable professional standards. The researcher, then, should disclose to participants the various financial, academic, political, and public health interests that have a stake in the research, so that participants can, of their own initiative, consider possible conflicts and agendas.

6. The researcher should actively solicit participant questions in a number of settings (clinical, community, focus groups, and other settings that are comfortable for the participant) and develop education materials based on key themes and questions that emerge from those discussions. This can result in the creation of an evolving "concordance" that can assist in the ongoing process of participant education.

7. The researcher should attempt to understand the individual circumstantial factors that will condition an individual's enrollment in and continuing involvement in research, and assist the participant in managing those factors. The oft-cited example of married women participating in research when husbands have ambiguous or negative feelings about such participation remains a good example. It may be a foreign value to posit that such women ought to simply assert their independence under conditions of inequality. An inadequate understanding of women's lives under local conditions may underestimate research risks to participants, may complicate and confuse issues related to enrollment and retention, and ultimately, undermine research sampling protocols themselves.

8. The researcher could develop, for research staff and research participant use, an open-ended lexicon of key terms and descriptions that regularly surface in researcher-participant conversations about the research; research staff can, in preparation, be trained to observe word usage and common points of confusion about terminology. This can help facilitate shared understandings of commonly-used terms, and avoid confusion or misinterpretation about meanings. Such a lexicon, in a comprehensible format, can be made available to all research participants.

9. The researcher should recognize that individual responses to research participation are conditioned not merely by historical legacies, but by strictly individual variations in literacy, comprehension, learning styles, mental health, disease status, and a host of other factors—and develop educational

materials and strategies, and participant retention support materials, that take those factors into account.

10. The researcher can solicit varied media life and health narratives from participants, encouraging individuals to orally, visually, artistically, or in written form to share their stories with researchers and, as appropriate, other research participants. The solicitation of narrative that is broader and deeper than standard clinical histories will enable researchers to contextualize disease and research involvement in the lives of research participants, and will assist participants in the process of voicing their realities, concerns, expectations, and hopes.

11. Conversely, researchers should utilize varied media tools to create their own narratives and share them with research participants. There are two ways in which this makes sense. First, mutual sharing of narratives recognizes that the research relationship is not merely abstractly "professional," but social as well, and such sociability is important to the construction of trust-based relationships between researchers and participants. And second, a relationship in which one party is expected to disclose private information, but the other party is somehow viewed as exempt, is likely to be seen as nonreciprocal, and inherently unequal.

12. Researchers should encourage research participants to "voice their complaints" not merely as individuals interacting with researchers, but as social and political actors participating in a larger social narrative about health and illness, power and powerlessness, agency and control, individuals rights and responsibilities in collective contexts, and justice. Encourage participants to voice their realities in other ways. Research participants may or may not be aware of advocacy initiatives such as participate

13. And finally, researchers should make every effort to involve host-country researcher in the design and monitoring of ethics protocols to govern the research. This would tend, all other things being equal, to add additional sensitivity to the ethics-related needs of research participants, since host-country researchers are more likely to be familiar with host country conditions and cultures sharing individual lives. However, relatively few scientists and practitioners from developing countries have participated on the larger dialogue related to ethics of clinical research (Thomas, 1998).

Researcher Obligation to Control Arm Participants

The African 076 trials discussed elsewhere in the volume raise critical questions about the obligation of researchers to control group participants in clinical trials. A growing consensus in favor of the "highest ethical standard" would suggest that placebo-controls are highly problematic in the case of HIV/AIDS research, and that control group participants should have equal access to medical care and monitoring and post-trial access to tested treatments, as those participants who are not in the control group. Minimally, the researcher's specific duties with regard to control group participants must be spelled out in the protocol, and thoroughly communicated to participants, so that full informed consent is achieved.

Researcher Obligations to Potential Participants who are Excluded

Too often overlooked is the researcher's obligation to individuals wishing to participate in a study, but who are excluded for categorically appropriate reasons (i.e., men who want to participate in a study related to treatment effects on women). In societies characterized by extreme health care resource limitations, it is entirely understandable that HIV + individuals would desire access to any available resource—including clinical trials, which almost always include care and monitoring during the course of the trial—that had the potential to extend individual quality and length of life. It is also entirely understandable that individuals who are excluded from trials by virtue of justifiable categorical exclusions may feel disappointed, frustrated, angry, and any number of other possible emotions.

The researcher has the obligation, in informing a potential participant of an exclusion decision, to fully inform the individual of the basis for that exclusion, and the options and alternatives available to the individual. The researcher also has the obligation to listen thoroughly to an individual's complaint about exclusion, and if reasonable, make adjustments to current research or future research protocols to address that complaint. Trials to test the efficacy of vaginal microbicides, for example, have been challenged by individuals and communities who have complained that an exclusive focus on intravaginal product development has effaced gay sexual identity and practices, including the prevalence of anal intercourse among men who have sex with men. This has led to the allocation of additional resources for the development of an effective rectal microbicide as a parallel research track along vaginal microbicide development.

At the least, the study should clearly spell out, as with control participants, the rights of those who are excluded from the study, and should outline how the research team will manage participant concerns and help the individual resolve the challenge of accessing care in resource-limited settings. Excluded potential participants deserve to be fully informed of these rights, and should fully understand their grievance options under such conditions. In addition—as outlined elsewhere—researchers continue to be positively bound by the duty to work toward intra-national and transnational universal access to health care resources, as defined based on the highest clinically-established standards of care now being promulgated.

References

AIDS Vaccine Advocacy Coalition. (2005). *Finding Your Way: A Guide to Understanding Ethical Issues Related to Participation in Clinical Trials for Preventive HIV Vaccines.* Last modified July, 2005; Last accessed January 1, 2007; Available at http://www.avac.org/pdf/reports/finding_way.pdf

Angell, M. (1997). The ethics of clinical research in the Third World. *New England Journal of Medicine, 337(12),* 847–849.

Baker, R. (1998). A theory of international bioethics: The negotiable and the non-negotiable. *Kennedy Institute of Ethics Journal, 8(3),* 233–273.

Baylies, C. (2004). HIV/AIDS and older women in Zambia. In Poku, N.K. & Whiteside, A. (Eds.). *Global Health and Governance: HIV/AIDS* (pp. 161–185). New York, New York: Palgrave MacMillan.

Booth, K.M. (2004). *Local Women, Global Science: Fighting AIDS in Kenya.* Indianapolis: Indiana University Press.

D'Adesky, C. (2004). *Moving Mountains: The Race to Treat Global AIDS.* London: Verso.

Decosas, J. (1996). HIV and development. *AIDS, 10(suppl. 3),* S69-S94.

Edejer, T.T.T. (1999). North-South research partnerships: The ethics of carrying out research in developing countries. *British Medical Journal, 319,* 438–441.

Godwin, J. & Csele, J. (2005). *Community Action Kit: HIV Vaccines and Human Rights.* Canadian HIV/AIDS Legal Network. Last accessed January 1, 2007 Available at http://www.aidslaw.ca/publications/interfaces/downloadFile.php?ref = 350

International Council of AIDS Service Organizations (ICASO). (2006). *Community Involvement in HIV Vaccine Research: Making it Work.* Last modified July, 2006; Last accessed January 1, 2007; Available at http://www.icaso.org

International Council of AIDS Service Organizations (ICASO). (2002). *Developing Vaccines for HIV and AIDS: Preparedness: An Introduction for Community Groups.* Last accessed January 1, 2007; Available at http://www.icaso.org

Kim, R. (1998). Letter to the Editor. *New England Journal of Medicine, 338(12),* 838.

Kramer, L. (2003). Angels on my shoulders. Remarks made to the Association of Nurses in AIDS Care, New York Hilton, November 20, 2003. Reprinted in *Journal of the Association of Nurses in AIDS Care, 15 (5),* 25–29.

Lurie, P. & Wolfe, S.M. (1997). Unethical trials of interventions to reduce perinatal transmission of the human immunodeficiency virus in developing countries. *New England Journal of Medicine, 337(12),* 853–856.

Nairn, T.A. (1993). The use of Zairian children in HIV vaccine experimentation: A cross-cultural study in medical ethics. *The Annals of the Society of Christian Ethics, 13,* 223–243.

Parker, R.G., Easton, D., & Klein, C.H. (2000). Structural barriers and facilitators in HIV prevention: A review of international research. *AIDS, 14(suppl. 1),* S22-S32.

Tan, K.C. (1997). Kantian ethics and social justice. *Social Theory and Practice, 23(1),* 53–73.

Thomas, J. (1998). Ethical challenges of HIV trials in developing countries. *Bioethics, 12(4),* 320–327.

Varmus, H. & Satcher, D. (1997). Ethical complexities of conducting research in developing countries. *New England Journal of Medicine, 337(12),* 1003–1005).

Wolf, L.E. & Lo, B. (1999). What about the ethics? *Culture and Medicine, 171,* 365–367.

Zion, D. (2004). HIV/AIDS clinical research, and the claims of beneficence, justice, and integrity. *Cambridge Quarterly of Healthcare Ethics, 13,* 404–413.

Chapter 6
Researcher-Community Relations

Introduction

This chapter explores the ethical dimensions of the relationship between HIV/AIDS researchers and participant communities. It is, by necessity, a broad discussion; a thorough exploration of the relationship between HIV/AIDS researchers and each of the specific communities now participating in HIV/AIDS research would require volumes, not a single chapter.

And "community," in this discussion, is certainly not limited to traditional definitions, which are often bounded by geography. "Community" herein implies any group of individuals, however physically close or distant, that self-recognizes and affirms common beliefs, values, identities, practices and behaviors, or aspirations. While, for example, there is extraordinary diversity among African Americans in United States, there is nevertheless a commonly-recognized identity as an "African American community" in many places, and those local African American communities often have histories and characteristics that are unique and describable. It is those "communities" that constitute the subject of this chapter.

A *New York Times* article in the summer of 2005 documented how multi-layered and contentious ethical debates can become. A report posted on the internet—without any documenting names, or supporting proof—contended that New York City child welfare officials were guilty of abuse in allowing HIV + children in the New York City foster care system to participate in clinical trials. Further, the internet posting charged overt racism, noting that most of the children participating in the clinical trials were African American or Latino. The accusations rapidly migrated from internet site to internet site, eventually making their way into a story in the *New York Post*, and then into a BBC documentary. Eventually, City Council hearings found themselves packed with angry community members and defensive child welfare officials—and the "truth" of the situation seemed increasingly elusive.

Subsequent investigations found no credence to the claim that foster care officials were guilty of abuse, or had inappropriately offered consent for participation of HIV + foster children in the study. Even further, reviewers later noted that the original internet article had been posted by an "independent journalist" who himself pre-subscribed to the notion that HIV was not the causative agent for AIDS, and that AIDS "treatment" was therefore inherently injurious (a belief that many have termed "HIV denialism"). But it was too late: the damage had been done, and the relationship between "the community"—itself a fluid and dynamic configuration composed of sometimes overlapping and sometimes separate entities, such as racial communities, foster parents, and the alternative press—and foster care officials was, seemingly, damaged beyond repair. The long-term outcome, though not tested, is predictable: the next researcher who comes along, and who wants to conduct research using HIV + foster children as participants, will, even if the research meets the highest ethical standards, and is being proposed entirely based on the interests of the children and communities involved, be met with at least suspicion, if not outright hostility (Scott and Kaufman, 2005).

This example is not provided in order to point the finger at one or the other party as the ultimate culprit in the controversy, but to illustrate how easily controversy can erupt. Additionally, it underscores the degree to which "community" may include a wide spectrum of individuals and interests whose actions lie beyond anyone's control, two factors which, together, highlight both the complexity and necessity of researcher-community relations. Benatar (2002) makes a special and simultaneously more universal case for the obligations of researchers in relation to participants communities by locating HIV/AIDS within a context of economically- and ideologically-supported resource disparity:

The world at the beginning of the 21st century is thus characterized by widening economic and health disparities between rich and poor (within and between countries), and by suffering, conflict, and alienation associated with pervasive social forces. This scenario provides a strong case for viewing the emergence of new diseases such as AIDS that afflict predominantly those marginalized by poverty (80% of HIV positive persons live in the poorest countries of the world), as directly related to the ecological niches created and sustained by the nature of the global political economy and its ideology (Benatar, 2002: 1132).

For Benatar, the subsequent obligation is clear: "In the same way that racism, paternalism, gender discrimination and interpersonal abuse have become discredited, so too should autocratic/unaccountable institutions and exploitation be actively contested" (Benatar, 2002: 1132).

To a certain extent, mechanisms for community involvement in the planning and implementation of HIV/AIDS care, services, and research are more structurally (if imperfectly) embedded in the United States. The community planning requirements related to allocation and monitoring of Ryan White CARE Act funds is a good example. But in even this, a legislatively-mandated process of community involvement, there have been enormous problems in practice. In an empirical observation of the HIV community planning process carried out in Michigan, Dearing and colleagues (1998) identified considerable shortcomings in the CDC *Guidance* for such planning efforts, concluding that "considerable reinvention" is required to make community planning work effectively. Specifically, they concluded that planning council efforts required both *information-seeking* and *decision-making* skills and capacities, and that planning councils composed of representative cross-sections of community stakeholders could not be fairly expected to shoulder the burden of the former task (Dearing et al., 1998). If community planning, in this case narrowly focused on allocation of care resources (without consideration of research or prevention resources), is often contentious, even though typically supported by staffing, meeting resources (food, meeting space, transportation to and from meetings sites, and the like), and the compulsion of legislative mandate—if, under such conditions, community planning is often unsuccessful, how can community planning, or the maintenance of good relationships between researchers and participants, be expected to succeed in resource-poor settings in developing countries, where such supports or mandates rarely exist?

General Goals

Kilmarx and colleagues (2001) have made a number of overall recommendations to enhance community participation in research, including (1) community consultation, such as might occur through a representative community advisory group, (2) community review of the study protocol, with opportunities for suggestions, and (3) community review of consent forms and informational materials. This is a start, but hardly adequate. Before protocols related to specific and narrow ethical dilemmas are outlined, researchers should endeavor to understand, as completely as possible, the context in which research will take place—the preconditions affecting the characteristics of the community in which the research is to be conducted. In formulating guidelines for biomedical research in developing countries, Benatar (2002) also suggests that special consideration should include the conditions in developing countries; the research agenda of the industrialized world; informed consent; and justice in the distribution of knowledge and resources flowing from research. Thomas (1998) makes a similar point when he argues for a maximalist perspective on the bioethics of HIV clinical trials that would consider such issues as continued access to advanced treatment and technologies, equity, the nature of intersectoral collaboration, and human rights of research participants.

These are perspectives that travel far beyond the limited boundary of individual consent, and include an analysis of the relations and conditions in which research takes place, as well as an explicit consideration of research gains as a shared resource. And they are not universal perspectives; the debate continues. Brody (2002), for instance, suggests that many of the types of community-wide requirements that have been suggested should be viewed as moral aspirations only, with the focus of attention to remain on the research participants themselves.

Recognition of the community context in which research is to take place must face, head on, the stakes for local communities: in some cases, community and cultural survival are implicated. Research endeavors in international settings have profound implications and potential impacts, both for the very individuals and communities now alive. The urgency to fairly distribute knowledge and resources, as Benatar counsels, is immediate and palpable. What is often noted as uncommon about HIV/AIDS is the degree to which emotionally-charged public debate about the epidemic has given rise to the need to address deeper, meta-ethical questions, to re-examine foundational theory in light of postmodern, multicultural, feminist, and queer theory. At its worst these explorations could rapidly devolve into exercises in intellectual parsing, but they cannot be allowed to do so: lives, whole countries, are at stake.

More specifically, WHO/CIOMS guidelines have recognized the need to conduct research in developing countries, and subsequently sought to expand on the *application* of bioethical principles. Guidelines for such application address the urgency for research within communities in which disease is endemic; the need to solicit not only personal consent but the active engagement of community leadership; the need to work in collaboration with local public health authorities; the need to educate the community fully about the aims of the research and potential hazards or inconveniences; the need for multi-disciplinary review and auditing that involves community participants; and the need to fully address the ethical dimensions of externally-sponsored research. (Council of International Organization of Medical Sciences, 1993). These are the broad considerations which the researcher must address, but there are still precious few models to utilize in field practice.

In recent years one way that communities have addressed the systemic nature of health challenges is through the formation of community health partnerships (CHPs). CHPs are defined as "voluntary collaborations of diverse community organizations, which have joined forces to pursue a shared interest in improving community health" (Mitchell and Shortell, 2000: 242). "Partnerships," in this sense, can include coalitions, alliances, consortia, councils, and other organizational forms. These bodies almost always certainly include formal entities, but can include non-attached individuals who are also affected by health issues— such as consumers of HIV/AIDS services.

Despite the wisdom of the approach, however, CHPs have, in practice, not always achieved measurable results (Cheadle, Berry, Wagner, et. al., 1997; Knocke and Wood, 1981; Wandersman, Goodfman, and Butterfoss, 1993). In a multidisciplinary analysis drawing from the literature on community organization, social work, business strategy, organizational theory, transaction cost economics, and public health, Mitchell and Shortell (2000) hypothesize that CHPs with a high degree of external and internal alignment, and a high degree of centrality, will be better able to demonstrate long-run sustainable performance, long-run ability to secure needed resources, and long-run ability to use those resources effectively and efficiently. By "external alignment" the authors mean the correspondence between the composition of the CHP or partnership, and the breadth or scope of problems addressed; by "internal alignment" the authors mean the number of different services or initiatives the CHP has undertaken, the level of decisionmaking, and mechanisms of coordination and integration. By "centrality," they mean the importance and influence of the CHP or partnership within the power structure and organizational ecology of its community. Put more simply, community health partnerships, Mitchell and Mitchell suggest, are more likely to be successful if they adopt strategies that are realistic in relation to the external environment, utilize governance models that embrace heterogeneity and clear decisionmaking, and occupy a meaningful place within local decision-making structures and processes.

If HIV/AIDS research is embedded in and at least partially realized through a community partnership process—and the present volume argues that it ought to be—then there is much to recommend in this hypothesis in the design and management of community structures to inform and guide ethical and proce-dural practice in research. The very concept of a community health partnership, in the first place, implies an entity more durable and comprehensive than a mere advisory committee. Mitchell's and Shortell's hypothesis indicates that the part-nership should (a) play a discernable community role, not merely an advisory role to investigators, (b) pay particular attention to the partnership's diversity of composition and governance structures, so that they nurture cohesion and involvement, and (c) articulate realistic strategic goals. Thus, instead of estab-lishing time-limited, study-specific consultative bodies that advise and consent, we might examine the possibility of creating more lasting, non-study-dependent research partnerships with specific characteristics that develop long-term goals and strategy.

The establishment of a community coalition or alliance for research can be an effective tool, but we must still outline specific tools and strategies that such coalitions can adopt, and that can facilitate effective relationships between researchers and local communities. The following is a brief discussion of but some of those strategies and tools, and the concerns and issues they may raise.

Research Committees and IRBs

Research committees are charged with the tasks of (1) evaluating research proposals; (2) educating and assisting faculty, researchers, and the community in understanding and appreciating research ethics; and (3) monitoring and auditing research, and providing accountability to the public. In the conduct of research in international settings, researcher-community relationships can be strengthened by (1) ensuring representative community participation in research committees; (2) conducting research committee deliberations as openly and transparently as possible; (3) developing and consistently following protocols for reporting back to the community the research committee's deliberations, conclusions, and concerns; and (4) ensuring that the research committee has developed lines of communica-tion with all key community constituents and stakeholders, including health care providers, public health officials, other academic and research institutions, government, the media, non-governmental organizations working to provide HIV/AIDS treatment and prevention education, other civil society organizations, and faith-based institutions.

Particularly important—and highly problematic—is researcher relationships with host country IRBs and standing research and ethics committees, since they can play a significant role in authorizing the proposed research. The challenge is problematic because of the reality of poorly-resourced developing country IRBs, and because of the historical tendency of Western science to adopt paternalistic stances in laudable efforts to improve the capacity of developing-country science. In a study of research ethics committees (RECs) in fifteen African countries, Milford, Wassenaar, and Slack (2006) requested that a sample of African RECs rated their own capacity to review HIV vaccine protocols as "moderate to limited." Based on their findings from this study, the authors recommend additional training and support around issues of informed consent, adding that "while there is a need for ethics training applied to HIV vaccine trials, generic ethics training needs could also be addressed by ethics initiatives and sponsors of ethics training programs" (Milford, Wassennaar, and Slack, 2006: 8). This can be viewed as a problematic statement at face value: it seems to presuppose that local RECs are ethically deficient, when, in fact, they may not be as thoroughly schooled in Western tools of ethical analysis and reasoning as their Western counterparts. It would perhaps be more inclusive and culturally respectful to recommend that "generic ethics training" should be a bilateral enterprise, in which Western and non-Western ethicists—whether they are professionally deemed as such—seek to come to a greater and more adept understanding of how "an ethics" is created, and applied, in local context.

Community and Opinion Leaders

Every community has designated (i.e. formal) and informal leaders, individuals who by virtue of an office, position, education, experience, age, or personal charisma enjoy elevated status or influence over others. Indeed, the utilization of so-called "Population Opinion Leaders" to impact community norms about HIV risk and risk reduction behaviors has been shown to be an effective tool in the United States, where the conscious mobilization of such leaders serves as the centerpiece of some prevention initiatives.

Identification of community and opinion leaders, especially outside of the researcher's own cultural community, can be a complex task. Formal leadership may, in some circumstances, possess little authority, and may not be capable of adequately representing community concerns, viewpoints, and needs. Nevertheless, it will be helpful to the researcher to identify community leaders who can fairly and accurately reflect community sentiments, and who can also communicate, with reasonable authority, the research team's goals, intentions, motivations, strategies, and needs. Such leadership can be identified through semi-structured polling of community members. One must be careful, however, to frame questions carefully: asking, "who is the leader of the community?" or some variation thereof may generate the names of official community leadership, but such individuals may or may not enjoy the respect and trust of the community. Asking, on the other hand, "Who, even without a title or office, knows how to get things done in the community?" may identify individuals who are viewed as self-absorbed, fiercely competitive, or unethical, but who nevertheless, because of sheer persistence or force of will, get things done—without, again, enjoying the trust and respect of the community. We propose other, more carefully-worded queries: Who do most people view as fair, even-handed? Who would most people select as someone who could do a good job describing what's happening in the community, or parts of the community? Who, among people you know, can talk about or explain community issues so that others can understand? Who do others trust and respect, even if they don't always agree with that person?

It is unlikely, given the complexity of even small communities, that a single individual will be identified through such a process. It is much more likely that a number of individuals, each with different leadership capacities, and capable of voicing certain aspects of community life, emerge out of focused queries.

Once identified, community leadership can be engaged, formally or informally, to gain support for research objectives, and to serve as a conduit through which community concerns can be relayed to researchers. Dorothy Mbori-Ngacha (2001) offers a good example—through failure to do so—of the importance of engaging directly with communities and community leadership:

Over half the women accept testing, but less than a third of those who test positive come back for the interventions. We are trying to understand this. Why would you not return, after going through the whole process, to benefit from what we promised in the beginning? When women come to the antenatal clinic, their agenda isn't to learn their HIV status. They want antenatal care. They may get tested, but if it comes out positive, many aren't ready to deal with that. Many are afraid.

I think we did it backwards, in a sense. We should have mobilized the communities so they would support a woman in using antiretrovirals for preventing transmission or for not breast-feeding her baby. Right now there isn't enough support. Her mother-in-law will ask and visitors will ask, and it will be very difficult for her to justify why she's not breastfeeding (International AIDS Vaccine Initiative, 2001).

Educational Materials and Campaigns

Representative community leadership can serve as key informants in the development of community-focused educational materials and campaigns, which will attempt to communicate the nature of the proposed research, its short- and long-term goals and objectives, its specific protocols, its eligibility criteria and exclusions, its ethical architecture, and its sponsorship, financing, the credentials of the research team, and other key facts back to the community. The essential questions are, What do members of the community want and deserve to know, and how do they want to learn it? There are a wide variety of formats for such education, including written materials, oral presentations, community media, plays and skits, and a number of social marketing initiatives such as billboards. Educational materials should be translated into local languages and dialects (and back-translated to ensure accuracy), and subject to continual testing with the intended audience.

Community Decisionmaking and Research Ownership

Researchers would do well to establish formal or recognizable structures, such as the kind of CHP coalitions or alliances mentioned above, that can offer opportunities for community-decision about the deign and implementation of research, and ownership of the research process and its findings. Researchers should explicitly clarify the limits of decisionmaking and ownership in such structures, and the rationale for those restrictions. Meetings of coalitions, alliances, or comities should be open, with regular reporting back to the community in a variety of formats, and should receive adequate staffing support in the form of secretarial assistance, technical support, and needed equipment. The vexing question of who "owns" research results should be forthrightly faced, and complex ethical questions, such as which standard of care to adopt in the provision of medical care to trial participants, or the appropriate and fair construction of controls, should be subject to open discussion. Such coalitions, alliances, or committees cannot be viewed simply as mechanisms that will facilitate researcher-managed ratification of research decisions. Rather, they should be regarded more broadly, as venues in which members of the community and the research team can, together, (1) genuinely grapple with the unresolved questions of research sponsorship and objectives, protocols, and ethics; and (2) bilaterally exchange knowledge and wisdom about the conduct of science and the life of the local community and its residents.

Termination Aftershocks and the Next Study

The conclusion of a particular research study will invariably (if no additional studies are scheduled) result in disruption. Research dollars that have been flowing into the community, and that have circulated through the local economy, will cease; the provision of medical care to research participants may be discontinued or reduced; the mobilization of the community toward the development of solutions to the local challenge of HIV/AIDS may suffer because of suspended technical and staffing support for coalitions, alliances, or committees; and opportunities for local skill development through research employment and education may dry up. Worse, developing world and domestic minority community perceptions that research consists of something akin to "drive-by" operations, in which researchers set up shop, extract data, and then depart, may be reinforced, thus increasing mistrust, hopelessness, and adversarial relations.

Research studies are often typically multi-year initiatives. This affords researchers the opportunity, *from the very outset*, to work in partnership with the community to forecast what could happen at the study's conclusion, and plan proactively for management of any anticipated negative consequences. Even more importantly, researchers should integrate such forecasts into the research plan; it would be unethical, for example, to provide medical care for study participants during the course of the study, and then, at its conclusion, simply cease provision of such services, and in a similar manner, it can be viewed as unethical to alter local economies, disrupt employment and learning prospects, and suspend support for emergent political/advocacy structures (committees, alliances, and coalitions) without thought of potential local aftershocks.

Two other post-study issues, related to researcher-community relations, deserve mention. The first is the ethical responsibility of the researcher (discussed elsewhere in this volume) to advocate for additional resources that can help rectify the inherent inequality of HIV/AIDS health care resource allocations that now exists. This should take the form not only of working to fund continuing research, which supports the self-interest of researchers, but also advocating for additional medical, ancillary service, and community support funding that is not based on research. Such additional services can simply help improve the lives of members of the community and the community's capacity for survival in the face of the epidemic. The researcher-community relationship is an ideal setting from which to advocate or mobilize for such resources.

The second has to do with who is credited with "authorship" of research findings. Too often, no mention is made of the local community's contributions in research publications. It is as though the local community were effaced or interchangeable with any number of other "community-subjects." Explicit acknowledgement of the community's contribution will help engender ownership of the research and facilitate the smooth and respectful implementation of future research. That findings should be reported back not merely in peer-reviewed journals, but directly, to the community through open forums and other venues, should be taken as a given. But when that reporting occurs, it usually functions as the sole setting in which thanks and appreciations are offered; such acknowledgement needs to make its way into the scientific literature as well.

References

Benatar, S.R. (2002). Reflections and recommendations on research ethics in developing countries. *Social Science and Medicine, 54*, 1131–1141.

Brody, B.A. (2002). Ethical issues in trials in developing countries. *Statistics in Medicine, 21*, 2853–2858.

Cheadle, A., Berry, W., Wagner, E., et. al. (1997). Conference report: Community-based health promotion—state of the art and recommendations for the future. *American Journal of Preventative Medicine, 13*, 240–243.

Council of International Organization of Medical Sciences. (1993). *International Guidelines for Biomedical Research Involving Human Subjects.* Geneva, Switzerland: Author.

Dearing, J.W., Larson, R.S., Randall, L.M., & Pope, R.S. (1998). Local reinvention of the CDC HIV prevention community planning initiative." *Journal of Community Health, 23(2),* 113–126.

International AIDS Vaccine Initiative. (2001). Women and HIV in Kenya: An interview with Dorothy Mbori-Ngacha. IAVI Report: The Newsletter on International AIDS Vaccine Research, 5(8), 7–8, 23; Last modified September 2001; Last accessed January 1, 2007; Available at http://www.iavireport.org/Issues/0901/72001.pdf.

Kilmarx, P.H, Ramjee, G., Kitayaporn, D., & Kunasol, P. (2001). Protection of human subjects' rights in HIV-preventive clinical trials in Africa and Asia: Experiences and recommendations. *AIDS, 15 (suppl 5),* S73-S79.

Knocke, D. & Wood, J.R. (1981). *Organized for Action: Commitment in Voluntary Organizations.* New Brunswick, New Jersey: Rutgers University Press.

Milford, C., Wassennaar, D., & Slack, C. (2006). Resources and needs of ethics committees in Africa: Preparations for vaccine trials. *IRB: Ethics & Human Research, 28(2),* 1–9.

Mitchell, S.M & Shortell, S.M. (2000). The governance and management of effective community health partnerships: A typology for research, policy, and practice. *The Milbank Quarterly, 78(2),* 241–289.

Scott, J. & Kaufman, L. (2005). Belated charge ignites furor over AIDS drug trial. *New York Times,* July 17, Sec. 1, Col. 1, Metropolitan Desk, p. 1.

Thomas, J. (1998). Ethical challenges of HIV trials in developing countries. *Bioethics, 12(4),* 320–327.

Wandersman, A., Goodfman, R.M., & Butterfoss, F.D. (1993). Understanding coalitions and how they operate." In M. Minker (Ed.).*Community Organizing and Community Building for Health* (pp. 261–277). New Brunswick, New Jersey: Rutgers University Press.

Chapter 7
Recruiting for HIV-Related Research

Introduction

The concept of recruitment refers to the strategies that are utilized to approach individuals who may be eligible to participate. Ethical issues may arise during this process that relate to the extent to which detailed information about the study is provided, the voluntary nature of the interaction with the research team (Roberts, Warner, Anderson, Smithpeter, and Rogers, 2004), and the risks and benefits that may be associated with the recruitment process, as distinct from study participation. Although the process of recruitment is often perceived as being seamless with that of enrollment and informed consent (Roberts, Warner, Anderson, Smithpeter, and Rogers, 2004), the ethical issues may be distinct.

Providing Information

Federal regulations clearly specify the information that is to be provided to study participants in the context of the informed consent process. This includes

1. A statement that the study involves research, an explanation of the purposes of the research and the expected duration of the subject's participation, a description of the procedures to be followed, and identification of any procedures which are experimental;
2. A description of any reasonably foreseeable risks or discomforts to the subject;
3. A description of any benefits to the subject or to others which may reasonably be expected from the research;
4. A disclosure of appropriate alternative procedures or courses of treatment, if any, that might be advantageous to the subject;
5. A statement describing the extent, if any, to which confidentiality of records identifying the subject will be maintained;
6. For research involving more than minimal risk, an explanation as to whether any compensation and an explanation as to whether any medical treatments are available if injury occurs and, if so, what they consist of, or where further information may be obtained;

7. An explanation of whom to contact for answers to pertinent questions about the research and research subjects' rights, and whom to contact in the event of a research-related injury to the subject; and

8. A statement that participation is voluntary, refusal to participate will involve no penalty or loss of benefits to which the subject is otherwise entitled, and the subject may discontinue participation at any time without penalty or loss of benefits to which the subject is otherwise entitled (United States Department of Health and Human Services, 2006).

The regulations suggest that the following additional information be provided to study participants, as appropriate:

1. A statement that the particular treatment or procedure may involve risks to the subject (or to the embryo or fetus, if the subject is or may become pregnant) which are currently unforeseeable;

2. Anticipated circumstances under which the subject's participation may be terminated by the investigator without regard to the subject's consent;

3. Any additional costs to the subject that may result from participation in the research;

4. The consequences of a subject's decision to withdraw from the research and procedures for orderly termination of participation by the subject;

5. A statement that significant new findings developed during the course of the research which may relate to the subject's willingness to continue participation will be provided to the subject; and

6. The approximate number of subjects involved in the study (United States Department of Health and Human Services, 2006).

Clearly, the withholding of critical information may obviate any consent that is given for participation in a study. Many of the abuses that have occurred in the context of health-related research have resulted from the intentional omission of critical details from the consent process (United States Department of Health, Education, and Welfare, 1973). However, ethically and legally, it is unclear to what extent specific details about the study must be included in the initial *recruitment* materials, such as flyers, radio announcements, or advertisements, as distinct from the provision of this information during the consent process. Neither international guidelines for the ethical conduct of research nor the federal regulations address this specific issue, although some institutional review boards appear to demand that these details be furnished in recruitment materials.

Consider, as an example, a scenario in which investigators wish to conduct research to formulate a culturally-appropriate HIV prevention intervention to reduce sexually risky behavior among women in a relatively closed community that is experiencing an increasing risk of HIV infection. The provision of all such details in recruitment materials potentially stigmatizes individuals who are associated with the participants, including even those who are not HIV-infected. The participating women may be perceived as promiscuous by others in their community and may be ostracized as a result. Recruitment at a site that is designated specifically and solely for the purpose of HIV research may also lead to the same consequence if participants' presence at the site becomes known to their community. Contrast these approaches with the conduct of a study that states in its recruitment materials that it seeks to improve the health of women in the context of their relationships and is conducted through a clinic or social service agency that provides a panoply of needed services to the community, so that entrance into the building is not automatically associated with HIV status and/or HIV-related, socially undesirable behaviors, such as multiple sexual partners and/or substance use.

Various international guidelines recognize the risk of stigmatization and marginalization that may result or be associated with participation in health-related research. The *International Guidelines for Biomedical Research Involving Human Subjects* (Council for International Organizations of Medical Sciences, 2002) notes in the commentary to Guideline 8, which addresses the benefits and risks of study participation, that

research in certain fields, such as epidemiology, genetics, or sociology, may present risks to the interests of communities, societies, or racially or ethnically defined groups. Information might be published that could stigmatize a group or expose its members to discrimination. Such information, for example, could indicate, rightly or wrongly, that the group has a higher than average prevalence of alcoholism, mental illness or sexually transmitted disease, or is particularly susceptible to certain genetic disorders. Plans to conduct such research should be sensitive to such consideration, to the need to maintain confidentiality during and after the study, and for the need to publish the resulting data in a manner that is respectful of the interests of all concerned, or in certain circumstances not to publish them. The ethical review committee should ensure that the interests of all concerned are given due consideration; often it will be advisable to have individual consent supplemented by community consultation.

Guidelines 19 and 21 of the *International Guidelines for the Ethical Review of Epidemiological Studies* (Council for International Organizations of Medical Sciences, 1991, 2005) also caution investigators to be aware of this potential risk and to protect research participants from such risk to the extent possible. Guideline 19 provides that "ethical review must always assess the risk of subjects or groups suffering stigmatization, prejudice, loss of prestige or self-esteem, or economic loss as a result of taking part in a study . . . ," while Guideline 21 notes that

Epidemiological studies may inadvertently expose groups as well as individuals to harm, such as economic loss, stigmatization, blame, or withdrawal of services. Investigators who find sensitive information that may put a group at risk of adverse criticism or treatment should be discreet in communicating and explaining their findings. When the location or circumstances of a study are important to understanding the results, the investigators will explain by what means they propose to protect the group from harm or disadvantage; such means include provisions for confidentiality and the use of language that does not imply moral criticism of subjects' behaviour.

The duty of the researcher to be cognizant of and to minimize such risks to the research participants arises from the ethical principles of beneficence and non-maleficence. *Beneficence* refers to the "ethical obligation to maximize possible benefits and to minimize possible harms and wrongs," while the principle of *nonmaleficence* counsels researchers to protect participants from avoidable harms (Council for International Organizations of Medical Sciences, 1991).

These same guidelines should be equally applicable to the recruitment process. As seen from the above example, inadequate attention to such concerns during the recruitment process could inadvertently result in the stigmatization and ostracism of study participants. There may be methodological implications, as well. Enrollment may move more slowly than originally anticipated due to individuals' reluctance to be associated with the study and risk the potential adverse social and economic consequences.

Ensuring Voluntariness

Numerous reasons have been identified for individuals' decisions to participate in research studies. These include a sense of altruism (Baker, Studies, Lavender, and Tincello, 2005; Elbourne, 1987; Fry and Dwyer, 2001; Jenkins, Chinaworapong, Morgan, Ruangyuttikarn, Sontirat, Chiu et al., 1998; Stanford et al., 2003), a desire for enhanced medical care (Baker, Studies, Lavender, and Tincello, 2005; Stanford et al., 2003), assurances of confidentiality and privacy (Stanford et al., 2003), a sense of personal satisfaction (Fry and Dwyer, 2001), an expression of citizenship (Fry and Dwyer, 2001), activism (Fry and Dwyer, 2001), the availability of additional information or assistance (Fry and Dwyer, 2001), and economic gain (Fry and Dwyer, 2001). Yet other individuals may agree to participate in a specific research study out of fear that they will alienate or disappoint their health care provider who has approached them for their participation (Warner, Roberts, and Nguyen, 2003). In the context of HIV research specifically, several motivations for participating have been identified: altruism or the desire to help others (Hays and Kegeles, 1999; Harro et al., 2004; Leonard, Lester, Rotheram-Borus, Mattes, Gwadz, and Ferns, 2003; Reeder, Davison, Gipson, and Hesson-McInnis, 2001; Thapinta et al., 2004), a desire to understand or to end the AIDS epidemic (Hays and Kegeles, 1999; Reeder, Davison, Gipson, and Hesson-McInnis, 2001), a desire to obtain needed services or increase health awareness (Leonard, Lester, Rotheram-Borus, Mattes, Gwadz, and Ferns, 2003; Moreno-Black, Shor-Posner, Miguez, Burbano, O'Mellan, and Yovanoff, 2004), and/or a wish to reduce the likelihood of becoming HIV-infected (Hays and Kegeles, 1999). The most common reasons for refusal to participate in studies include a lack of interest (Cooley et al., 2003), a distrust of research and/or researchers (Baker, Studies, Lavender, and Tincello, 2005; Thompson, Neighbors, Munday, and Jackson, 1996), health limitations (Cooley et al., 2003), inconvenience (Baker, Studies, Lavender, and Tincello, 2005; Cooley et al., 2003), and a sense of disempowerment associated with the research process (Baker, Studies, Lavender, and Tincello, 2005).

Several of these scenarios raise concerns regarding the extent to which the decision to participate is actually voluntary. For instance, if a prospective participant is unable to obtain adequate medical care outside of the context of the research study, one must question the extent to which the individual perceives his or her decision as voluntary. If the response to recruitment efforts is premised on economic gain, the voluntary nature of the resulting consent to participate may be suspect if the gain constitutes a significant portion of an individual's income or assets at the time it is offered. An individual who is approached by his or her physician for participation in a study may feel that he or she has no real choice and must participate in order to avoid potential unwelcome consequences. In each case, one must ask whether the underlying motivation is merely an incentive to participate and an acceptable form of persuasion or whether it constitutes a form of undue influence or coercion that would negate voluntariness and obviate informed consent.

Persuasion, Incentives, Inducement, and Coercion Defined

Persuasion, which is seen as an acceptable form of influence that is compatible with informed consent, "is the successful and intentional use of reason to convince a person to willingly accept the beliefs, choices, or decisions favored by the persuader" (Erlen, Sauder, and Mellors, 1999: 85–86). An incentive has been defined as "that which influences or encourages to action; motive; spurs; stimulus" (McKechnie, 1976: 921). In contrast, an inducement is considered to be "undue" if it is so "attractive that [it can] blind prospective subjects to potential risks or impair their ability to exercise their proper judgment . . . ." (Office of Protection from Research Risks, 1993). Coercion represents an even more forceful form of influence: "an extreme form of influence by another person that completely controls a person's decision . . . depriv[ing] the person of autonomous choice, and thus is incompatible with informed consent" (Faden and Beauchamp, 1986: 338–339). Coercion has been understood "to involve a threat of physical, psychological, or social harm in order to compel an individual to do something, such as participate in research" (Grady, 2005: 1683). What is offered as an incentive may constitute either an acceptable form of persuasion or a form of coercion, depending upon the nature of the incentive and the context in which it is offered (Erlen, Sauder, and Mellors, 1999). And, according to some scholars, incentives may be morally suspect even if they are not coercive if they succeed in convincing individuals to do something that they would not ordinarily agree to do. (Steinbock, 1995).

Medical Care as an Incentive

It may be difficult to understand how the provision of medical care in the context of a study about a medical condition might be sufficient to persuade an otherwise reluctant and possibly generally unwilling individual to sign on as a research participant. One scholar with significant experience providing medical care in

Mozambique, where the local provincial health service budget was approximately $3 US per capita, noted:

This is the sort of health service where every clinician finds him or herself from time to time looking at the pharmacy cupboard and wondering how to divide the remaining three vials of penicillin between the five patients in the ward who need it. (Whether to give starting doses to everyone in the hope that the promised new supplies will arrive, or just give it to one seriously ill child, for whom at least it represents a curative course.) From that perspective, enrolling patients in a clinical trial will always look attractive, no matter how unethical that research may turn out to be (Loff and Black, 2000: 293).

Consider, as well, the situation of HIV-infected members of U.S. racial/ethnic minority groups with respect to health care. HIV has been found to be one of the largest contributors to the gap in life expectancy that exists between blacks and non-Hispanic whites in the U.S., due to both disproportionately higher HIV infection rates and death rates, which persist despite the advent of HAART (Wong et al., 2002). Minority individuals experience increased difficulty obtaining needed HIV care and are less likely to receive the medications needed to treat their infection (Moore et al., 1994; Weissman et al., 1994; Mor et al., 1992; Stone et al., 1998; Turner et al., 2000). Compared to non-Hispanic whites, African Americans and Hispanics are more likely to experience a delay in the receipt of HIV care following a diagnosis (Turner et al., 2000), and the care that they receive is less likely to adhere to recommended treatment protocols (Shapiro et al., 1999; Stone et al., 1998). When one considers the relative lack of health insurance among African Americans and Hispanics in comparison with non-Hispanic whites (Hall, Collins, and Glied, 1999; Henry J. Kaiser Family Foundation, 2003; Mills and Bhandari, 2003) in conjunction with the documented delays in receiving care and the quality of that care, it is not surprising that some individuals may feel that they have no real choice but to participate.

The question, though, is whether the provision of medical care constitutes an incentive, spurring an individual towards action, or an inducement, blinding the person to the risks associated with study participation. And, a failure or refusal to provide medical care associated with the condition under study would also raise ethical questions. For instance, it is not unreasonable that a study participant receive the medical care necessary to ensure the safe conduct of the research or treatment for an adverse reaction to a drug that is the subject of the study in which they are participating. Indeed, a failure to provide such care arguably contravenes the provisions of the Nuremberg Code. Additionally, the *International Ethical Guidelines for Biomedical Research Involving Human Subjects* of the Council for International Organizations of Medical Sciences (2005) states in Guideline 21:

External sponsors are ethically obliged to ensure the availability of: health-care services that are essential to the safe conduct of the research; treatment for subjects who suffer injury as a consequence of research interventions; and, services that are a necessary part of the commitment of a sponsor to make a beneficial intervention or product developed as a result of the research reasonably available to the population or community concerned.

Monetary Incentives

The idea of an incentive is often confused with the concept of recruitment. While recruitment refers to the overall strategy to identify, interest, and provide information to potentially eligible persons, incentives may comprise only one aspect of this overall strategy and may be offered for a variety of reasons (Dickert and Grady, 1999; Rice and Broome, 2004). Incentives may also be offered to further retention in addition to the initial recruitment of participants.

Grady (2005) has identified a number of reasons that underlie the provision of a monetary incentive to recruit research participants: (1) to assure the recruitment of an adequate number of participants for a study; (2) as a mechanism to overcome opportunity costs, inertia, and distrust and recruit hard-to-reach populations; (3) as reimbursement for lost wages and other participation-related expenses, such as transportation; and (4) as fair compensation or remuneration for the expended time and associated inconvenience. Studies that have examined the attitudes of unpaid research participants towards the provision of incentives have reported that the participants approve of the use of incentives to improve problematic recruitment, to reimburse participants for their study-associated costs, and to recognize participants for their investment of time and effort (Russell, Moralejo, and Burgess, 2000). Although a monetary incentive may be a motivation for some individuals to participate in HIV-related research, it appears to be a relatively minor reason (Hays and Kegeles, 1999).

Although Grady (2005) recognizes that payment may be perceived as a form of coercion, she rejects the validity of that notion, arguing that an offer of money in return for participation cannot be coercive because it does not involve a threat of harm. In analyzing whether the provision of a payment constitutes an undue inducement, Grady (2005) acknowledges that scholars have questioned whether payment can impair potential participants' judgment, compromise the voluntary nature of their decisionmaking, or unduly influence the participants to misrepresent themselves in order to appear eligible for enrollment into a particular study. She counters such concerns by asserting that

voluntary decisions are motivated by various factors, sometimes including money, and are not necessarily motivated by altruism alone. When people are choosing a job, making purchases, or making other voluntary decisions, they often consider the money aspects of their choice in the form of salary, benefits, or sales price. Decisions are generally complex and multifaceted, however, and are rarely based solely on monetary considerations. Similarly, people participate in clinical research for multiple reasons, and money may be one among those reasons or even the main reason. Limited data suggest that the offer of money is one factor in the decision making of some, but not all, potential participants (Grady, 2005: 1683).

Grady further argues that more careful screening can minimize the risk of misrepresentation by prospective participants, which could compromise both the safety of the particular participant and the validity of the study. Although few studies have examined this particular issue, the research that has been conducted suggests that the provision of a monetary incentive may, indeed, influence participants to conceal activities that would render them ineligible to participate in a study (Bentley and Thacker, 2006).

Research further suggests that, although the payment of a monetary incentive will not blind prospective participants' to the risks of participation, payment appears to increase individuals' willingness to participate regardless of the level of risk involved (Bentley and Thacker, 2006). Additionally, larger payments appear to result in increased willingness.

Concerns have also been voiced that the provision of payments may be more attractive to individuals of lower socioeconomic status, resulting in a disproportionate burden on this population in contravention of the principle of justice, as well as methodological difficulties resulting from reduced generalizability of the study findings (Grady, 2005). Grady (2005) has suggested that prorating payment for studies involving multiple visits may minimize the possibility that someone will volunteer to participate on the basis of payment alone.

Active Recruitment Strategies

A variety of mechanisms may be used to recruit individuals for enrollment into research studies. These strategies may be active, consisting of direct contact between the researcher and the potential participant, such as by telephone or personal contact, or they may be passive, whereby the participant must contact the research team after having been made aware of the study, such as through an advertisement or brochure (Amthauer, Gaglio, Glasgow, Dortch, and King, 2003; Coley et al., 2003; Milgrom, Hujoel, Weinstein, and Holborow, 1997). Active recruitment strategies may raise particular ethical concerns because they involve direct contact with the participant and, as such, carry the possibility of being coercive.

It is not unusual for individuals to be approached by their care providers for participation in clinical studies. Several research groups have noted the high esteem in which physicians are often held by their patients, and the concomitant reticence of patients to disappoint their providers by refusing their requests, patients' trust in their physicians to hold their best interests in mind, and patients' fear of repercussions if they should decline their provider's suggestion of participation (Pearn, 2001; Warner, Roberts, and Nguyen, 2003). Additionally, when approached for participation by providers, individuals may confuse the purposes of research with the provision of clinical care (Grady, 2005).

Researcher Conflict of Interest

Increasing attention has been focused on the payment of finders' fees, which are "offers of money to physicians, nurses, or other health professionals in reward for their referral of patients eligible for research participation . . . over and above reasonable remuneration for services rendered" (Lemmens and Miller, 2003: 399). It has been hypothesized that health professionals who receive such fees may "be more lenient with respect to informed consent procedures, they may convince themselves that research participation is in their patient's best interests, or they may be overly flexible with regard to the study inclusion and exclusion criteria" (Lemmens and Miller, 2003: 401). Unwittingly, they may exert increased pressure on individuals to agree to participation. Should this occur, the participants who are the focus of recruitment efforts are potentially at increased risk of harm, and the validity of the study is questionable. And, although the American Medical Association has condemned such payments as unethical (American Medical Association, 1994, 2000) and IRBs are charged with the responsibility to review the methods used to recruit study participants (Food and Drug Administration, 1998), there is no legal requirement that investigators disclose payments that they receive in exchange for recruitment efforts.

Risks and Benefits in the Recruitment Process

Relatively little attention has been focused on an examination of the risks and benefits associated with the recruitment process, as distinct from participation in the study itself. However, the manner in which recruitment is conducted may itself raise such concerns.

The direct recruitment of patients by their providers into research studies may bring about harm to the participants and the violation of the principle of nonmaleficence, in addition to those issues related to voluntariness, discussed above. In one study involving patients, more than half indicated that they would find an invitation from their care provider to enter a clinical trial upsetting. Two-thirds of the patients who thought that such an invitation would affect their recovery believed that it would make them worse (Corbett, Oldham, and Lilford, 1996). Researchers surmised from these findings that patients expect their physicians to focus on their best interests and their clinical care in the context of providing that care (Habiba and Evans, 2002).

The indirect recruitment of additional participants through already-enrolled participants has been utilized frequently as a recruitment strategy in the context of HIV research, such as research designed to understand the transmission patterns of HIV infection through sexual and drug-using networks (Margolis, 2000). However, this approach carries the potential to harm current and prospective participants, in contravention of the ethical principle of nonmaleficence. Already-enrolled research participants may be asked for a listing of their sexual contacts during a specified time period. Margolis (2000) has decried the use of such a recruitment strategy, arguing that the practice carries with it significant risks. First, this approach violates the basic need for privacy. Second, it may result in the inadvertent deductive identification of the index participant who does not wish to be identified to the potential participant. Third, the disclosure of the identity of a sexual partner by the index participant may undermine the trust between those persons. Finally, individuals within the community may refuse to participate in research studies if they believe that information provided to researchers will be disclosed to others or they believe that they will be asked to violate trust. The practice of recruiting new participants through existing participants also raises issues of voluntariness in the informed consent process, because the newly recruited individuals may be reticent to refuse to participate for fear of injuring their relationship with the participant(s) who provided their names.

Increasingly, heterosexual couples are recruited into HIV-related research to test the effectiveness of HIV prevention strategies (McMahon, Tortu, Torres, Pouget, and Hamid, 2003; Witte, El-Bassel, Gilbert, Wu, Chang, and Steinglass, 2004). Numerous ethical issues may arise in this context. Depending upon the dynamics within a relationship, a woman may be pressured by her male partner to participate when she does not want to, or blamed if the couple does not meet study eligibility criteria (McMahon et al., 2003). There is also a danger that information provided by one partner that has been withheld from the other may be inadvertently disclosed or revealed by a staff member during the course of the study. Depending upon the nature of that information and the dynamics within the relationship, such a disclosure could leave the partner vulnerable to abuse.

Concerns have been voiced with respect to the use of active recruitment strategies in the context of studies with bereaved families, whereby members of bereaved families are contacted directly by members of the research team who become aware of their situation through their care providers (Cohen, Davis, Hunter, Carp, Geromanos, and Sunkle, 1997; Steeves, Kahn, Ropka, and Wise, 2001). The bereaved family members may feel that their privacy is being violated by the physicians or other ancillary care providers, such as clergy, who facilitate these communications. And, although active recruitment strategies appear to result in higher response rates from prospective participants, many IRBs often refuse to permit their use (Nelson et al., 2002).

Indeed, research suggests that such concerns are well-founded. In one study focusing on attitudes about recruitment approaches, the majority of the 498 respondents indicated that they would mind if their names were given to researchers by their doctor without their permission (Hull, Glanz, Steffen, and Wilfond, 2004). However, the majority also indicated that they would be willing to give such permission if asked to do so.

Researchers are increasingly relying on the internet as a mechanism for recruitment into HIV studies (Bowen, Williams, and Horvath, 2004; Bull, Lloyd, Rietmeijer, and McFarlane, 2004; Fernandez et al., 2004; Fernandez et al., 2005), raising numerous ethical and methodological challenges. First, it is incumbent upon the researcher recruiting through chatrooms to divulge the nature of his/her participation in those communications. It would be easy, for instance, for the researcher to withhold the true purpose of his or her participation and steer other chatroom participants to a research study. And, while such deception is not prohibited, international ethical guidelines strongly discourage its use (Council for International Organizations of Medical Sciences, 1991, 2005). Reliance on such tactics may result in feelings of betrayal among the individuals targeted and a community-wide distrust of scientists and their trade.

Second, web-based recruitment strategies that require individuals to reveal information about themselves on-line in order to facilitate future contact must incorporate sufficient safeguards to protect the confidentiality of those data from intending hackers. A failure to do so could lead to adverse social, economic, and legal consequences to respondents.

Recruitment and Justice

The ethical principle of justice has been interpreted as demanding distributive justice: that the benefits and the burdens of research participation be available across populations (Council for International Organizations of Medical Sciences, 2002, 2005). However, until relatively recently, biomedical research studies were often conducted without female participants (Ramasubbu, Gurm, and Litaker, 2001; Vidader, Lafleur, Tong, Bradshaw, and Marts, 2000). This omission rises not only ethical issues, but methodological concerns as well. A sufficient number of females must be included in studies in order to ensure the generalizability of the research findings. For instance, the therapeutic actions and side effects of drugs and drug dosages may vary between men and women due to differences in body composition, hormones, and metabolism (Mastroianni, Faden, and Federman, 1994). Gender and sex differences may also exist with respect to psychological, social, behavioral, or epidemiological factors.

In the United States, the 1993 NIH Revitalization Act mandated the inclusion of women in all relevant clinical research (Federal Register, 1994). At the same time, the Food and Drug Administration revised its policy to permit women of childbearing potential to participate in early drug trials and emphasized the need for analyses of research data by sex. Despite these policy changes and the increased enrollment of women as participants in clinical studies, relatively few research groups conduct analyses of the resulting data by sex (General Accounting Office, 2000). A similar situation exists with respect to the enrollment of women and analysis of data in research conducted in Canada (Marrocco and Stewart, 2001; Stewart, Cheung, Layne, and Evis, 2000).

Minorities have also been excluded from participation in clinical research until relatively recently (Armistead et al., 2004; Daunt, 2003). Various factors are believed to be responsible for this lack of participation, including stigmatization, employment obligations that permit little time off from work, lesser financial resources to pay for child care or transportation expenses associated with study participation, and poor literacy skills that reduce individuals' ability to participate in survey- or questionnaire-based research (Armistead et al., 2004; Family Health Research Group, 1998).

The attitudes and perceptions of providers may be key to the underrepresentation of both women and minorities in HIV/AIDS clinical trials. One study found that providers believed that African Americans, Latinos, and Haitians were less likely than non-Hispanic whites to be interested in HIV clinical trials (Stone, Mauch, and Steger, 1998). Similar judgments were made about women. A proportion of the providers further reported that, based on these perceptions, they were less likely to inform Latinos and Haitians about the possibility of clinical trial participation.

While the principle of justice argues for increased inclusion of women and minority group members, the ethical principles of respect for persons and nonmaleficence may require that persons be excluded from participation in research studies if they are vulnerable and if those studies do not directly address their needs or if other, nonvulnerable populations can be relied on for the conduct of the particular study (Council for International Organizations of Medical Sciences, 2002). Vulnerable participants are persons who are relatively or absolutely unable to protect their own interests because "they have insufficient power, prowess, intelligence, resources, strength, or other needed attributes to protect their own interests through negotiation for informed consent" (Levine, 1988: 72). It has been argued that individuals suffering from incurable or fatal diseases, such as HIV/AIDS, may constitute a vulnerable group, in part because of their relationship with their physician and confusion between clinical care and participation in clinical research (the "therapeutic misconception") (Grady, 2005). The vulnerability of such persons may be further compounded in the presence of other factors, including poverty, illiteracy, and stigmatization associated with minority status (Glantz, Annas, Grodin, and Mariner, 2001). Indeed, in view of the demographic characteristics of populations most impacted by HIV infection in developing countries, it is likely that many participants in phase III HIV vaccine trials will be drawn in those countries from black communities that may be at increased vulnerability due to these factors (Barsdorf and Wassenaar, 2005; Lindegger, Wassenaar, and Slack, 2001).

Some might argue that individuals characterized by this compound vulnerability be excluded from research studies because of an increased likelihood that their decision to participate will be less than completely voluntary due to their economic and social circumstances. The few research studies that have been conducted on minority groups' perceptions of research suggest that, at least in some contexts, individuals from historically persecuted minority groups are less likely than others to perceive participation in research as voluntary (Barsdorf and Wassenaar, 2005).

However, the exclusion of such persons from research due to their vulnerability would be reflective of paternalism resulting from an inappropriate overemphasis on the principle of nonmaleficence and de-emphasis on the principle of justice, which dictates that the benefits and burdens of research be accessible across populations (Beauchamp, Jennings, Kinney, and Levine, 2002; Erlen, Sauder, & Mellors, 1999; Roberts, Geppert, & Brody, 2004; Stanley, Stanley, Lautin, Kane, and Schwartz, 1981). Instead, such persons should be afforded the opportunity to participate in research protocols that have been structured to afford special protection to participant groups that may be particularly vulnerable (Council for International Organizations of Medical Sciences, 2002, Guideline 13).

References

American Medical Association, Council on Ethical and Judicial Affairs. (2000). Fee splitting: Referrals to health care facilities (Opinion E-603). In American Medical Association Council on Ethical and Judicial Affairs, *Code of Medical Ethics: Current Opinions*. Chicago, Illinois: Author.

American Medical Association, Council on Ethical and Judicial Affairs. (1994). *Finder's Fees: Payment for the Referral of Patients to Clinical Research Studies* (report #65). Chicago, Illinois: Author.

Amthauer, H., Gaglio, B., Glasgow, R.E., Dortch, W., & King, D.K. (2003). Lessons learned: Patient strategies for a type 2 diabetes intervention in a primary care setting. *The Diabetes Educator, 29(4)*, 673–681.

Armistead, L.P., Clark, H., Barber, C.N., Dorsey, S., Hughley, J., Favors, M., & Wyckoff, S.C. (2004). Participant retention in the Parents Matter! Program: Strategies and outcome. *Journal of Child and Family Studies, 13(1)*, 67–80.

Baker, L., Studies, D., Lavender, T., & Tincello, D. (2005). Factors that influence women's decisions about whether to participate in research: An exploratory study. *Birth, 32(1)*, 60–66.

Barsdorf, N.W. & Wassenaar, D.R. (2005). Racial differences in public perceptions of voluntariness of medical research participants in South Africa. *Social Science & Medicine, 60*, 1087–1098.

Beauchamp, T.L., Jennings, B., Kinney, E.D., & Levine, R.J. (2002). Pharmaceutical research involving the homeless. *Journal of Medicine and Philosophy, 27(5)*, 547–564.

Beecher, H.K. (1966). Ethics and clinical research. *New England Journal of Medicine, 274*, 1354–1360.

Bentley, J.P. & Thacker, P.G. (2006). The influence of risk and monetary payment on the research participation decision making process. *Journal of Medical Ethics, 30*, 293–298.

Bowen, A., Williams, M., & Horvath, K. (2004). Using the internet to recruit rural MSM for HIV risk assessment: Sampling issues. *AIDS and Behavior, 8(3)*, 311–319.

Bull, S.S., Lloyd, L., Rietmeijer, C., & McFarlane, M. (2004). Recruitment and retention of an online sample for HIV prevention intervention targeting men who have sex with men: The Smart Sex Quest Project. *AIDS Care, 16(8)*, 931–943.

Cohen, H.L., Davis, L., Hunter, J., Carp, D., Geromanos, K., & Sunkle, S. (1997). Coordinating a large multicentered HIV research project. *JANAC, 8(1)*, 41–50.

Cooley, M.E., Sarna, L., Brown, J.K., Williams, R.D., Chernecky, C., Padilla, G. et al. (2003). Challenges of recruitment and retention in multisite clinical research. *Cancer Nursing, 26(5)*, 376–384.

Corbett, F., Oldham, J., & Lilford, R. (1996). Offering patients entry in clinical trials: Preliminary study of the views of prospective participants. *Journal of Medical Ethics, 22*, 227–231.

Council of Organizations for Medical Sciences. (2002). *International Ethical Guidelines for the Conduct of Biomedical Research Involving Human Beings*. Geneva, Switzerland: Author.

Council of Organizations for Medical Sciences. (1991). *International Guidelines for Ethical Review of Epidemiological Studies*. Geneva, Switzerland: Author.

Council of Organizations for Medical Sciences. (2005). *International Guidelines for Ethical Review of Epidemiological Studies*. Geneva, Switzerland: Author.

Daunt, D.J. (2003). Ethnicity and recruitment rates in clinical research studies. *Applied Nursing Research, 16(3)*, 189–195.

Dickert, N. & Grady, C. (1999). What's the price of a research subject? Approaches to payment for research participation. *New England Journal of Medicine, 341*, 198–202.

Elbourne, D. (1987). Subjects' views about participation in a randomized controlled trial. *Journal of Reproduction and Infant Psychology, 5*, 3–8.

Erlen, J.A., Sauder, R.J., & Mellors, M.P. (1999). Incentives in research: Ethical issues. *Orthopedic Nursing, 18(2)*, 84–87.

Faden, R.R. & Beauchamp, T.L. (1986). *A History and Theory of Informed Consent*. New York: Oxford.

Family Health Research Group. (1998). The Family Health Project: A multidisciplinary longitudinal investigation of children whose mothers are HIV-infected. *Clinical Psychology Review, 18*, 839–856.

Fernandez, M.I., Perrino, T., Collazo, J.B. et al. (2005). Surfing new territory: Club-drug use and risky sex among Hispanic men who have sex with men recruited on the internet. *Journal of Urban Health, 82(1 Supp. 1)*, i79-i88.

Fernandez, M.I., Varga, L.M., Perrino, T. et al. (2004). The Internet as recruitment tool for HIV studies: Viable strategy for reaching at-risk Hispanic MSM in Miami? *AIDS Care, 16(8)*, 953–963.

Food and Drug Administration. (1998). *Information Sheets: Guidance for Institutional Review Boards and Clinical Investigators*. Rockville, Maryland: Author.

Fry, C., & Dwyer R. (2001). For love or money? An exploratory study of why injecting drug users participate in research. *Addiction, 96*, 1319–1325.

General Accounting Office. (2000). NIH has increased its efforts to include women in research [Pub. GAO/HEHES-00–96]. Washington, D.C.: Government Printing Office.

Glantz, L.H., Annas, G.J., Grodin, M.A., & Mariner, W.K. (2001). Research in developing countries: Taking "benefit" seriously. In W. Teays & L. Purdy (Eds.), *Bioethics, Justice, and Health Care* (pp. 261–267). Belmont, California: Wadsworth.

Grady, C. (2005). Payment of clinical research subjects. *Journal of Clinical Investigation, 115(7)*, 1681–1687.

Habiba, M. & Evans, M. (2002). The inter-role confidentiality conflict in recruitment for clinical research. *Journal of Medicine and Philosophy, 27(5)*, 565–587.

Hall, A.G., Collins, K.S., & Glied, S. (1999). *Employer-Sponsored Health Insurance: Implications for Minority Workers*. New York: The Commonwealth Fund.

Harro, C.D., Judson, F.N., Gorse, G.J. et al. (2004). Recruitment and baseline epidemiologic profile of participants in first phase 3 HIV vaccine efficacy trial. *Journal of Acquired Immune Deficiency Syndromes, 37(3)*, 1385–1392.

Hays, R.B. & Kegeles, S.M. (1999). Factors related to the willingness of young gay men to participate in preventive HIV vaccine trials. *Journal of Acquired Immune Deficiency Syndromes and Human Retrovirology, 20*, 164–171.

Henry J. Kaiser Family Foundation. (2003). Key facts: African Americans and HIV/AIDS. Available at http://www.kff.org/hivaids/hiv6090chartbook.cfm. Last accessed June 8, 2006.

Hull, S.C., Glanz, K., Steffen, A., & Wilfond, B.S. (2004). Recruitment approaches for family studies: Attitudes of index patients and their relatives. *IRB: Ethics & Human Research, 26(4)*, 12–18.

Jenkins, R.A., Chinaworapong, S., Morgan, P.A., Ruangyuttikarn, C., Sontirat, A., Chiu, J. et al. (1998). Motivation, recruitment, and screening of volunteers for a phase I/II HIV preventive vaccine trial in Thailand. *Journal of Acquired Immune Deficiency Syndromes and Human Retrovirology, 18*, 171–177.

Katz, J. (1972). *Experimentation with Human Beings*. New York: Russell Sage Foundation.

Levine, R. (1988). *Ethics and the Regulation of Clinical Research*. New Haven, Connecticut: Yale University Press.

Lemmens, T. & Miller, P.B. (2003). The human subjects trade: Ethical and legal issues surrounding recruitment incentives. *Journal of Law, Medicine & Ethics, 31*, 398–418.

Leonard, N.R., Lester, P., Rotheram-Borus, M.J., Mattes, K., Gwadz, M., & Ferns, B. (2003). Successful recruitment and retention of participants in longitudinal behavioral research. *AIDS Education and Prevention, 15(3)*, 269–281.

Lindegger, G.C., Wassenaar, D.R., & Slack, C.M. (2001). HIV vaccine trials in South Africa: Some ethical considerations. *Grace & Truth, 18(3)*, 20–30.

Loff, B. & Black, J. (2000). The Declaration of Helsinki and research in vulnerable populations. *Medical Journal of Australia, 172*, 292–295.

Margolis, L.H. (2000). Taking names: The ethics of indirect recruitment in research on sexual networks. *Journal of Law, Medicine & Ethics, 28*, 159–164.

Marrocco, A. & Stewart, D.E. (2001). We've come a long way, maybe: Recruitment of women and analysis of results by sex in clinical research. *Journal of Women's Health & Gender-Based Medicine, 10(2)*, 175–179.

McKechnie, J. (1976). *Webster's New Twentieth Century Dictionary*, 2nd ed. New York: Collins World.

McMahon, J.M., Tortu, S., Torres, L., Pouget, E.R., & Hamid, R. (2003). Recruitment of heterosexual couples in public health research: A study protocol. *BMC Medical Research Methodology, 3*:24. Available at http://www.biomedcentral.com/1471–2288/3/24.

Milgrom, P.M., Hujoel, P.P., Weinstein, P., & Holborow, D.W. (1997). Subject recruitment, retention, and compliance in clinical trials in periodontics. *Annals of Periodontology, 2*, 67–74.

Mills, R.J. & Bhandari, S. (2003). *Health Insurance Coverage in the United States: 2002* [Current Population Reports P60–223]. Washington, D.C.: United States Department of Commerce

Moore, R., Stanton, D., Gopalan, R. et al. (1994). Racial differences in the use of drug therapy for HIV disease in an urban community. *New England Journal of Medicine, 330*, 763–768.

Mor, V., Fleishman, J.A., Dresser, M. et al. (1992). Variations in health service among HIV infected patients. *Medical Care, 30*, 17–29.

Moreno-Black, G., Shor-Posner, G., Miguez, M.J., Burbano, X., O'Mellan, S., & Yovanoff, P. (2004). "I will miss the study, God bless you all": Participation in a nutritional chemoprevention trial. *Ethnicity & Disease, 14*, 469–475.

National Institutes of Health. (1994). NIH guidelines on the inclusion of women and minorities as subjects in clinical research. *Federal Register, 59*, 14508–14513.

Nelson, K., Garcia, R.E., Brown, J., Mangione, C.M., Louis, T.A., Keeler, E., & Cretin, S. (2002). Do patient consent procedures affect participation rates in health services research? *Medical Care, 40(4)*, 283–288.

Pearn, J.H.A (2001). The ethics of recruitment. *Medical Journal of Australia, 174(10)*, 542–543.

Ramasubbu, K., Gurm, H., & Litaker, D. (2001). Gender bias in clinical trials: Do double standards still apply? *Journal of Women's Health and Gender-Based Medicine, 10*, 757–764.

Reeder, G.D., Davison, D.M., Gipson, K.L., & Hesson-McInnis, M.S. (2001). Identifying the motivations of African American volunteers working to prevent HIV/AIDS. *AIDS Education and Prevention, 13(4)*, 343–354.

Rice, M. & Broome, M.E. (2004). Incentives for children in research. *Journal of Nursing Scholarship, 36(2)*, 167–172.

Roberts, L.W., Geppert, C.M.A., & Brody, J.L. (2004). A framework for considering the ethical aspects of psychiatric research protocols. *Comprehensive Psychiatry, 42(5)*, 351–363.

Roberts, L.W., Warner, T.D., Anderson, C.T., Smithpeter, M.V., & Rogers, M.K. (2004). Schizophrenia research participants' responses to protocol safeguards: Recruitment, consent, and debriefing. *Schizophrenia Research, 67*, 283–291.

Russell, M.L., Moralejo, D.G., & Burgess, E.D. (2000). Paying research subjects: Participants' perspectives. *Journal of Medical Ethics, 26*, 126–130.

Shapiro, M.F., Morton, S.C., McCaffrey, D.F. et al. (1999). Variations in the care of HIV-infected patients in the United States. *Journal of the American Medical Association, 281*, 2305–2315.

Stanford, P.D., Monte, D.A., Briggs F.M., Flynn, P.M., Tanney, M., Ellenberg, J.H. et al. (2003). Recruitment and retention of adolescent participants in HIV research: Findings from the REACH (Reaching for Excellence in Adolescent Care and Health) Project. *Journal of Adolescent Health, 32*, 192–203.

Stanley, B., Stanley, M., Lautin, A., Kane, J., & Schwartz, N. (1981). Preliminary findings on psychiatric patients as research participants: A population at risk? *American Journal of Psychiatry, 138(5),* 669–671.

Steeves, R., Kahn, D., Ropka, M.E., & Wise, C. (2001). Ethical considerations in research with bereaved families. *Community Health, 23(4),* 75–83.

Steinbock, B. (1995). Coercion and long-term contraception. *Hastings Center Report, 25(1),* S19-S22.

Stewart, D.E., Cheung, A.M.W., Layne, D., & Evis, M. (2000). Are we there yet? The representation of women in clinical research populations in Canada. *Annals of the Royal College of Physicians and Surgeons of Canada, 33,* 229–231.

Stone, V.E., Mauch, M.Y., & Steger, K.A. (1998). Provider attitudes regarding participation of women and persons of color in AIDS clinical trials. *Journal of Acquired Immune Deficiency Syndromes, 19,* 245–253.

Stone, V.E., Steger, K.A., Hirschorn, L.R. et al. (1998). Access to treatment with protease inhibitor containing regimens: Is it equal for all? Presented at the 12th International AIDS Conference, Geneva, Switzerland, June 28-July 3 [abstract no. 42305].

Thapinta, D., Jenkins, J.A., Morgan, P.A. et al. (2002). Recruiting volunteers for a multiphase I/II HIV preventive vaccine trial in Thailand. *Journal of Acquired Immune Deficiency Syndromes, 30,* 503–513.

Thompson, E.E., Neighbors, H.W., Munday, C., & Jackson, J.S. (1996). Recruitment and retention of African American patients for clinical research: An exploration of response rates in an urban psychiatric hospital. *Journal of Consulting and Clinical Psychology, 64(5),* 861–867.

Turner, B.J., Cunningham, W.E., Duan, N. et al. (2000). Delayed medical care after diagnosis in a US national probability sample of persons infected with HIV. *Archives of Internal Medicine, 160,* 2614–2622.

United States Department of Health, Education, and Welfare. (1973). *Final Report of the Tuskegee Syphilis Study Ad Hoc Advisory Panel.* Washington, D.C.: Tuskegee Syphilis Study Ad Hoc Advisory Panel.

United States Department of Health and Human Services, Office for Protection from Research Risks. (1993). IRB guidebook. Washington, D.C.: United States Department of Health and Human Services. Available at http://www.hhs.gov/ohrp/irb/irb_guidebook.htm; last accessed May 17, 2006.

Vidader, R.M., Lafleur, B., Tong, C., Bradshaw, R., & Marts, S.A. (2000). Women participants in NIH-funded clinical research literature: Lack of progress in both representation and analysis by sex. *Journal of Women's Health and Gender-Based Medicine, 9,* 495–504.

Warner, T.D., Roberts, L.W., & Nguyen, K. (2003). Do psychiatrists understand research-related experiences, attitudes, and motivations of schizophrenia study patients? *Comprehensive Psychiatry, 44(3),* 227–233.

Weissman, J.S., Mkadon, H.J., Seage, G.R. et al. (1994). Changes in insurance status and access to care by persons with AIDS in the Boston Health Study. *American Journal of Public Health, 84,* 1997–2000.

Witte, S.S., El-Bassel, N., Gilbert, L., Wu, E., Chang, M., & Steinglass, P. (2004). Recruitment of minority women and their main sexual partners in an HIV/STI prevention trial. *Journal of Women's Health, 13(10),* 1137–1147.

Wong, M.D., Shapiro, M.F., Boscardin, W.J. et al. (2002). Contributions of major diseases to disparities in mortality. *New England Journal of Medicine, 347,* 585–592.

Case Study One
Ethical Issues in Internet-Based HIV Primary Prevention Research

B. R. Simon Rosser, PhD, MPH, and Keith Horvath, PhD

Behavioral Intervention Trials

Behavioral intervention trials use similar methods as drug intervention trials when rigorously testing the efficacy of an intervention. However, unlike drug trials, behavior is the primary outcome of interest (in this case, reduction in unsafe sex behavior) and masking the condition to which participants are randomized is often not possible. In behavioral randomized controlled trials (RCTs), participants are prospectively assigned to either the new intervention (e.g., counseling, group intervention, community level awareness) or to a control condition (usually a comparison "tested" intervention). Baseline and follow-up surveys are used to assess self-reported sexual or drug using behavior over a preceding time period, typically 2–3 months. At baseline, provided the randomization was effective, there should be no significant differences between those in the treatment or control conditions on risk factors or co-factors. At follow-up (typically 3-months for short-term interventions or 12–18 months for long-term HIV prevention), any differences in knowledge, attitudes and behavior are attributed to the intervention.

Men who have sex with men (MSM) are among the most infected and affected populations with HIV, both nationally and globally. While the reasons why MSM are at at increased risk for HIV are multifactorial, identified barriers to HIV prevention for this population include less access to education, greater experience of discrimination, lack of self-acceptance of sexual orientation and greater stress (US Surgeon General, 2001). Even online, accessing good information is difficult. Some government websites often use well-meaning but vague terms like "always use a condom", while other websites provide different, and at times, inaccurate or unhelpful information about homosexuality. For these reasons, the need for effective HIV prevention that addresses this community's sexual health concerns is particularly pressing.

Introducing the Study

The potential of the Internet to reach at-risk and stigmatized populations on a global scale makes researching Internet-based interventions of great significance to HIV prevention. The leading challenge of HIV primary prevention is demonstrating whether online interventions can effectively promote health by reducing risk behavior, especially among those at greatest risk of acquiring or transmitting disease.

The *Men's INTernet Sex II (MINTS-II)* study is one of the first NIH-funded HIV prevention trials of its kind with three aims:

1. To investigate risk behavior with sex partners met online among Men who use the Internet to seek Sex with Men (MISM);
2. To develop a highly-interactive, Internet-based HIV prevention intervention that is theoretically-sound and state-of-the-art in both HIV prevention and e-learning; and
3. To conduct an RCT to test the effects of the new Internet-based HIV prevention intervention for MISM on participants' risk behavior.

This study has been ethically challenging for multiple reasons. First, the field of study is highly innovative. As discussed below, there are frequently ethical dimensions to practical questions, and thus we find ourselves adapting principles. Second, some ethical challenges are secondary to technological innovation. The Internet is complex: it may be viewed simultaneously as a *tool* or medium by which to impart information; an *environment* where individuals join together to become virtual communities; and a *culture* with its own language, rules, and styles of communication (Rosser et al., 2007). The scope of the Internet brings challenges. While it is global in reach (i.e., the world wide web), it is both a public space (people feel free to surf anywhere on the net) and a private medium (for example chat rooms, web cams, and email where people may engage in highly personal, private communications). Third, sexual behavior is intimately tied to the Internet, which may be viewed as the largest sex venue in the world. The Internet is a widely used tool for individuals to get their sexual information, obtain sexually explicit material, and meet other people both socially and for sex. As an example, Gay.com currently has 3.0 million members in the US, which numerically represents a gay virtual city approximately the size of Chicago. Sexual minorities including MSM were early adopters of the Internet, and many (14% in an earlier study we conducted) report only meeting men for sex online (not conventionally). All these characteristics have made ethical decisionmaking more challenging.

The Ethical Dilemmas

In this paper we highlight five ethical dilemmas that we have encountered.

The first ethical dilemma was *who to target*? In HIV prevention, there is frequently a trade-off: the more targeted or personalized an intervention, the more effective it is, but the more broad or impersonal an intervention, the more people may find it relevant. So we had to decide whether the first highly interactive HIV prevention online intervention should attempt to target all MISM, or only a subgroup? To make the dilemma more real, we asked the question this way: Is it more important to address the HIV prevention needs of an 18-year old, HIV negative white student living in North Dakota who is just coming out, becoming sexually active, and likely needs education about his sexuality and information on HIV prevention, or is it more important to target the 45-year old HIV positive African American living in downtown New York who has already received a lot of education about HIV? Is there a way of addressing their common prevention needs, or if their needs are too different, who deserves HIV prevention more?

The second ethical dilemma we encountered was *how explicit to be*? For example, when talking about sexual acts, should we use the word "intercourse" (highly professional but very clinical and an advanced reading level), "sex" (a middle ground term that can be ambiguous), or "fuck" (a clear term but also offensive to some participants)? Sexual explicitness presents a difficult dilemma, because MISM are a diverse group with opinions and values ranging from very traditional to highly non-traditional. The language with which one man may most easily identify to protect his health, may offend another. How do you chose between promoting good and doing as little harm as possible (causing offense and turning people off to prevention)? In face-to-face individualized interventions, HIV preventionists can circumvent this problem by using the language a client uses or prefers. In group interventions and public pamphlets, language typically is "toned down" to that needed to convey the information while trying not to offend. However, while MSM may be diverse, most gay men talk about sex differently and far more openly than other groups. Similarly, online communication is typically much more direct, explicit and visual than offline communication (Gurak and Lannon, 2003). Consider as well, the clearer and more direct the language, the greater the chance that the message will be received as relevant and understood. The more visually explicit (e.g., watching a condom being placed on a penis), the greater the chance that men will mirror the behavior, but also that others may judge the site as pornographic.

The third dilemma was *how to make the new intervention most effective*? The best online learning environments keep content fun, alive and entertaining and ultimately improves learning and retention (Allen, 2003). However, some people may consider keeping sex fun, alive and entertaining to be inadvertently encouraging men to have more sex, thereby possibly increasing HIV risk. We also considered that in a highly conservative political environment, interventions that gain greater public attention may be at greater risk for being targeted by groups opposed to programs that benefit MSM.

The fourth dilemma was *how participant-to-participant interactive to make the site*? In traditional offline interventions, the presenter has some control over the participants' interactions. Creating a forum for participants to interact with one another during an online intervention has the advantage of promoting virtual community and possibly establishing safer sex group norms. However, because of the anonymous nature of the Internet, online communication may turn hostile (i.e., "flaming") and, in some cases, participants may promote unsafe sex values. While in a principle of community-education is trust of the community. But if trusting the community destroys the purpose of the intervention, how interactive should we make this intervention?

The fifth dilemma was in the *choice of control group*. In a drug study, the typical design is a double-blind placebo-controlled trial. In a behavioral trial, the typical control is state-of-treatment, offered both so that researchers can test whether the new intervention is more effective than what is already available, and to address the ethical necessity that everyone gets at least some information (usually the standard-of-care) on how to lower their HIV risk behavior. But on the internet, what is a standard-of-care? Thus, we asked ourselves whether we should (a) use a null control and just let people search for themselves, (b) require that they read an online brochure about HIV risk reduction, as is sometimes used in offline studies, or (c) send them to specific sites to assess whether the new intervention is superior to existing sites but have not been rigorously tested?

Our Solutions and Why We Chose Them

In choosing *who to target,* ethically we reasoned there is no way to value one life over another. Both men, in our example, have a right to education, but with limited resources, how can we do the greatest good? To address the dilemma of whose needs to prioritize, we took an empirical approach. In order to make the best informed decision, we collected baseline data on risk behavior. A priori, we decided that if a subgroup of MISM reported significantly higher risk than other subgroups, we would target them first *because targeting prevention to the highest risk group should reduce the greatest amount of HIV transmission.*

We collected formative research data on from 2,718 MISM living in the US to inform our decision, but the data turned out to be more complex than anticipated. Across demographics, most of the sample reported no unprotected anal intercourse in the last 90 days with male partners either met online (n = 2059 of 2710 or 76%) or offline (n = 2237 of 2709 or 83%). African American MISM did report significantly more (almost twice as much) unprotected anal intercourse (UAI) than other groups, however, as shown in Table 7.1, it was a small number of men that (those in the top 10th percentile) who reported UAI with large numbers of men, that accounted for the differences in risk found across race and ethnicity. Thus, based on the evidence, we decided to target the 10–18% of men across all races and ethnicities who engage in sex with the most partners.

Another important demographic characteristics was HIV status. HIV positive MISM (n = 119 or 5%) were almost 4 times more likely to report UAI with multiple men than non-HIV positive MISM. The dilemma became do we target a large group of HIV negative men to help them avoid becoming infected or a small number of HIV positive men who appear at great risk of infecting large numbers

TABLE 7.1 Risk of engaging in unprotected anal intercourse, last 90 days, by the race/ethnicity of men who use the internet to seek sex with other men (N = 2,883 MISM, aged 18 years or old, US residents)

Race/ethnicity	N	# UAI partners		p50	Percentiles			Relative risk		
		Mean	SD		p90	p95	p99	Max	RR*	p
White-American	728	0.96	3.14	0	2	5	12	45	–	–
Latino-American	683	1.42	6.80	0	3	5	19	150	1.5	.09
Asian-American	512	0.74	3.53	0	2	3	6	70	0.9	.34
African-American	445	2.11	9.86	0	2	4	8	60	1.9	.005
Other/multiracial	348	1.12	3.76	0	0	5	19	150	1.5	.35

*Negative binomial regression, $\chi^2(3) = 10.2$ p = .01.

of men, but need specialized intervention to address prevention from an HIV positive's perspective. We decided both were important, yet our resources were insufficient to respond to both. Thus, because HIV-positive men constituted such a small percentage of our assessment sample, we prioritized prevention for uninfected or HIV status unknown MISM. Concurrently, we are conducting formative research to address the online needs of HIV-positive persons.

The second ethical dilemma was *how explicit to be*? The decision to target the men engaging in the most risk behavior helped us determine that the language had to be highly explicit and direct (to be credible). We prioritized providing explicit information that may save men from becoming infected as more important than possibly offending someone coming across our website or someone at lower risk. But how explicit is explicit enough or too explicit (to the target population)? What about racial, cultural, age and other differences in what is considered too sexually explicit? And could we protect the study from those opposed to such education shutting it down?

We decided the answers to these questions needed to come from the target population. Here, the principle is that all communities have the right to be consulted on how they are portrayed and respect in how health professionals address them. Hence as part of the survey, we conducted a needs assessment of what MSM would find acceptable in an online sexual health intervention. To illustrate, we asked questions such as, "How acceptable to you is . . ." sexually explicit language, pictures of nude men, images of male-male sex, and audio of men speaking sexually. Over 90% of 2,718 MISM rated sexually explicit words, images, audio, and video helpful and acceptable. Only minor differences in acceptability were found across age, HIV status, race/ethnicity, religion, religiosity and other demographics. Nevertheless, a minority of men may find some materials, notably images of male-female sex, less acceptable. Hence, as we develop the intervention modules, we are exploring whether we can give these men options in choosing less explicit versions of the curriculum (although time and budget constraints make this prohibitive in Version 1.0).

Building on the first two decisions, that these men are at extreme risk and report wanting realistic sexually explicit education addressing a wide range of sexual health topics, the third dilemma (*how to make the new intervention most effective*) became moot. There is no scientific evidence that sexually explicit health interventions increase risk behavior. Further, "toning the intervention down," we reasoned would lead us to deliberately create a barrier to effective HIV prevention. Even though some might consider it politically expedient or prudent to "tone down" the intervention, this misuses public funds given implicitly to produce the strongest, not the weakest, intervention.

The fourth dilemma (*how participant-to-participant interactive to make the site*) pits values of community trust against having an intervention of integrity. We considered running facilitated chat interventions, or asynchronous chat where we could intervene if necessary. But such options are unlikely to be sustainable in an intervention that we hope ultimately will be automated and live, 24/7. Here, a middle option became appealing: developing modules where the community can participate and contribute (e.g., bulletin boards on favorite sexual terminology), provide referrals to experts to answer specific questions (e.g., email the experts), while keeping core elements of HIV prevention focused on risk reduction to prevent flaming.

The final dilemma, choice of control group, was the most difficult, and one on which our team could not agree even after substantial discussion. Some argued, based on the Tuskegee syphilis study, that a null control was ethically unacceptable. Since we know our subjects are engaging in high risk behavior, at a minimum those in the control should receive information warning them of their risk. To study these men for 18 months while they continue to engage in risk behavior would be to know of substantial risk yet do nothing. This part of the team argued the ethical principle, "first, do no harm." Another part of the team argued for a null control. They noted that the greatest good is served by knowing whether we can lower HIV risk online. As one of the first studies of this type, if we do not find it effective – and providing education in the control makes this less likely – then a whole area of science may incorrectly conclude online interventions are ineffective. This, in turn, could cause immeasurable harm to all of public health. This part of the team noted also that, by definition, all our participants are already online, and hence could search on "HIV prevention" and easily find information. Also, they were concerned that requiring people to read websites would come across as artificial. Finally, some argued to send subjects to other HIV prevention websites, thus controlling for time and effort. But others argued this would threaten the scientific validity. Technically, it would change the study to an intervention-comparison study where the other websites have not been rigorously evaluated. If a difference was found, we would not know if our intervention was successful or whether the other websites were harmful. Conversely, if no difference was found we would not know if both or neither sites were effective. Science, and the whole point of the study, would be compromised.

Since as a team we could not agree between the first two options, we took two approaches. First, we wrote an application for additional funds so that we could study both control conditions – a null condition and a minimal ethical standard condition – which would answer what the right control should be. As this was not funded, we had to make a choice. So we decided to identify five experts in the field (who did not have an investment in the outcome), to present the three options to them in a neutral way, and to abide by whatever the majority decided. Four of the five consultants recommended choosing a null control condition. They argued that until the effects of other types of interventions (e.g., reading online pamphlets or going to other sites) is known, the risk of introducing contamination into the study outweighed the benefits. Since MISM by definition can easily access other sites, and may be more likely to do so as a result of participating, we have not prevented them from accessing other HIV prevention information (an important difference to the Tuskegee experiment), and indeed provide a more real life control condition. The team abided by their decision.

Summary and Conclusions

There are at least seven ethical lessons we have learned from this study:

1. When working in a new area (e.g., online interventions), many if not most decisions include an ethical component. It is important to anticipate ethical dilemmas.
2. When ethical issues are identified, whenever practical, we bring them to the team for discussion. Since our team is multidisciplinary, and includes experts in ethics, this frequently leads to many points of view and spirited discussions.
3. When agreement cannot be reached, all deferring to a third party or method of resolution appears both effective and respectful.
4. Behavioral intervention trials can be complex, as ethics, values, politics, and personal morality all interplay.
5. It is not enough to focus on the needs of the study, the public, or on what is acceptable or prudent in order to identify what is right. In health and in science, the rights and needs of minorities and those most vulnerable must be prioritized.
6. At times, ethical principles such as "do good" and "do no harm" can end up seeming in opposition.
7. When confronting an ethical dilemma, it is helpful to ask, "what do we need to make the best *informed* decision?" In this study, answers included consulting the target population, formative evaluation research, external consultation, expert advice, group discussion, and seeking other ways to address equally important priorities.

Acknowledgment. We would like to acknowledge the team of investigators on MINTS, all of whom have helped identify and resolve the ethical dilemmas identified above. We also express thanks to the external consultants and to the staff of our university's IRB with whom we consult regularly and consider a national leader in human subjects' considerations in Internet-based studies. *MINTS-II* is a NIMH-funded study to develop and test the effects of "next generation" online, highly interactive HIV prevention interventions that at time of writing is still in progress. For more information about this study, go to www.mints.umn.edu.

References

Allen, M.W. (2003). *Michael Allen's Guide to E-Learning: Building Interactive, Fun, and Effective Learning Programs for any Company.* Hoboken, NJ: John Wiley & Sons, Inc.

Gurak, L.J. & Lannon, J.M. (2004). *A Concise Guide to Technical Communication.* New York, NY: Pearson Longman.

Hooper, S., Rosser, B.R.S., Horvath, K.J., Remafedi, G., Naumann, C. and the Men's INTernet Study II (MINTS-II) Team. An evidence-based approach to designing appropriate and effective Internet-based HIV prevention interventions: Results of the Men's INTernet Sex (MINTS-II) Study needs assessment. *Abstracts of the XVI International AIDS Conference, Toronto, Canada, August 17, 2006, #ThPe0451.*

Office of the Surgeon General [of the United States] (2001). *Surgeon General's Call to Action to Promote Sexual Health and Responsible Sexual Behavior. At:* www.surgeongeneral.gov/library/sexualhealth/call.htm.

Pequegnat, W., Rosser, B.R.S., Bowen, A.M., Bull, S.S., DiClemente, R.J., Bockting, W.O. et al. (2007). Internet-based HIV/STD prevention survey lessons: Practical challenges and opportunities. *AIDS & Behavior, 11(4),* 505-521.

Rosser, B.R.S., Oakes, J.M., Konstan, J., Remafedi, G., Zamboni, B., and the Men's INTernet Sex II (MINTS-II) Team. Demographics Characteristics of HIV Sexual Risk Behavior by Men who use the Internet to seek Sex with Men: Results of the Men's INTernet Sex Study-II. *Abstracts of the XVI International AIDS Conference, Toronto, Canada, August 15, 2006, #TuPE0615.*

Rosser, B.R.S., Bockting, W.O., Ross, M.W., Miner, M.H., & Coleman, E. (2006). The relationship between homosexuality, internalized homonegativity and mental health in men who have sex with men. *Journal of Homosexuality,* in press.

Chapter 8
Clinical Trials

Designing Clinical Trials

Types of Trials

There are various types of clinical trials. The specific procedures in a clinical trial depend upon the type of trial that is to be conducted. Trials may be

- *Treatment trials*, to test experimental drugs, combinations of drugs, or surgical or radiological interventions to treat a disease or condition;
- *Prevention trials*, to prevent the initial development or recurrence of a disease or condition;
- *Diagnostic trials*, to find new methods for the diagnosis of a disease or condition;
- *Screening trials*, to examine new ways to detect a disease or condition; or
- *Quality of life trials*, to develop interventions to improve the quality of life for individuals suffering from chronic diseases (National Institutes of Health, 2005).

Phases of Trials

Phase I: Assessing Toxicity and Dosage

Clinical trials can be designed to be Phase I, II, or III trials. Phase I trials are designed to evaluate the safety, tolerability, pharmacokinetics, and pharmacody-namics of a therapy. Phase I trials often include dose-ranging studies that permit the refinement of doses for clinical use; the range of doses to be tested are usually a small fraction of the dose that causes harm in animal testing. Phase I trials usually involve between 20 and 80 participants (Chow and Liu, 2003; Meinert, 1986) and may last as long as 8 to 12 months (Canadian HIV/AIDS Legal Network, 2002). Although this phase frequently relies on healthy volunteers to participate, HIV drug trials and cancer trials rely on individuals with the disease that is the focus of the trial (Chow and Liu, 2003; Meinert, 1986).

Phase I trials may be formulated as single ascending dose studies (SAD) or multiple ascending dose studies (MAD). In SAD studies, groups of three or six patients are given a small dose of the drug and then are observed for a specific period of time. If they do not exhibit any adverse side effects, a new group of patients is then given a higher dose. This is continued until intolerable side effects begin to appear. At this point, the drug is said to have reached the maximum tolerated dose (MTD) (Chow and Liu, 2003; Meinert, 1986).

MAD studies are conducted to better understand the pharmacokinetics and pharmacodynamics of the drug being tested. In these studies, participants are administered a low dose of the drug and the dose is then increased up to a prede-termined level. Blood and other samples are collected at various points in time and are analyzed to understand how the drug is processed within the body.

Phases II and III: Clinical Efficacy

Phase II trials are conducted only after the initial safety of the therapy has been confirmed in Phase I trials. Phase II trials, which typically involve several hundred participants, are designed to assess clinical efficacy of the therapy. Phase II trials may also continue Phase I assessments in a larger group of volunteers and patients. Phase II may be designed to evaluate the toxicity and efficacy of the new treatment, or may be divided into Phase IIA and Phase IIB. Phase IIA is designed to assess dosing requirements, while Phase IIB focuses on treatment efficacy (Chow and Liu, 2003; Meinert, 1986). Phase II trials usually involve 50 to 500 volunteers and may require 18 to 24 months to complete (Canadian HIV/AIDS Legal Network, 2002).

Ethically, clinical trials to assess the efficacy of a treatment are not to be initiated unless *clinical equipoise* exists with respect to the experimental treatment and the treatment to which it is being compared. For instance, at the commencement of the 076 protocol to assess the efficacy of zidovudine (AZT), it was unknown whether the AZT regimen was more efficacious, less efficacious, or of the same level of efficacy as a placebo. [Efficacy refers to "the extent to which a specific intervention, procedure, regimen or service produces a beneficial result under ideal conditions" (Last, 1988: 41). It is often confused with the concept of effectiveness, which is "the extent to which a specific intervention, procedure, regimen, or service, when deployed in the field, does what it is intended to do for a defined population" (Last, 1988: 41).] "The aim of the study must be to disturb equipoise and, thus, alter clinical practice . . . (Crouch and Arras, 1998: 27).

In contrast to the relatively small size of Phase I and II studies, Phase III stud-ies are much larger and typically involve 1,000 to 3,000 participants. They are often conducted as large double-blind randomized controlled trials to assess the efficacy of the new intervention. For the purpose of this discussion, *double-blind randomized controlled trial* refers to drug trials in which the study participants, the investigators, and the outcome assessors do not know which participants are receiving the experimental drug and which are receiving the comparison drug or the placebo (Heckerling, 2005). In contrast, a *single-blind randomized controlled trial* involves the withholding of treatment assignment to only the study partici-pants; the investigators and the outcome assessors, who are often one and the same, know which participants are receiving which treatment exposure. In both cases, participants are randomized (assigned) to any of two or more study arms, each of which involves the administration of a different treatment or treatments.

One of the arms may be a placebo, so that the study is a *single- or double-blind randomized placebo controlled trial.*

Each of these designs is accompanied by various advantages and disadvantages. Double blinding increases scientific validity by protecting against ascertainment bias, whereby a provider's knowledge of the individual's treatment affects the outcome assessment, and expectation bias, whereby a participant's knowledge of his or her treatment affects his/her response (Heckerling, 2005). The protection against ascertainment bias is lost with a single blind trial. A potential benefit of single blind trials over double blind trials is the ability of the investigator to monitor directly any adverse treatment effects in the study participants. This may be particularly important in small single center trials (Heckerling, 2005).

In some cases, individuals suffering from serious or life-threatening conditions, such as HIV/AIDS, may be able to access an experimental drug on a nonrandomized basis while a clinical trial is being conducted. This is known as a treatment IND (investigational new drug) protocol (Veatch, 1989). Various objections have been raised to this practice, premised on both ethical and methodological concerns. First, it has been argued that if the intervention is scarce or expensive, society has an interest in making sure that it is used efficiently in order to answer the relevant scientific questions. Second, investigators should not be forced to provide medications in ways that violate their personal or corporate conscience, that is, they should not be required to provide medications in situations that they believe entail morally unacceptable risks. It has also been argued, however, that the principle of autonomy gives individuals not only the right to receive the standard treatment, but also the right to access the experimental intervention.

In contrast to a randomized controlled clinical trial, in which the participant does not know to which study arm he or she has been assigned, a drug trial that utilizes a *quasi-experimental design* allows participants to decide to which of the study arms they would like to be assigned. The study of didanosine (ddI) conducted by the Medical Research Council (U.K.) in HIV-symptomatic zidovudine (AZT)-intolerant individuals provides an example of this type of design. The study was designed with two arms; participants could elect the arm in which they wanted to participate. If the individual selected Arm A, he or she would be randomized to high dose, low dose, or placebo, while those individuals selecting Arm B would be randomized to either high dose or low dose didanosine (Institute of Medical Ethics Working Party on the Ethical Implications of AIDS, 1992).

In contrast to placebo-controlled trials, *equivalency trials*, in which the experimental treatment or intervention is compared to a standard treatment for the same condition, may be undertaken to compare the efficacy of the new treatment or to assess other aspects of the treatment, such as undesired side effects. The *International Ethical Guidelines for Biomedical Research Involving Human Subjects* (CIOMS, 2002), explains in Guideline 11 how equivalency trials might be used in countries lacking an effective intervention for a particular condition:

An equivalency trial in a country in which no established effective intervention is available is not designed to determine whether the investigational intervention is superior to an established effective intervention currently used somewhere in the world; its purpose is, rather, to determine whether the investigational intervention is, in effectiveness and safety, equivalent to, or almost equivalent to, the established effective intervention. It would be hazardous to conclude, however, that an intervention demonstrated to be equivalent, or almost equivalent, to an established effective intervention is better than nothing or superior to whatever intervention is available in the country; there may be substantial differences between the results of superficially identical clinical trials carried out in different countries. If there are such differences, it would be scientifically acceptable and ethically preferable to conduct such 'equivalency' trials in countries in which an established effective intervention is already available.

If a drug is demonstrated to be efficacious in Phase III trials, the trial results are compiled into a document that contains a description of the methodology utilized, the results of human and animal studies, the manufacturing procedures, the formulation details, and the shelf life (Chow and Liu, 2003). This document comprises the regulatory submission to be made to the relevant regulatory authority in the country in which approval for marketing is to be sought. In the United States, the submission is made to the Food and drug Administration (FDA) (Chow and Liu, 2003).

Phase IV: Post-Marketing Surveillance

Phase IV trials involve safety surveillance and ongoing technical support of a drug following its introduction into the market. These studies are designed to detect adverse effects that may occur with long-term use of the drug or in combined use with other drugs. Such studies may be initiated by the manufacturer on a voluntary basis or may be required by the regulatory agency that granted marketing approval (Chow and Liu, 2003).

Ethical Issues

Access to Participation

In recruiting and enrolling individuals for participation in a clinical trial, prospective participants will be evaluated for their eligibility based upon a pre-formulated listing of inclusion and exclusion criteria. Inclusion generally requires that individuals possess a specified set of characteristics or a specified disease or condition. Often, certain demographic characteristics will also be specified where it is relevant to the underlying condition or treatment under study. For instance, if a condition manifests primarily in women, a study might be restricted to female participants. Exclusion criteria are developed to exclude from participation individuals who might have conditions that would render it more difficult to understand the effect of the intervention, or who might be at increased risk of harm during participation as a result of the condition. As an example, it would be harder to evaluate the effect of an experimental treatment on the neurological effects of HIV/AIDS if participants had a co-occurring non-HIV-related neurological condition whose symptoms were similar to the neurological manifestations of HIV. This restriction of enrollment to individuals with specified characteristics who lack other enumerated characteristics, together with randomization, helps to ensure a more homogenous participant sample, which reduces the number of individuals needed to participate in a study, facilitates the statistical analysis, and reduces the costs associated with conducting the clinical trial. (It is beyond the scope of this book to examine the statistical implications of restriction and randomization. Interested readers are urged to consult the listed references by Chow and Liu, Feinberg and Japour, and Meinert, below.)

Because of the methodological need to restrict enrollment to individuals possessing certain characteristics and not others, minorities and women have been excluded from participation in clinical research, including HIV-related clinical trials, until relatively recently (Armistead et al., 2004; Daunt, 2003; Kass, Taylor, and King, 1996). Pregnant women, in particular, are likely to be excluded from HIV-related trials focused on anything other than vertical transmission, due to fears of teratogenicity and the legal liability that might arise if a fetus were injured as a result of the mother's participation (Caschetta, Chavkin, and McGovern, 1993; Kass, Taylor, and King, 1996). Additional factors believed to be responsible for this lack of participation include stigmatization, employment obligations that permit little time off from work, lesser financial resources to pay for child care or transportation expenses associated with study participation, and poor literacy skills that reduce individuals' ability to participate in survey- or questionnaire-based research (Armistead et al., 2004; Family Health Research Group, 1998). Attitudes and perceptions of providers may also be key to the underrepresentation of women and minorities in HIV/AIDS clinical trials. One study found that providers believed that African Americans, Latinos, and Haitians were less likely than non-Hispanic whites to be interested in HIV clinical trials (Stone, Mauch, and Steger, 1998). A proportion of the providers reported that, based on these perceptions, they were less likely to inform Latinos and Haitians about the possibility of clinical trial participation.

This relative lack of access to clinical trials of minorities and women contravenes the ethical principle of justice, which provides for an equitable distribution of the burdens and benefits of research across groups. There may also be legal issues that arise as a result of this relative lack of access. It could be argued, first, that exclusionary practices or policies are unfair to specific classes of persons. If certain benefits, such as a higher standard of medical care, are available only through participation in research, then the exclusion of specific groups from that research results in disproportionate harm to those classes of persons (Kass, Taylor, and King, 1996). And, if a drug is prescribed to individuals within these excluded classes but has never been tested on them, the prescribing physician and/or the pharmaceutical manufacturer may be potentially liable for harm that befalls the woman or fetus as a result of its use (Kass, Taylor, and King, 1996).

Prisoners as a class are generally unable to participate in clinical trials due to significant federal restrictions on their participation. These restrictions were formulated because of prisoners' relative inability to act autonomously within the prison environment. Federal regulations currently provide that studies may not be conducted with prisoners unless they are

(a) studies of the possible causes, effects, and processes of incarceration and criminal behavior;
(b) studies of prisons as institutional structures or of prisoners as incarcerated persons;
(c) research on conditions particularly affecting prisoners as a class; or,
(d) research on innovative and accepted practices that have the "intent and reasonable probability of improving the health or well-being of the subject" (Code of Federal Regulations, 2006).

Although federal regulations severely curtail inmates' access to clinical trials, they do not prohibit their participation in trials that do not contain a placebo or no treatment arm. It is theoretically possible, then, for inmates to participate in HIV-related clinical trials. This may be critical due to the high proportion of inmates in correctional facilities infected with HIV. For example, studies conducted in New Jersey in 1991 and 1992 indicated that almost 9% of adult male inmates and more than 14% of adult female inmates were HIV-seropositive. Among those with histories of injection drug use, the proportion increased to 40% of the men and almost 43% of the women (Stein and Headley, 1996).

However, additional restrictions on prisoner participation in clinical trials are often imposed by the states themselves. Of 30 state correctional systems surveyed in 1994, 9 had policies that permitted prisoners to participate in HIV-related trials, 19 specifically prohibited their participation, and 2 had ambiguous policies relating to individual participation in research (Collins and Baumgartner, 1995). At the time of the study, 8 of the 9 prison systems permitting participation reported that a total of 185 prisoners were enrolled in research.

It has been argued that the severe restrictions on prisoner participation in clinical trials deprives individuals of the benefits as well as the burdens of research, thereby violating the principle of justice, and fails to permit individuals to exercise their autonomy, in contravention of the principle of respect for persons (Pasquerella, 2002; Stein and Headley, 1996). And, although prison systems may prohibit inmate participation in clinical trials due to a fear of lawsuits (Collins and Baumgartner, 1995), they may be equally susceptible to lawsuit for refusal to permit such access. In New Jersey, for instance, prisoners initiated a class action lawsuit challenging their lack of access to drug trials which resulted in their ability to assess for themselves the risks and benefits associated with their enrollment in HIV/AIDS-related clinical trials (DeVesa, 1993; *Roe v. Fauver*, 1992).

Randomization

Several scholars have asserted that the very structure of clinical trials interferes with individual autonomy and, therefore, the ethical principle of respect for persons. Kodish, Lantos, and Siegler (1990) argue, for instance, that "patient autonomy in RCT is completely safeguarded only if the patient is free to choose (without agreeing to participate in the RCT) any therapy which they might have received by participating in the RCT and is equally free to choose the randomization alternative." This assumes, however, that individuals have, or should have, the right to choose an intervention before its safety and efficacy have been established (Freedman, 1991), which often occurs following a series of clinical trials. And, because clinical trials are not to be conducted unless clinical equipoise exists, there is an underlying assumption, as well, that any treatment must be better than no treatment (placebo) (Freedman, 1991).

This perspective seems to suggest that trials must rely on quasi-experimental designs, which would permit individuals to choose the treatment they are to receive. However, if an insufficient number of individuals selected participation in the placebo arm of the trial, there would be inadequate statistical power to conduct analyses of the results.

A modified consent process, known as the Zelen design, provides prospective participants with a seemingly increased ability to choose. This process involves the randomization of individuals prior to seeking their consent (Zelen, 1979, 1990). In a "single consent" design, participants randomized to the control arm are not asked for their consent; they receive the standard treatment without mention of the trial. Persons randomized to the experimental arm are asked for their consent; if they refuse, they receive the standard of care. This process has been deemed suitable when the experimental treatment is available only in the context of the clinical trial and when it is being compared to the standard of care (Homer, 2002). The "double consent" process asks both those randomized to the experimental arm and those allocated to the standard treatment for their consent. Those in the standard care arm who decline to participate are provided with the experimental or other treatment. Individuals randomized to the experimental arm who decline to participate are provided with the standard treatment. In both the single and double consent processes, the investigators know who is approached for enrollment into which arm.

Although the Zelen design may facilitate recruitment because individuals know whether they will be receiving the experimental intervention, there are serious ethical questions associated with its use. Methodologically, there may be inadequate statistical power to analyze the results depending upon how many individuals have ultimately selected the experimental versus the standard treatment. Second, the procedure requires the collection of data from individuals who have declined to participate and others who do not know that their data are being used for research. This use of data contrary to the wishes of the individuals violates the basic ethical principle of respect for persons and the requirement for informed consent. Ultimately, the use of this procedure to make individuals unwitting participants in research could undermine public confidence in research and researchers.

Standard of Care and Participant Safety

Standard of Care

Significant attention has been focused on the standard of care to be provided to individuals who are randomized to the control arm of a clinical trial. Although this should be a concern regardless of the disease that is the focus of a clinical trial, the issue has been most acutely raised in the context of treatment and prevention trials related to HIV/AIDS. This section briefly reviews the ethical concerns and arguments raised that relate to the standard of care to be provided in the context of clinical trials. The case study by Whalen that follows this chapters explores these issues in greater detail.

The 076 Treatment Trial. In 1994, the Pediatric AIDS Clinical Trials Group (PACTG) conducted a large randomized controlled trial to assess the efficacy of a zidovudine (AZT) regimen in reducing vertical transmission of HIV from pregnant women to their unborn infants. This protocol required the initiation of oral AZT after 14 weeks' gestation, intravenous AZT during labor and delivery, and AZT for the infant for six weeks following delivery (Connor et al., 1994). The study was halted following the findings from preliminary data, which indicated that the transmission rate among women receiving the zidovudine was 8.3%, compared to 25.5% among the women receiving the placebo. The course of treatment could reduce HIV transmission by as much as 68%.

Although of great importance in reducing HIV transmission, the drug regimen was not unproblematic. The estimated cost per patient for this treatment was estimated at $800 for zidovudine alone, apart from the cost of establishing and maintaining the infrastructure necessary to provide prophylactic care, including the intravenous delivery of the drug during labor and delivery (Varmus and Satcher, 1997). This expense was unaffordable in developing countries, where maternal-infant HIV transmission was significant.

These trials generated intense, acrimonious debate within the scientific and bioethics communities. In 1997, Lurie and Wolfe identified 15 placebo-controlled HIV-related clinical trials being conducted in developing countries by the U.S. and other nations. Lurie and Wolfe claimed that these trials were in contravention of international guidelines for the conduct of biomedical research (Lurie and Wolfe, 1997). The then-existing version of the Helsinki Declaration provided:

The potential benefits, hazards and discomfort of a new method should be weighed against the advantages of the best current diagnostic and therapeutic methods.

Various commentators argued vociferously against the conduct of these trials, asserting that, following the findings of the 076 study, investigators were ethically mandated to provide research participants in the control arm of the study with the standard of care that is prevalent in the West (Angell, 1997; Lurie and Wolfe, 1997). This would represent an equivalence trial, whereby the efficacy of the reduced course of AZT provided to participants in the experimental arm would be compared to that of the 076 protocol. Proponents of this position have remained adamant (Annas and Grodin, 1998), despite the proffered arguments relating to the affordability or sustainability of the longer course of treatment in the participants' own country (Varmus and Satcher, 1997); the desire of the country to find a less expensive, albeit somewhat less efficacious treatment that could be provided; or the possibility that the provision of such a level of care would effectively create a two-tier system of care, as well inadvertently serve to coerce individuals to participate in the trial in order to obtain that care (Muthuswamy, 2005).

Other researchers argued in favor of an alternative design that would utilize three arms: a short-course AZT arm, an 076 regimen arm, and a placebo arm comprised of HIV-seropositive pregnant women in the community who were not actively recruited into the third arm of the study (Edi-Osagie and Edi-Osagie, 1998). It is unclear, however, how researchers would have access to data pertaining to the HIV status of these women without their enrollment into the study (cf. Halsey, Sommer, and Black, 1998), which would then, according to the underlying premise, trigger an obligation to provide them with some form of treatment other than placebo.

Proponents of a placebo-controlled randomized trial design offered the following arguments to support their position.

1. Comparing an unknown, more affordable intervention with that of the known 076 intervention, which is affordable only in developed countries, would not provide useful information for the country in which the trial is conducted. If the affordable intervention is found to be of greater benefit, the information will be of no use because the intervention is not affordable. It would still be unclear as to whether the unknown intervention provides any benefit because it would have been compared against the known intervention, which is believed *a priori* to be better (Varmus and Satcher, 1997). In such a scenario, the research is less relevant to the population that is the focus of the study; paradoxically, the potential for exploitation is increased, rather than decreased (Killen, Grady, Folkers, and Fauci, 2002). This perspective found support in a report from a Bangkok study, which found that 15% of the pregnant women in the AZT study arm transmitted HIV to their infants, compared to 16% of the pregnant women in the placebo arm (Kaiser, 1997). Absent the placebo, researchers might have erroneously concluded that the AZT was producing significant results as measured against the background transmission rate of 24% to 33% (Kaiser, 1997).

2. In countries in which there is no intervention, participation in the placebo arm of the trial does not carry any greater risk than would be faced with standard practice in that locale (Varmus and Satcher, 1997).

3. The placebo-controlled studies were approved following rigorous review by ethics and scientific review committees in both the sponsor and the host countries (Varmus and Satcher, 1997).

Yet other scholars concluded that "trials must be designed to provide at least the highest standard of care practically attainable in the host country in which the trial is being done" (Writing Committee, 1999).

The debate prompted the re-examination and revision of the relevant provisions of the Declaration of Helsinki. Paragraph 29 now provides:

The benefits, risks, burdens and effectiveness of a new method should be tested against those of the best current prophylactic, diagnostic, and therapeutic methods. This does not exclude the use of placebo, or no treatment, in studies where no proven prophylactic, diagnostic or therapeutic method exists (World Medical Association, 1996).

And, in 2002, the World Medical Association issued a clarification of paragraph 29, cautioning investigators regarding the use of placebo controls. This cautionary note provides:

The WMA hereby reaffirms its position that extreme care must be taken in making use of a placebo-controlled trial and that in general this methodology should only be used in the absence of existing proven therapy. However, a placebo-controlled trial may be ethically acceptable, even if proven therapy is available, under the following circumstances:

– Where for compelling and scientifically sound methodological reasons its use is necessary to determine the efficacy or safety of a prophylactic, diagnostic or therapeutic method; or
– Where a prophylactic, diagnostic or therapeutic method is being investigated for a minor condition and the patients who receive placebo will not be subject to any additional risk of serious or irreversible harm.

All other provisions of the Declaration of Helsinki must be adhered to, especially the need for appropriate ethical and scientific review.

Currently, no consensus exists with respect to the standard of care to be provided to participants in the control arm of a randomized trial (Bayer, 1998). Even if there were consensus in principle, identification of the specific equivalent to be utilized within a specific trial could be difficult due to differing standards of medical practice and/or the absence of a "gold standard" of treatment (Porter, Forrest, and Kennedy, 1992). Some scholars have noted that, although researchers may be required to provide participants randomized to the control arm of a trial with the highest standard of care, regardless of where it may be found, no such requirement exists with respect to participants randomized to the experimental arm, who are to be administered a drug that may or may not be equally efficacious as that provided to those in the control arm. And, if all trial participants were to be provided with an equivalent standard of care, clinical trials to assess the efficacy of new drugs could no longer be undertaken (Bloom, 1998).

The most recent version of the *International Ethical Guidelines for Biomedical Research Involving Human Subjects* acknowledges the existing lack of consensus with respect to the conduct of placebo-controlled trials and the standard of care to be provided to clinical trial participants (CIOMS, 2002). Guideline 11 provides:

As a general rule, research subjects in the control group of a trial of a diagnostic, therapeutic, or preventive intervention should receive an established effective intervention. In some circumstances it may be ethically acceptable to use an alternative comparator, such as placebo or "no treatment".

Placebo may be used:

- when there is no established effective intervention;
- when withholding an established effective intervention would expose subjects to, at most, temporary discomfort or delay in relief of symptoms;
- when use of an established effective intervention as comparator would not yield scientifically reliable results and use of placebo would not add any risk of serious or irreversible harm to the subjects.

A recent web-based survey of subscribers to two listservs for individuals interested in international health research ethics was utilized to test the range of beliefs relating to the conduct of research in resource-poor settings (Kent, Mwamburi, Cash, Rabin, and Bennish, 2003). The researchers developed a hypothetical scenario in which they planned to test an HIV treatment known to be less effective than highly active antiretroviral therapy (HAART), and queried respondents as to whether the various trials designs were or were not ethical.

A total of 215 individuals from 47 countries responded to the survey. Researchers reported that 97% endorsed testing the new treatment in HIV-infected patients, without HAART, and 86% endorsed testing the experimental treatment against placebo. Sixty-eight percent of the respondents endorsed a "standard of care" for the study participants that reflected the local standard for treatment, rather than a universal standard (Kent, Mwamburi, Cash, Rabin, and Bennish, 2003).

Beyond Protocol 076. Issues relating to the standard of care have been raised in other contexts as well. Loue, Lurie, and Lloyd (1995) explored the ethical issues that would arise in the context of a trial to evaluate the efficacy of needle exchange programs. They concluded that the use of a placebo for the comparison arm of such a trial would contravene the principles of respect for autonomy, which mandates the formulation of additional protections for vulnerable participants, and of nonmaleficence, which imposes the obligation on researchers to refrain from actions that would cause harm.

Ethical challenges have also been identified in trials designed to test the efficacy of vaginal microbicides in preventing HIV transmission. The ethical principles of beneficence and nonmaleficence require that the investigator prevent harm to the participants. Additionally, they must provide the "most appropriate currently established" intervention to the trial participants. Accordingly, participants in both the treatment and comparison arms of a microbicide trial should be provided with male latex condoms, since they are currently available and are known to reduce the risk of HIV transmission (de Zoysa, Elias, and Bentley, 1998; Faden and Kass, 1998; Muthuswamy, 2005). However, a high level of condom use will reduce the investigators' ability to ascertain the protective effect of the microbicide, because a relatively small proportion of sex acts will have been protected through the use of microbicide alone. Similarly, the provision of treatment of existing sexually transmitted infections is ethically mandated since such treatment has been demonstrated to reduce the risk of HIV transmission (Grosskurth et al., 1995).

Participant Safety

Closely related to the issue of standard of care, but not identical, is the issue of participant safety. This issue has been raised most frequently in discussions relating to HIV preventative or vaccine trials. (Additional discussion regarding the multiple ethical and legal issues related to participant safety in the context of vaccine trials may be found in Stein, 1994).

As an example, concerns were raised in public meetings in Melbourne and Sydney, Australia in 1993 regarding the recruitment of gay men to HIV prophylactic trials. The concern was that participation in such trials would create a false sense of security, resulting in a decrease in condom use and an increased rate of HIV transmission, which would ultimately harm the individuals, the larger community, and the fate of the trial (Zion, 1995). Zion has posed the question: "Should the good of the gay community be put before the individual autonomy of gay men wishing to participate in a vaccine trial, and the potential good for future generations that will come about through the development of an efficacious vaccine?" (Zion, 1995: 518). The resolution of this dilemma necessarily rests on the moral agency of a particular community vis-à-vis the individual claims, and the power relationships and structures of advocacy that exist within a specific community, however it is defined. (For a discussion of various approaches to defining community, see chapter 4's discussion of community advisory boards.)

Participation in HIV vaccine trials may entail other potential threats to participants' safety, as well, which may manifest in a variety of forms. Serodiagnostic testing is used to detect antibody production after becoming infected with HIV; the development of antibody to HIV antigens after immunization indicates an immune response to a potentially effective vaccine (Frey, 2003). The induction of antibodies to HIV vaccines could, in the United States, result in a denial of health, life or disability insurance; the denial of employment, promotion, or assignment; and/or restrictions on the ability to travel internationally. Participants in HIV vaccine trials have encountered negative reactions from family, friends, and coworkers as a result of their participation (Allen et al., 2001; Sheon et al., 1998).

Uganda prepared for the possibility that some vaccine trial participants might suffer harms despite the best efforts of the investigators to prevent such an occurrence. The vaccine manufacturer arranged insurance for trial participants who experienced injuries related to the vaccine (Mugerwa, Kaleebu, Mugyenyi et al., 2002). In addition, individuals who tested positive for HIV antibodies as a result of the vaccine were to be provided with "special participation cards" that would presumably protect them from discrimination on the basis of their presumed HIV status. (Although this may have protected the participants from harms resulting from their participation, the investigators do not discuss the potential adverse effect on those who are HIV-infected that such a practice might create. The issuance of these cards to differentiate the truly infected from the vaccine-induced false positives may result in the creation of two classes of individuals who are seemingly HIV-positive, and an enhanced ability, through the use of the cards, to recognize and discriminate more easily against the true positives.)

It has been recommended that researchers work closely with communities in all phases of vaccine trial development and implementation in order to minimize the potential harm that could result from undue optimism about the vaccine (Canadian HIV/AIDS Legal Network, 2002). The establishment of a relationship with the community will facilitate the development of linkages with service providers, whose assistance may be needed to address the concerns of potential trial participants and/or ameliorate any harms.

Although various documents indicate that vaccine trial participants should receive care if they suffer a vaccine-related injury (Canadian HIV/AIDS Legal Network, 2002; UNAIDS, 2000), the quality of that care has not been delineated. UNAIDS (2000) has recommended that prior to initiating a vaccine trial, the host, the community, and the trial sponsor reach a consensus regarding the level of care to be provided and that discussion consider the level of care and treatment available in the country sponsoring the trial; the highest level of care available in the host country; the highest level of treatment available in the host country, including the availability of antiretroviral therapy outside of the study in the host country; the availability of an infrastructure in the host country that is able to provide care and treatment in the context of research; and the potential duration and sustainability of care and treatment for the trial participant.

It has also been suggested that participants who suffer trial-related injuries are entitled to receive reasonable compensation (Canadian HIV/AIDS Legal Network, 2002; UNAIDS, 2000). Additional questions have been raised regarding who is responsible for the payment of the compensation: the industry sponsor, an insurer, the investigators and/or their institution, the government(s) funding the research, the government(s) approving the research?

Balancing Risks and Benefits

The ethical principle of beneficence has been explained as follows:

Persons are treated in an ethical manner not only by respecting their decisions and protecting them from harm [the ethical principle of respect for persons], but also by making efforts to secure their well-being. . . . Two general rules have been formulated as complementary expressions of beneficent actions in this sense: (1) do no harm and (2) maximize possible benefits and minimize possible harms (National Commission for the Protection of Human Subjects, 1979: 4).

The issue, then, "is to decide when it is justifiable to seek certain benefits despite the risks involved, and when the benefits should be foregone because of the risks (National Commission for the Protection of Human Subjects, 1979: 4).

This calculation may be quite difficult. For instance, volunteers in an HIV vaccine trial may not receive direct benefit, but may experience significant risks as a result of their participation. These risks include social discrimination; discrimination in employment, travel, and health care; and seroconversion (Moodley, 2002). However, their service is invaluable to the larger society because without their participation, a vaccine could never developed. How then can the risks to the individuals be balanced against the benefit of the study, which will inure to the larger society?

Prior to the public debate regarding the use of placebo in the vertical transmission studies, Macklin and Friedland (1986) explored the question of how much more than minimal risk should be permitted in conducting a clinical trial. They concluded:

The respect-for-persons principle allows human subjects to assume the risks involved in the research they agree to participate in, even if those risks are greater than minimal. The principle requires that subjects' consent to participate be fully informed, but that consent does not preclude the use of placebos. As long as the features of a randomized controlled trial are explained to the subjects, in language they can understand, it is not unethical to conduct clinical investigations that pose more than minimal risk. Although subjects will not be informed which substance they will receive, the requirements of informed consent are met if they are told they will be randomly assigned to receive either the experimental drug or the inactive substance (Macklin and Friedland, 1986: 277).

This does not, however, imply that knowledge, understanding, and voluntariness on the part of the participant are sufficient to justify any level of risk, but only that these are necessary prerequisites. Regardless of the willingness of the individual to assume increasingly grave levels of risk in the hope of obtaining some individual benefit from participation, the investigator remains bound by the provisions of the Nuremberg Code, which establish a limit to the risks that may be imposed:

The experiment should be so conducted as to avoid all unnecessary physical and mental suffering and injury.

No experiment should be conducted where there is an *a priori* reason to believe that death or disabling injury will occur; except, perhaps, in those experiments where the experimental physicians also serve as subjects.

The degree of risk to be taken should never exceed that determined by the humanitarian importance of the problem to be solved by the experiment (*Trials of War Criminals*, n.d.).

The application of these precepts in a specific situation may not, however, be easily determined. One scholar has argued that in the case of diseases with a grave prognosis, such as rabies, it would be unethical to conduct a placebo-controlled trial of a new therapy if there existed the possibility that the new therapy might be successful in preventing this outcome in even a small number of cases (Levine, 1985). This position establishes two criteria to be assessed in evaluating the risk-benefit ratio: lethality and the likelihood of success. This does not provide guidance, however, as to the degree of success or the nature of success that is necessary to justify increasingly elevated levels of risk. In the case of HIV, for instance, which is a fatal disease, can success be measured by a delay in the most serious manifestations of disease? A decrease in the viral load? An improvement in immune functioning? And then, how much of a decrease in viral load or improvement in immune functioning?

Obtaining Informed Consent

Authority to Consent

Several questions arise related to informed consent to participate in HIV-related trials, although they are not specific to HIV research and they may arise in other contexts as well. (For a more detailed discussion of informed consent and related issues in the context of HIV vaccine trials, readers are urged to consult Lurie et al., 1994).

First, the issue is raised as to who may give valid consent to participate in a clinical trial. This was a critical question, for instance, in the maternal-child vertical transmission studies. At least one scholar questioned whether women should be asked for their consent, or be permitted to give their consent, to participate in placebo controlled trials because to do so would require that they "waive the interests of their future children" (de Zulueta, 2001). This perspective rests on a number of implicit assumptions, none of which are certainties: (1) that receipt of the placebo will necessarily result in harm; (2) that the experimental treatment, whatever it may be in a specific trial, is necessarily preferable to a placebo and will yield greater benefit and produce less harm; (3) that all pregnancies of HIV-seropositive women result in viral transmission to their children who are to be born; and (4) that the interests of the mother are not congruent, and are likely in opposition to, those of the fetus.

A similar question also arises in the context of cultures in which members of households may be required to obtain permission from a designated individual prior to entering into any form of an agreement. Such might be the case, for instance, in communities in which the eldest male member of the household is responsible for all decisions affecting family members (Loue, Okello, and Kawuma, 1996). In still other situations, the consent of a community leader may be necessary prior to seeking the consent of individual prospective participants. These issues are addressed in the context of international studies, in Chapter 15.

Voluntariness

A number of scholars have questioned whether individuals who are infected with HIV and who do not have access to necessary medical care can provide valid informed consent to participate in a clinical trial, due to the progressive and fatal course of the disease and a sense that there may exist no other mechanism apart from clinical trial participation by which to obtain the necessary treatment (de Zulueta, 2001; Schüklenk, 1998). Such circumstances, it has been argued, may preclude the possibility of truly voluntary consent because individuals may be willing to participate in research, regardless of the risk-benefit ratio, in the hope or false belief that participation may prolong their lives.

Continuing Consent

It is possible that, during the course of a clinical trial, new information will become known from sources outside the trial that may affect participants' willingness to continue with their participation and/or that may modify the risk-benefit ratio that existed at the inception of the investigation. As an example, consider the following situation.

The Veterans' Administration was conducting a long-term clinical trial to compare early AZT therapy with later-initiated AZT therapy for individuals with symptomatic HIV infection (Simberkoff et al., 1993). During the course of the trial, two important events occurred. First, the AIDS Clinical Trials Groups (ACTG) terminated a similar trial based on its finding of clear benefits for those participants in the AZT arm of the study. An interim analysis of the VA data at this time did not reveal such a difference. Second, the Food and Drug Administration (FDA) in 1990 announced approval of the earlier use of AZT, a decision that was relevant to the VA study participants.

The investigators instituted a series of procedures to ensure that the autonomy of the VA study participants was not compromised. Participants were asked to reaffirm their consent to participate, after being informed of the ACTG findings, the FDA-approved revised treatment recommendations, and the rationale for continuing the VA trial. The investigators did not, however, advise the participants of the results of the interim analyses conducted with the VA study data. Participants were also advised that continued blinded participation in the trial was optional, and they could elect to receive unblended treatment and follow-up (Simberkoff et al., 1993).

When the Study is Over: Access and Sustainability

A fundamental issue relates to the conduct of clinical trials in developing countries that do not have access to adequate health care services. The participation of their populations in clinical trials in the absence of plans to make the treatments available should they be found to be efficacious has been characterized as exploitative of the population and violative of the ethical principle of justice. (For a detailed analysis of the concept of exploitation in the context of biomedical research, readers are referred to Resnik, 2003.) In such situations, absent the intent and a mechanism to make the treatments available, the participants become a means to an end (Annas and Grodin, 1998). This same issue, however, is relevant to U.S. populations that may lack access to care due to limited finances, lack of health care insurance, or other factors.

The *International Ethical Guidelines for Biomedical Research Involving Human Subjects* urge that the products of research be made reasonably available to local populations following the conclusion of the investigations (Council for International Organizations of Medical Sciences, 2002). This view has been echoed by numerous commentators, many of whom have indicated that the unavailability of the treatment or intervention would reframe the investigation as having been exploitative of that population (Crouch and Arras, 1998).

The parameters of "reasonable availability" have not, however, been delineated and it is not easy to do so. Crouch and Arras (1998) have questioned the legitimacy of relying on political boundaries as a criterion, noting that the nonparticipating segments of a population residing in the distant reaches of the country in which the study takes place may be less similar to those who participated than individuals immediately across the national border from the study site.

It has been suggested that the creation of a post-clinical trial drug fund may provide one mechanism for continued access to treatment following the conclusion of a clinical trial (Ananworanich et al., 2004). This can be accomplished through the development of rollover protocols and the imposition of a requirement on trial sponsors that they provide at least a two-year supply of the trial drug to participants either in the form of fixed funds or drug (Ananworanich et al., 2004). Income for the drug fund can also be derived from revenues derived from the research studies (overhead costs) and profits from related activities, such as registration fees from symposia and training courses that are offered.

Conflict of Interest

Conflicts in Provider-Investigator Roles

It is not uncommon for clinicians who are also investigators to recruit research participants from among their patients, based on the information that they have acquired during the course of providing the patients with clinical care. Some scholars have argued that, because patients see their physicians as care providers only, the dual provider-investigator role represents the moral equivalent of sharing the patient's information with another individual who would not otherwise have access to that information (Habiba and Evans, 2002). This "inter-role breach of confidentiality" is impermissible, they assert, because clinical care is distinct from activity associated with recruitment and enrollment into clinical trials; trials are not intended to benefit patients directly and so cannot be considered a legitimate clinical interest; the goals and obligations of researchers and clinicians are quite different and sometimes in conflict; patients provide their physicians information with the understanding that it will not be shared with others, such as between physician-as-provider and physician-as-researcher; and patients experience difficulty refusing the requests of their providers.

However, valid reasons may exist for a provider to approach his or her patient to participate in a clinical trial. First, the clinician may have a good reason to believe that participation in the trial may help a particular patient (Iltis, 2005). This does not necessarily violate the principle of clinical equipoise, which is a prerequisite for the conduct of a clinical trial. (See discussion of equipoise, above.) Rather, the physician may know specific details about the treatment to be tested and the condition of the particular patient that makes him or her more likely to believe that the particular treatment might work for a specific patient. Additionally, in situations in which there is no known treatment for the patient's condition, or the patient has been unable to obtain benefit from any of the existing treatments, and the treatment under study is unavailable outside of the investigation, the physician may correctly believe that participation in the trial provides the only possibility of conferring benefit on the patient (Iltis, 2005).

In order to avoid the potential conflict for the patient that is created by the dual physician-investigator role, it has been suggested that the physician explain his or her multiple roles to the patient, inform the patient of his or her right to decline to participate in research, and time the recruitment of the patient in a manner that is sensitive to the risks and burdens that the patient might experience as a participant (Iltis, 2005). It has also been recommended that, in general, where there exists an established therapy for a patient's condition, the patient should not be recruited into placebo-controlled trials for a new, but similar treatment.

Financial Conflicts of Interest

Conflicts of interest in biomedical research appear to be ubiquitous (Korn, 2000). To some extent, this has resulted from the increasing proportion of bio-medical research and clinical trials and their associated costs that is funded by pharmaceutical companies. For instance, in 1999, the top 10 pharmaceutical companies expended $22.7 billion primarily on clinical research, compared to NIH's $17.8 billion for basic research (DeAngelis, 2001). Additionally, there has been a shift away from academic centers to nonacademic research organizations for the conduct of clinical trials, resulting in increased control by industry of the trial design, access to and analysis of data, and the publication of research findings (Bodenheimer, 2000; Wadman, 1996).

Because of concerns relating to financial conflicts of interest on the part of researchers, federal regulations require the disclosure of such interests at the time of proposal submission, as new investigators are added to the project, and as investigators' financial interests change. The state of California has instituted even more stringent requirements, mandating investigator disclosure of such potential conflicts at the close-out of the investigation, in addition to the fore-going. (For a detailed and relatively recent review of federal and state law relat-ing to the regulation of investigator financial conflict of interest, the reader is referred to Henderson and Smith, 2002). And, as the result of a court decision in a lawsuit by a patient against his physician-investigator, physicians in California are obligated to disclose to patients "any personal interests unrelated to [the] patient's health, whether research or economic, that may affect the physician's professional judgment" (*Moore v. Regents of the University of California*, 1990: 483).

There has also been an increased reliance by industry on the use of finders' fees, "offers of money to physicians, nurses, or other health professionals in reward for their referral of patients eligible for research participation . . . over and above reasonable remuneration for services rendered" (Lemmens and Miller, 2003: 399). It has been suggested that the receipt of such fees may encourage health professionals to "be more lenient with respect to informed consent proce-dures, they may convince themselves that research participation is in their patient's best interests, or they may be overly flexible with regard to the study inclusion and exclusion criteria" (Lemmens and Miller, 2003: 401). This could potentially result in their placing increased pressure on individuals to agree to participation, inadvertently placing the recruited participants at increased risk of harm and compromising the validity of the study. IRBs are charged with the responsibility to review the methods used to recruit study participants (Food and Drug Administration, 1998), but there is no legal requirement that investigators disclose payments that they receive in exchange for recruitment efforts. The American Medical Association has condemned such payments as unethical (American Medical Association, 1994, 2000).

Despite these concerns, and the existence of federal regulations mandating the disclosure by investigators of significant financial interests in companies that might reasonably appear to be affected by the research, university policies governing faculty and research staff conflicts of interest vary widely (Lo, Wolf, and Berkeley, 2000). It has been suggested as a remedy that university investigators and research staff be prohibited from holding stock, stock options, or decision-making positions in companies that may reasonably appear to be affected by the results of their clinical research (Lo, Wolf, and Berkeley, 2000).

Other Conflicts of Interest

In April 1991, the *Chicago Tribune* carried a series of articles that focused on the deaths in Paris of three individuals who had participated in an AIDS vaccine experiment by Dr. Daniel Zagury, an immunologist at the Institut Jean Godinot of the Université Piere et Marie Curie (Nairn, 1993). The deaths resulted from acute necrosis, a vaccinia disease, that appeared to have been caused by Dr. Zagury's Phase I experiment with 9 HIV seronegative children in Zaire. The experiment involved the insertion of HIV envelope proteins into the recombinant vaccinia strain. Because the vaccinia recombinant contained only part of the HIV virus, the children would not have contracted HIV, but it was hypothesized that their immune systems might generate antibodies to the protein produced by the inserted gene and would, as a result, be protected from later contracting the infection.

The Phase I vaccine trial had been approved by a review committee in Zaire, but not in France. An investigation into the study and the deaths raised questions about the validity of the informed consent furnished by the children's mothers, who were infected with HIV; the children's fathers had all died from the disease. Zagury maintained that he had conducted the trial and enrolled the children out of compassion for their situation and the pleas of their mothers for help in safeguarding their children from the infection (Nairn, 1993). Altruism can be the source of a conflict of interest, such as when a provider-researcher ignores a randomization scheme in order to ensure that a severely ill patient, for whom every treatment has failed, receives whatever benefit may be gained from an experimental intervention. However, Zagury's motives were questioned by both scholars and those investigating the deaths (Nairn, 1993). It has been noted, for instance, that the experiment exposed healthy children to significant risk without any direct benefit. This seems incongruent with altruism and prompts questions about the existence of other underlying motives.

References

Allen, M., Israel, H., Rybczyk, K. et al. (2001). Trial-related discrimination in HIV vaccine clinical trials. *AIDS Research and Human Retroviruses, 17,* 667–674.

American Medical Association, Council on Ethical and Judicial Affairs. (2000). Fee splitting: Referrals to health care facilities (Opinion E-603). In American Medical Association Council on Ethical and Judicial Affairs, *Code of Medical Ethics: Current Opinions.* Chicago, Illinois: Author.

American Medical Association, Council on Ethical and Judicial Affairs. (1994). *Finder's Fees: Payment for the Referral of Patients to Clinical Research Studies* (report #65). Chicago, Illinois: Author.

Ananworanich, J., Cheunyam, T., Teeratakulpisarn, S., et al. (2004). Creation of a drug fund for post-clinical trial access to antiretrovirals. *Lancet, 364,* 101–102.

Angell, M. (1997). The ethics of clinical research in the Third World. *New England Journal of Medicine, 337,* 847–849.

Annas, G.J. & Grodin, M.A. (1998). Human rights and maternal-fetal HIV transmission prevention trials in Africa. *American Journal of Public Health, 88(4),* 560–563.

Armistead, L.P., Clark, H., Barber, C.N., Dorsey, S., Hughley, J., Favors, M., & Wyckoff, S.C. (2004). Participant retention in the Parents Matter! Program: Strategies and outcome. *Journal of Child and Family Studies, 13(1),* 67–80.

Bayer, R. (1998). The debate over maternal-fetal HIV transmission prevention trials in Africa, Asia, and the Caribbean: Racist exploitation or exploitation of racism? *American Journal of Public Health, 88(4),* 567–570.

Bloom, B.R. (1998). The highest attainable standard: Ethical issues in AIDS vaccines. *Science, 279 (5348),* 186–188.

Bodenheimer, T. (2000). Uneasy alliance: Clinical investigators and the pharmaceutical industry. *New England Journal of Medicine, 342,* 1539–1543.

Canadian HIV/AIDS Legal Network. (2002). *HIV Vaccines in Canada: Legal and Ethical Issues. An Overview.* Toronto, Canada: Author.

Caschetta, M.B., Chavkin, W., & McGovern, T. (1993). FDA policy on women in drug trials [correspondence]. *New England Journal of Medicine, 329,* 1815.

Chow, S.C. & Liu, J-P. (2003). *Design and Analysis of Clinical Trials: Concepts and Methodologies.* Hoboken, New Jersey: John Wiley & Sons, Inc.

Collins, A. & Baumgartner, D. (1995). U.S. prisoners' access to experimental HIV therapies. *Minnesota Medicine, 78(11),* 45–48.

Connor, E.M., Sperling, R.S., Gelber, R., et al. (1994). Reduction of maternal-infant transmission of HIV type I with ZDV treatment. *New England Journal of Medicine, 331,* 1173–1180.

Council of International Organizations for Medical Sciences. (2002). *International Ethical Guidelines for Biomedical Research Involving Human Subjects.* Geneva, Switzerland: Author.

Crouch, R.A. & Arras, J.D. (1998). AZT trials and tribulations. *Hastings Center Report, 28(6),* 26–34.

Daunt, D.J. (2003). Ethnicity and recruitment rates in clinical research studies. *Applied Nursing Research, 16(3),* 189–195.

De Zoysa, I., Elias, C.J., & Bentley, M.E. (1998). Ethical challenges in efficacy trials of vaginal microbicides for HIV prevention. *American Journal of Public Health, 88(4),* 571–575.

De Zulueta, P. (2001). Randomised placebo-controlled trials in HIV-infected pregnant women in developing countries. Ethical imperialism or unethical exploitation? *Bioethics, 15(4)*, 289–311.

DeAngelis, C.D. (2001). Conflict of interest and the public trust [editorial]. *Journal of the American Medical Association, 284*, 2237–2238.

DeVesa, F. (1993). Opinion letter from Attorney General of New Jersey to Commissioner W.H. Fauver, New Jersey Department of Corrections, November 30. Cited in G.L. Stein & L.D. Headley. (1996). Forum on prisoners' access to clinical trials: Summary of recommendations. *AIDS & Public Policy Journal, 11(1)*, 3–20.

Edi-Osagie, E.C.O. & Edi-Osagie, N.E. (1998). Placebo trials are unethical for established, untested treatments. *British Medical Journal, 316*, 625.

Faden, R. & Kass, N. (1998). HIV research, ethics, and the developing world [editorial]. *American Journal of Public Health, 88(4)*, 548–550.

Family Health Research Group. (1998). The Family Health Project: A multidisciplinary longitudinal investigation of children whose mothers are HIV-infected. *Clinical Psychology Review, 18*, 839–856.

Feinberg, J. & Japour, A.J. (2003). Scientific and ethical considerations in trial design for investigational agents for the treatment of human immunodeficiency virus. *Clinical Infectious Diseases, 36*, 201–206.

Food and Drug Administration. (1998). *Information Sheets: Guidance for Institutional Review Boards and Clinical Investigators.* Rockville, Maryland: Author.

Freedman, B. (1991). Suspended judgment: AIDS and the ethics of clinical trials: Learning the right lessons. *Controlled Clinical Trials, 13(1)*, 1–5.

Frey, S.E. (2003). Unique risks to volunteers in HIV vaccine trials. *Journal of Investigative Medicine, 51 (Supp. 1)*, S18-S20.

Grosskurth, H., Mosha, F., Todd, J. et al. (1995). Impact of improved treatment of sexually transmitted diseases on HIV infection in rural Tanzania: Randomized controlled trial. *Lancet, 346*, 530–536.

Habiba, M. & Evans, M. (2002). The inter-role confidentiality conflict in recruitment for clinical research. *The Journal of Medicine and Philosophy, 27(5)*, 565–587.

Halsey, N.A., Sommer, A., & Black, R.E. (1998). Authors' reply. *British Medical Journal, 316*, 627.

Heckerling, P.S. (2005). The ethics of single-blind trials. *IRB: Ethics & Human Research, 27(4)*, 12–16.

Henderson, J.A. & Smith, J.J. (2002). Financial conflict of interest in medical research: Overview and analysis of federal and state controls. *Food and Drug Law Journal, March*, 1–11.

Homer, C.S.E. (2002). Using the Zelen design in randomized controlled trials: Debates and controversies. *Journal of Advanced Nursing, 38(2)*, 200–207.

Iltis, A.S. (2005). Timing invitations to participate in clinical research: Preliminary versus informed consent. *Journal of Medicine and Philosophy, 30*, 89–106.

Institute of Medical Ethics Working Party on the Ethical Implications of AIDS. (1992). *British Medical Journal, 305*, 699–701.

Kaiser, J. (1997). Bangkok study adds fuel to AIDS ethics debate. *Science, 278 (5343)*, 1553.

Kass, N.E., Taylor, H.A., & King, P.A. (1996). Harms of excluding pregnant women from clinical research: The case of HIV-infected pregnant women. *Journal of Law, Medicine & Ethics, 24*, 36–46.

Kent, D.M., Mwamburi, M., Cash, R.A., Tabin, T.L., & Bennish, M.L. (2003). Testing therapies less effective than the best current standard: Ethical beliefs in an international sample of researchers. *American Journal of Bioethics, 3(2),* W28-W33.

Killen, J., Grady, C., Folkers, G.K., & Fauci, A.S. (2002). Ethics of clinical research in the developing world. *Nature Reviews, 2,* 210–215.

Kodish, E., Lantos, J.D., & Siegler, M. (1990). Ethical considerations in randomized controlled clinical trials. *Cancer, 65 (Suppl.),* 2400–2404.

Korn, D. (2000). Conflicts of interest in biomedical research. *Journal of the American Medical Association, 284,* 2234–2237.

Last, J.M. (Ed.) (1988). *A Dictionary of Epidemiology,* 2nd ed. New York: Oxford University Press.

Lemmens, T. & Miller, P.B. (2003). The human subjects trade: Ethical and legal issues surrounding recruitment incentives. *Journal of Law, Medicine & Ethics, 31,* 398–418.

Levine, R.J. (1985). The use of placebos in randomized controlled clinical trials. *IRB: A Review of Human Subjects Research, 7,* 1–6.

Lo, B., Wolf, L.E., & Berkeley, A. (2000). Conflict-of-interest policies for investigators in clinical trials. *New England Journal of Medicine, 343,* 1616–1620.

Loue, S., Lurie, P., & Lloyd, L.S. (1995). Ethical issues raised by needle exchange programs. *Journal of Law, Medicine, and Ethics, 23(4),* 382–388.

Loue, S., Okello, D., & Kawuma, M. (1996). Research bioethics in the Ugandan context. *Journal of Law, Medicine, and Ethics, 24(1),* 47–53.

Lurie, P., Bishaw, M., Chesney, M.A. et al. (1994). Ethical, behavioral, and social aspects of HIV vaccine trials in developing countries. *Journal of the American Medical Association, 271(4),* 295–301.

Lurie, P. & Wolfe, S.M. (1997). Unethical trials of interventions to reduce perinatal transmission of human immunodeficiency virus in developing countries. *New England Journal of Medicine, 337,* 853–856.

Macklin, R. & Friedland, G. (1986). AIDS research: The ethics of clinical trials. *Law, Medicine & Health Care, 14(5–6),* 273–280.

Meinert, C.L. (1986). *Clinical Trials: Design, Conduct, and Analysis.* New York: Oxford University Press.

Moodley, K. (2002). HIV vaccine trial participation in South Africa—An ethical assessment. *Journal of Medicine and Philosophy, 27(2),* 197–215.

Moore v. Regents of the University of California. (1990). 793 P.2d 479, *cert. denied,* 499 U.S. 936 (1991).

Mugerwa, R.D., Kaleebu, P., Mugyenyi, P. et al. (2002). First trial of the HIV-1 vaccine in Africa: Ugandan experience. *British Medical Journal, 324,* 226–229.

Muthuswamy, V. (2005). Ethical issues in HIV/AIDS research. *Indian Journal of Medical Research, 121,* 601–610.

Nairn, T.A. (1993). The use of Zairian children in HIV vaccine experimentation: A cross-cultural study in medical ethics. *The Annual of the Society of Christian Ethics, 13,* 223–243.

National Commission for the Protection of Human Subjects. (1979). *The Belmont Report.* Washington, D.C.: United States Department of Health and Human Services.

National Institutes of Health. (2005). An Introduction to Clinical trials. Available at http://www.clinicaltrials.gov. Last accessed August 14, 2006.

Pasquerella, L. (2002). Confining choices: Should inmates' participation in research be limited? *Theoretical Medicine, 23,* 519–536.

Porter, J.D.H., Forrest, B.D., & Kennedy, A.R. (1992). The ethics of placebos in AIDS drug trials. *Healthcare Ethics Committee Forum, 4(3)*, 155–162.

Resnik, D.B. (2003). Exploitation in biomedical research. *Theoretical Medicine, 24*, 233–259.

Roe v. Fauver. (1992). Consent decree. C.A. No. 88–1225-AET (D. New Jersey, March 3).

Schüklenk, U. (1998). Unethical perinatal HIV transmission trials establish bad precedent. *Bioethics, 12*, 312–319.

Sheon, A.R., Wagner, L., McElrath, M.J. et al. (1998). Preventing discrimination against volunteers in prophylactic HIV vaccine trials: Lessons learned from a phase II trial. *Journal of Acquired Immune Deficiency Syndromes and Human Retrovirology, 19*, 519–526,

Simberkoff, M.S., Hamilton, J.D., Gail, M. et al. (1993). Ethical dilemmas in continuing a zidovudine trial after early termination of similar trials. *Controlled Clinical Trials, 14(1)*, 6–18.

Stein, G.L. & Headley, L.D. (1996). Forum on prisoners' access to clinical trials: Summary of recommendations. *AIDS & Public Policy Journal, 11(1)*, 3–20.

Stein, R.E. (1994). Vaccine liability and participant compensation incentives in HIV vaccine trials. *AIDS Research and Human Retroviruses, 10(Supp. 2)*, S297-S300.

Stone, V.E., Steger, K.A., Hirschorn, L.R. et al. (1998). Access to treatment with protease inhibitor containing regimens: Is it equal for all? Presented at the 12th International AIDS Conference, Geneva, Switzerland, June 28-July 3 [abstract no. 42305].

Trials of War Criminals before the Nuremberg Military Tribunals under Control Council Law No. 10, Vol. 2 (pp. 181–182). Nuremberg October 1946 – April 1949, Washington. U.S. Government Printing Office (n.d.).

UNAIDS. (2000). *Ethical Considerations in HIV Preventive Vaccine Research: UNAIDS Guidance Document.* Geneva, Switzerland: Joint United Nations Programme on AIDS. Available at http://www.unaids.org.

Varmus, H. & Satcher, D. (1997). Ethical complexities of conducting research in developing countries. *New England Journal of Medicine, 337(14)*, 1003–1005.

Veatch, R.M. (1989). Drug research in humans: The ethics of nonrandomized access. *Clinical Pharmacy, 8(5)*, 366–370.

Wadman, M. (1996). Drug company "suppressed" publication of research. *Nature, 381(6577)*, 4.

World Medical Association. (1964). Declaration of Helsinki.

World Medical Association. (1996, 2002). Declaration of Helsinki.

Writing Committee. (1999). Science, ethics, and the future of research into maternal infant transmission of HIV-1. *Lancet, 353*, 832–835.

Zelen, M. (1979). A new design for randomized controlled trials. *New England Journal of Medicine, 300*, 1242–1245.

Zelen, M. (1990). Strategy and alternative designs for clinical trials: An update. *Statistics in Medicine, 9*, 645–656.

Case Study Two
Clinical Trials and International Ethics: Preventing Tuberculosis in Africa

C.C. Whalen, M.D., M.S.

Introduction

A randomized clinical trial is designed to determine whether one treatment is equivalent or superior to another treatment. When no standard therapy is recommended for a condition, a clinical trial may be designed with a placebo control arm so that comparisons between treated and untreated cases may be made. When a standard treatment for a condition has been established through peer-reviewed research or longstanding clinical practice and this treatment has been accepted as a standard of care, then a placebo-controlled clinical trial poses ethical concerns. In this setting, the clinical trial is designed to test equivalence between the standard therapy and the new treatment or intervention.

Although these principles are uniformly accepted by researchers, their application becomes complex and difficult when standards of care differ depending on the setting of the trial. This is best illustrated in international research in resource-limited countries where the local standard of care may differ from one established in the United States or another industrialized country.

Case Description

Tuberculosis is a disease caused by *Mycobacterium tuberculosis*. It is estimated that one-third of the world is infected with the organism. Six to seven million cases of tuberculosis disease develop each year and 2 million deaths are attributed to the disease (Dye, 2006).

There are two health states in the natural history of tuberculosis. In one state of health, an individual is infected with the organism but shows no signs of disease. Except for a reactive tuberculin skin test, the individual is healthy and not contagious. Most persons infected by *M. tuberculosis* never develop disease, and they are said to have latent tuberculosis infection. About 10 percent of persons with latent tuberculosis infection will develop tuberculosis disease at some point during their lifetime, sometimes decades after initial infection. It is pulmonary tuberculosis, the pneumonia, that poses the greatest threat to individual and public health because it is the most common form of disease, carries a significant risk for death, and is by far the most contagious.

Four strategies are used to control tuberculosis —passive case-finding and proper treatment, treatment of latent tuberculosis infection, BCG vaccination of children, environmental controls – but these public health interventions are not uniformly practiced throughout the world. BCG vaccination prevents disseminated and life-threatening forms of disease in children. It is the most widely used vaccine in the world and is given at birth as part of the World Health Organization's Expanded Programme on Immunization. In countries where tuberculosis is not endemic, BCG vaccine is not given because it may interfere with the interpretation of the tuberculin skin test.

The key strategy, however, is the first – the identification and treatment of infectious cases of tuberculosis. National tuberculosis control programs throughout the world, including the U.S., implement this approach, often using guidelines developed by the Centers for Disease Control and Prevention in the U.S. (Centers for Disease Control and Prevention, 2003) or the World Health Organization. Treatment of latent tuberculous infection is analogous to the treatment of high blood pressure; its main effect is to prevent the occurrence of tuberculosis disease. Treatment of latent tuberculosis infection is a central part of the tuberculosis control strategy in the U.S. but not in most countries where tuberculosis is endemic. Unlike a vaccine, treating latent infection does not prevent reinfection with another strain of *M. tuberculosis* and subsequent risks for disease. Where the risk of infection is high, such as in countries with a high prevalence of tuberculosis, treating latent infection is not given high priority by many ministries of health.

Since the mid-1980's, the global tuberculosis situation has been exacerbated by the HIV pandemic (De Cock, Soro, Coulibaly, and Lucas, 1992). HIV confers the greatest known risk for the development of tuberculosis. The annual incidence of tuberculosis in co-infected persons ranges from 3 to 12 percent, a risk that is 100 times greater than an HIV-seronegative person (O'Brien and Perriens, 1995). Moreover, tuberculosis may accelerate the natural history of HIV infection (Whalen et al., 1995; Whalen et al., 2000). These two organisms interact at the community level. In many developing countries of Africa, for example, 50 to 75 percent of tuberculosis cases are infected with HIV, whereas 10 to 15 percent of the population is infected with HIV. Thus, a small proportion of the population is giving rise to over 50 percent of the tuberculosis problem (De Cock and Chaisson, 1999). If tuberculosis disease could be prevented in this high risk group in the community, then it would in theory have a beneficial effect on tuberculosis control.

A series of clinical trials were designed and conducted in endemic areas of the world during the 1990s. Most of these studies were designed with a placebo group with the intent of understanding the true effect of the intervention and to measure the extent of side effects. One such study was conducted in Kampala, Uganda, from 1993 to 1998 (Whalen et al., 1997; Johnson et al., 2001) and will be the subject of the case discussion.

Study Description

This study was designed to assess whether three different regimens to treat latent tuberculosis infection were effective in reducing the risk of tuberculosis disease in HIV-infected adults. The study was designed as a randomized, placebo- controlled clinical trial in HIV-infected persons with either a reactive tuberculin skin test or cutaneous anergy. The trial compared placebo therapy for six months with isoniazid given for six months (6H), isoniazid and rifampin given for three months (3RH), and isoniazid, rifampin, and pyrazinamide for 3 months (3RHZ). The main outcome for the trial was the development of tuberculosis disease; safety and mortality were secondary outcomes. An interim analysis was performed after 84% of expected events had occurred in the study, 33 months after enrollment began. A statistically significant difference in the rates of incident tuberculosis between placebo and the 6H study arm was found so the intervention was stopped and study subjects receiving placebo were offered 6H. To evaluate the long-term benefit of the interventions, all study subjects were followed for up to 5 years and actively screened for tuberculosis. A final analysis of this cohort showed that the protective benefit of 6H was short-lived and that regimens containing rifampin might provide longer protection from tuberculosis disease.

The trial was conducted in Kampala,Uganda, under the auspices of the Uganda-Case Western Reserve University Research Collaboration and was funded by the Center for Disease Control and Prevention (ACET, cooperative agreement). The study protocol was approved by the AIDS Scientific Subcommittee at Makerere University in Uganda and by the Institutional Review Board at Case Western Reserve University.

Ethical Issues

Placebo Use and Standards of Care

When this study was first published, it was met with criticism because it included a placebo control arm (Angell, 1997). At the time this study began, there was no standard of practice for the prevention of tuberculosis in this high risk group, yet clinicians were seeking guidance in treatment. To this end, the Centers for Disease Control and Prevention published a set of guidelines for the treatment of latent tuberculosis infection in HIV-infected individuals (Centers for Disease Control and Prevention, 1990). These guidelines were based on results from several small observational studies in injection drug users from New York City that showed protective effect of isoniazid treatment (Selwyn et al., 1989), known effectiveness of isoniazid therapy in persons with intact immune systems (Ferebee, 1970), and consensus of expert opinion. Although these guidelines were adopted in the U.S. policies on tuberculosis control, they were not universally practiced as a standard of clinical care, especially in countries with limited economies and medical care systems.

As often happens, the sponsors of medical research are based in wealthy, industrialized countries but the research is carried out in a resource-limited developing country. In the design of an international study, which standard of care should be enforced? The standard in the country of the sponsor? Or the local practice in the host country of the study? Investigators for this study found themselves exactly in this position, as the study was sponsored by the Centers for Disease Control and Prevention and conducted in Uganda.

At the time of the study, the Helsinki Declaration advised: "In any medical study, every patient- including those of a control group, if any- should be assured of the best proven diagnostic and therapeutic method" (World Medical Organization, 1996). This principle opens a debate about what represents "the best proven diagnostic and therapeutic method." If one accepts the findings regarding treating latent tuberculosis infection from persons with intact immune systems as a reliable guide for how to manage latent infection in HIV disease, then a placebo arm in the case study would not be justified on the grounds that another proven therapy was available. For many, the guidelines of 1990 from the Centers for Disease Control and Prevention were sufficient to meet this criterion and establish the standard of care in HIV infection. For some, it established the standard of care not only for the U.S. but also for the rest of the world. But many experts in the field were more circumspect in accepting the guidelines at face value and evaluated the quality of evidence that supported them.

The guidelines relied on the consensus of expert opinion and how those experts evaluated the available evidence for treatment of latent infection in HIV disease. There was a large body of evidence on the effectiveness of treatment latent tuberculosis infection in healthy individuals with intact immune systems. Through well-designed randomized clinical trials, it had been established that treatment of latent tuberculosis infection with isoniazid for 6 to 9 months would reduce the risk of tuberculosis disease by 70 to 80% in persons with intact immune systems (Ferebee, 1970; International Union Against Tuberculosis Committee on Prophylaxis, 1982). Since HIV infection weakens the immune system's response to *M. tuberculosis* so severely (Barnes, Quoc, and Davidson, 1993), it was unknown whether a single drug like isoniazid would be effective in reducing the risk of tuberculosis disease in HIV-infected persons. There was evidence from observational studies in injection drug users that treatment with isoniazid reduced the incidence of tuberculosis in HIV-infected drug users (Selwyn et al., 1989; Selwyn et al., 1992), but experts in the field did not regard these data as definitive because of the restricted nature of the study population and lack of true controls in the study. Indeed, within the 1990 guidelines from the Centers for Disease Control, there was uncertainty about the stated guidelines as the document concludes with a call for further research. These developments set the stage for a series of international studies designed to evaluate the efficacy of treating latent tuberculosis infection in HIV-infected persons (Wilkinson, Squire, and Garner, 1998).

Apart from the available evidence in HIV disease, some experts contended that treatment of latent infection is not effective in high transmission settings when compared to low transmission settings. They believe that after completing a course of isoniazid therapy, an individual faces the same risk for infection as before treatment. Moreover, in the areas where strains of *M. tuberculosis* with isoniazid drug resistance circulate, the effectiveness of isoniazid therapy for infection is certainly not established. Even today, there is controversy about the effectiveness of isoniazid treatment of latent tuberculosis infection on the control of tuberculosis.

The international nature of the trial brought this controversy into clear focus. Uganda was chosen as a trial site because it was the epicenter of the HIV/AIDS epidemic in Africa in the early 1990s, and it was ravaged by an evolving tuberculosis epidemic that was tightly linked to the HIV/AIDS epidemic in the country (Eriki et al., 1991). At the time of the study, no information was available on drug susceptibility profiles for strains in Uganda. The Ugandan investigators questioned the benefit of isoniazid treatment in the planning stages of this trial and argued for the use of a true placebo arm. For the study to be informative to public health policy in Uganda, they needed to know whether a treatment regimen reduced the absolute risk of tuberculosis in HIV-infected persons, not whether it was equivalent to another regimen of unknown efficacy.

Although isoniazid treatment was generally safe in immunocompetent persons, it is associated with an age-dependent risk for hepatic toxicity, that can on rare occasion be fatal (Moulding, Redeker, and Kanel, 1989). It was unknown at the time whether HIV-infected persons would be more, or less, susceptible to the hepatic toxicity, or whether they would be at risk for other less common adverse effects of isoniazid. This concern was heightened by reports of severe and fatal cutaneous reactions to tuberculosis medication in HIV-infected individuals in Africa (Coopman, Johnson, Platt, and Stern, 1993; Pozniak, MacLeod, Mahari, Legg, and Weinberg, 1992). Although this severe toxicity was attributed to thiacetazone, one could not exclude isoniazid as a potential cause. Furthermore, HIV-infected persons were more susceptible to adverse cutaneous reactions to trimethoprim- sulfamethoxazole when used to prevent *Pneumocystis carinii* Pneumonia (Colebunders et al., 1987; Gordin, Simon, Wofsy, and Mills, 1984).

Investigators offered clinical equipoise as the basis for the decision to use the placebo group in this case study (Whalen, Johnson, Mugerwa, and Ellner, 1998). Clinical equipoise refers to a spectrum of opinion within a field about the therapeutic benefit of a treatment. It allows for some members of the field, even an investigator on a study, to believe that one intervention tested in a randomized clinical trial is more beneficial than other tested regimens, including a placebo, but that the field remains divided about the therapeutic benefit of intervention (Miller and Brody, 2003). Certainly, there was a spectrum of opinions in the medical field of tuberculosis control and prevention regarding the proper role of treating latent tuberculosis infection in HIV disease. For some in the field, the decision to use the placebo control group rested on the need to know the efficacy and safety of treatment for latent tuberculosis infection in HIV disease. For others in the field, a consensus statement from an authoritative body like the Center for Disease Control and Prevention was sufficient to establish standard of care in the U.S. Since the study was sponsored by the U.S. federal government, this standard of care applied.

Clinical equipoise is not necessarily static. Opinion about standards of care can change with time as the result of new research findings, opinion of leaders in the field, or new consensus statements from governments or specialty groups. In this case study, the clinical equipoise at the start of the trial was challenged when new research findings from a randomized clinical trial from Haiti showed that treating latent tuberculosis infection reduced the risk for tuberculosis in HIV disease by 80% (Pape, Jean, Ho, Hafner, and Johnson, 1993). With this new information, investigators reconsidered the need for a placebo group and appealed to an independent body with authority to decide about international health issues. In April 1994, a steering committee on therapy for mycobacterial infections at the World Health Organization reviewed the available data, both from all published and unpublished clinical trials of tuberculosis prevention in HIV disease. The expert panel concluded that clinical equipoise remained because the results from the published study were not sufficiently conclusive to inform public health policy regarding treatment of latent tuberculosis infection in HIV disease.

Since the completion of this study, a large body of literature has been published relating to standards of care in clinical trials, especially in the international setting. There is uniform agreement that ethical principles and guidelines articulated in the Belmont Report must be met regardless of the setting (Shapiro and Meslin, 2001). Revised versions of the Declaration of Helsinki (World Medical Association, 2000) and the *Council for International Organizations of Medical Sciences* (CIOMS) international ethical guidelines for biomedical research (Council for International Organizations of Medical Sciences, 2002) provide recommendations regarding standard of care that reflect current ethical thinking. In its revised version the Helsinki Declaration states: 'The benefits, risks, burdens, and effectiveness of a new method should be tested against those of the best current prophylactic, diagnostic, and therapeutic methods.' This principle states that the 'best current method' must be offered to study subjects, presumably regardless of whether this method is the standard of care in the country of the clinical trial.

This restated principle may limit a study to testing the relative benefit of an intervention instead of the absolute benefit, and possibly limit the information needed to make public policy in a country. This principle may lead to other ethical dilemmas by creating restricted access to medical care not typically available in a country. These ethical principles, however, are secondary to the central principle that study subjects must not be exploited for scientific objectives, clinical knowledge, or public health programs. Both the Helsinki Declaration and the CIOMS international ethical guidelines leave open the opportunity of a placebo-controlled trial when a proven therapy exists, but in these instances, 'extreme care' must be taken in justifying the study. Regardless, the proposed study must undergo proper ethical and scientific review and approval.

Clinical equipoise remains an important condition for the use of a placebo in a clinical trial (Emmanuel, Wendler, and Grady, 2000). But some believe that clinical equipoise is a flawed concept and should be abandoned (Miller et al., 2003). Miller and Brody argue that research and clinical care are conflated in randomized clinical trials. The main objective of a randomized clinical trial is to perform a research experiment that will inform treatment of future patients, patients and physicians often view the clinical trial as a way to deliver medical care to the study patients. Thus, a patient may enter a trial with the understanding that regardless of which therapy he or she receives, it is the best one for them as an individual. Clinical equipoise fails to separate the research requirements of the clinical trial from the clinical care of its participants. Alternative tests for the need of a placebo exist (Lilford, 2003), but they are not commonly practiced. The framing of research questions and the testing of new therapies is settings with heterogeneous standards of care remains a challenge in clinical research.

Providing Participants with Care

This case study raises another ethical principle relating to standards of care and clinical trials. This principle addresses what should happen when the study has ended. What is the obligation of the investigators to study participants? The Declaration of Helsinki states: 'At the conclusion of the study, every patient entered into the study should be assured of access to the best proven prophylactic, diagnostic, and therapeutic methods identified by the study'. In other words, if the study results indicate that a new treatment is effective, or more effective than a standard treatment, then all patients in the study should be offered this treatment. Indeed, this is what was done in the case study. As part of the study design, interim analyses were performed to determine when one intervention appeared superior to the placebo. When the isoniazid intervention was shown to reduce the incidence of tuberculosis, then subjects enrolled in the placebo arm were offered isoniazid treatment as indicated in the informed consent for the trial. Both the researchers and the study sponsor took responsibility for treating these study participants with isoniazid. The study investigators systematically identified and traced living subjects enrolled in the control arm of the study, excluded active tuberculosis, explained the findings of the study, and offered a six month course of isoniazid therapy. Of the surviving participants without tuberculosis, 85% were traced and offered treatment.

Disseminating Research Findings

A current area of controversy is whether the researchers have an ethical obligation to ensure that their findings are incorporated into public policy in the host country where the research was performed. Although the Helsinki Declaration specifically refers to "patients entered into the study," one may reasonably ask whether the benefits of the study intervention should not be made readily available to the entire population. In fact, the Helsinki Declaration stipulates: "Medical research is only justified if there is a reasonable likelihood that the populations in which the research is carried out stand to benefit from the results of the research." Thus, from the inception of a research project, investigators are required to establish that there is a "reasonable likelihood" that the proven treatment will be made available to the public. It is not clear, however, what represents a reasonable likelihood, or how this likelihood would be established. Presumably, investigators from the collaborating and host countries consult the local ministry of health, local health professional organizations, or community advisory boards in advance of the study and gain their support. Moreover, it is not clear whether "population" refers to the study participants or the community from which the study participants were drawn. The intent of the principle seems to require an assessment of how a new treatment or intervention might be made available to people not participating in the clinical trial. This step enters the realm of public policy, an area where researchers may not be qualified to participate. Nevertheless, they may be called upon to explain their findings and put them into context for politicians and public health officials who will convert the findings into public policy. In this way, researchers are called upon to be active participants in the process of public health policy.

As for the case study, the investigators from the study developed a working relationship with the Ugandan Ministry of Health in advance of the study. Indeed, the input from the policy makers in the country urged the use of a placebo control group to allow for the assessment of the absolute risk of tuberculosis. Because of the size and scope of the study, investigators also engaged local hospitals and clinics caring for HIV-infected patients. This interaction served to inform local health care professionals of the need to screen for active tuberculosis and offer treatment for latent infection, once the results were known. Investigators for the study also presented the findings at international consensus conferences and participated in drafting guidelines for the treatment of latent tuberculosis infection in HIV disease. These guidelines recommend treatment of HIV-infected persons with reactive tuberculin skin tests with isoniazid therapy for 6 to 9 month (Maher, Floyd, and Raviglione, 2002). The statement falls short of advocating isoniazid treatment of latent infection as a programmatic approach to tuberculosis control. Currently, community-based clinical trials of isoniazid therapy are being carried out in endemic settings to determine whether prevention of tuberculosis in HIV-infected persons improves tuberculosis control.

Summary

This case illustrates the myriad of ethical issues involved in the design and conduct of a randomized clinical trial of tuberculosis prevention in developing countries. In particular, it shows how clinical standards of care for tuberculosis in HIV infection were established and why they were challenged. The case shows how the concept of clinical equipoise was applied to justify the placebo arm of the study. Finally, it raises the question of how results from a single clinical trial can be used to inform public policy about tuberculosis control.

Acknowledgements. This work was supported, in part, by grants from the National Institutes of Allergy and Infectious Diseases, the National Institutes of Health, including the AIDS International Training and Research Program of the Fogarty International Center (TW-00011) and the Center for AIDS Research (AI 36219) at Case Western Reserve University. The study of preventive therapy for tuberculosis in HIV infection was supported by a cooperative agreement with the Centers for Disease Control and Prevention (ADEPT/HIV-Related Tuberculosis Demonstration Project, U78/CCU506716–04).

References

Angell, M. (1997). The Ethics of Clinical Research in the Third World. *N Engl J Med, 337*, 847–849.

Barnes, P.F., Quoc Le, H., & Davidson, P.T. (1993). Tuberculosis in patients with HIV infection. *Med Clin N Am, 77*, 1369–1390.

Centers for Disease Control and Prevention (1990). The use of preventive therapy for tuberculous infection in the United States. *MMWR, 39*, 9–12.

Centers for Disease Control and Prevention (2003). Treatment of tuberculosis, American Thoracic Society, CDC, and Infectious Diseases Society of America. *MMWR, 52*, 36–41.

Colebunders, R., Izaley, L., Bila, K., Kabumpangi, K., Melameka, N., Francis, H. et al. (1987). Cutaneous reactions to trimethoprim-sulfamethoxazole in African patients with the acquired immunodeficiency syndrome. *Ann.Intern.Med., 107*, 599–600.

Coopman, S.A., Johnson, J.R., Platt, R., & Stern, R.S. (1993). Cutaneous disease and drug reactions in HIV infection. *N Engl J Med, 328*, 1670–1674.

Council for International Organizations of Medical Sciences (2002). In *International Ethical Guidelines for Biomedical Research Involving Human Subjects*. Geneva, Switzerland: Author.

De Cock, K.M., Soro, B., Coulibaly, I.M., & Lucas, S.B. (1992). Tuberculosis and HIV infection in sub-Saharan Africa. *JAMA, 268*, 1581–1587.

De Cock, K.M. & Chaisson, R. E. (1999). Will DOTS do it? A reappraisal of tuberculosis control in countries with high rates of HIV infection. *Int. J. Tuberc. Lung Dis., 3*, 457–465.

Dye, C. (2006). Global epidemiology of tuberculosis. *Lancet, 367*, 938–940.

Emmanuel, E.J., Wendler, D., & Grady C (2000). What makes clinical research ethical? *JAMA, 283*, 2701–2711.

Eriki, P.P., Okwera, A., Aisu, T., Morrissey, A.B., Ellner, J.J., & Daniel, T.M. (1991). The influence of human immunodeficiency virus infection on tuberculosis in Kampala, Uganda. *Am Rev Respir Dis, 143*, 185–187.

Ferebee, S.H. (1970). Controlled chemoprophylaxis trials in tuberculosis: a general review. *Adv Tuberc Res, 17*, 28–106.

Gordin, F. M., Simon, G. L., Wofsy, C. B., & Mills, J. (1984). Adverse reactions to trimethoprim-sulfamethoxazole in patients with the acquired immunodeficiency syndrome. *Ann.Intern.Med., 100*, 495–499.

International Union Against Tuberculosis Committee on Prophylaxis (1982). Efficacy of various durations of isoniazid preventive therapy for tuberculosis: five years of follow-up in the IUAT trial. *Bull World Health Org, 60*, 555–564.

Johnson, J. L., Okwera, A., Hom, D. L., Mayanja, H., Mutuluuza, K. C., Nsubuga, P. et al. (2001). Duration of efficacy of treatment of latent tuberculosis infection in HIV-infected adults. *AIDS, 15*, 2137–2147.

Lilford, R.J. (2003). Ethics of clinical trials from a Bayesian and decision analytic perspective: whose equipoise is it anyway? *BMJ, 326*, 980–981.

Maher, D., Floyd, K., & Raviglione, M (2002). Interventions to control tuberculosis in high HIV prevalence populations. In *Strategic Framework to Decrease the Burden of TB/HIV* (pp. 17–22). Geneva, Switzerland: World Health Organization.

Miller, F.G. & Brody, H. (2003). A critique of clinical equipoise: therapeutic misconception in the ethics of clinical trials. *Hastings Center Report, 33,* 19–28.

Moulding, T.S., Redeker, A.G., & Kanel, G.C. (1989). Twenty isoniazid-associated deaths in one state. *Am Rev Respir Dis, 140,* 575–576.

O'Brien, R.J. & Perriens, J.H. (1995). Preventive therapy for tuberculosis in HIV infection: the promise and the reality. *AIDS, 9,* 665–673.

Pape, J.W., Jean, S.S., Ho, J.L., Hafner, A., & Johnson, W.D. (1993). Effect of isoniazid prophylaxis on incidence of active tuberculosis and progression of HIV infection. *Lancet, 342,* 268–272.

Pozniak, A.L., MacLeod, G.A., Mahari, M., Legg, W., & Weinberg, J. (1992). The influence of HIV status on single and multiple drug reactions to antituberculous therapy in Africa. *AIDS, 6,* 809–814.

Selwyn, P.A., Hartel, D., Lewis, V.A., Schoenbaum, E.E., Vermund, S.H., Klein, R.S. et al. (1989). A prospective study of the risk of tuberculosis among intravenous drug users with human immunodeficiency virus infection. *N Engl J Med, 320,* 545–550.

Selwyn, P.A., Sckell, B.M., Alcabes, P., Friedland, G.H., Klein, R.S., & Schoenbaum, E.E. (1992). High risk of active tuberculosis in HIV-infected drug users with cutaneous anergy. *JAMA, 268,* 504–509.

Shapiro, H.T. & Meslin, E.M. (2001). Ethical issues in the design and conduct of clinical trials in developing countries. *N Engl J Med, 345,* 139–141.

Whalen, C., Horsburgh, C.R., Hom, D., Lahart, C., Simberkoff, M., & Ellner, J. (1995). Accelerated course of human immunodeficiency virus infection after tuberculosis. *Am J Respir Crit Care Med, 151,* 129–135.

Whalen, C.C., Johnson, J.L., Mugerwa, R., & Ellner, J.J. (1998). Preventive Therapy Trial in Uganda. *N Engl J Med, 338,* 842.

Whalen, C.C., Johnson, J.L., Okwera, A., Hom, D.L., Huebner, R., Mugyenyi, P. et al. (1997). A Trial of Three Regimens to Prevent Tuberculosis in Ugandan Adults Infected with the Human Immunodeficiency Virus. *N Engl J Med, 337,* 801–808.

Whalen, C. C., Nsubuga, P., Okwera, A., Johnson, J. L., Hom, D. L., Michael, N. L. et al. (2000). Impact of pulmonary tuberculosis on survival of HIV-infected adults: a prospective epidemiologic study in Uganda. *AIDS, 14,* 1219–1228.

Wilkinson, D., Squire SB, & Garner, P. (1998). Effect of preventive treatment for tuberculosis in adults infected with HIV: systematic review of randomised placebo controlled trials. *BMJ, 317,* 625–629.

World Medical Association (2000). Declaration of Helsinki: Ethical principles for medical research involving human subjects. *JAMA, 284,* 3043–3045.

World Medical Organization (1996). Declaration of Helsinki. *British Medical Journal, 313,* 1448–1449.

Chapter 9
Observational Studies

It may be helpful to briefly review the various observational study designs and the data collection strategies that are often used in such studies prior to embarking on a discussion of the ethical issues associated with both the designs and the approaches to data collection.

Observational Study Designs

In general, three basic study designs exist for the conduct of observational studies: cohort, case-control, and cross-sectional. Hybrids that utilize various components are also possible, and the reader is referred to other sources for a discussion of these.

Cohort Studies

Cohort studies are utilized to determine the association between an exposure of interest and an outcome of interest. Compared to other study designs, cohort studies offer an important advantage in that they allow the investigator to assess the likelihood of various outcomes over time.

Cohort studies may be conducted retrospectively or prospectively. In either case, the investigator selects a study population based on their exposure or non-exposure to the factor of interest, and then follows them prospectively in time in order to observe the number of new cases of the disease under investigation that occurs in each group during a specified period of time (Kelsey, Thompson, and Evans, 1986).

As an example, suppose that we do not have any knowledge about whether the use of methamphetamine is related to increased HIV risk, but we have such a hypothesis. The investigator may select a group of individuals who have been using methamphetamine for a predetermined amount of time, or have used a predetermined quantity, as specified in the protocol, and follow them prospectively to see how many individuals in this group contract HIV. Concurrently, the investigator must identify a group of individuals who are at risk for the same exposure but who are not exposed and follow them prospectively for the same duration of time, to determine how many in this group become HIV-infected. Finally, the investigator will calculate the ratio of those who are infected in the exposed group and those who are infected in the unexposed group to determine if the individuals who used methamphetamine are at increased risk for contracting HIV, a statistic known as the risk ratio. The investigator will have to control for other factors that might be related to the increased HIV risk and may do this by excluding from study participation individuals with specified characteristics, or by controlling for these differences in the statistical analysis.

If we wished to conduct a retrospective cohort study, we could utilize the same design. However, rather than beginning our observation period today and following individuals into the future to see how many in each group develop HIV infection, we would start our observation period at a point of time in the past, and then follow individuals forward to today or beyond. The exposures in the past would have to be reconstructed through memory, prior medical records, etc., which may be extremely difficult to do.

Case-Control Studies

Like cohort studies, case-control studies are designed to examine the relationship between an exposure of interest and a disease or outcome of interest. In contrast to cohort studies, however, the investigator begins the study by identifying the cases, that is, a group of individuals who have already developed the disease or outcome of interest, and a group of controls, individuals who are "representative of the same base experience" as the cases but who did not develop the disease (Miettinen, 1985). The investigator will then look backwards in time for a predetermined period from the time that the disease was identified. The length of time will depend on the particular nature of the disease and the exposure (Kelsey, Thompson, and Evans, 1986).

As an example, consider our hypothesis that there may exist a relationship between the use of methamphetamine and an increased risk of HIV infection. If an investigator is using a case-control design, he or she will identify cases, individuals who are already HIV-seropositive, and will ask these individuals about their exposures to not only methamphetamine for "X" number of years, but also about their exposures to other substances that have been implicated in increased HIV risk (cf. Rothman, 1986). At the same time, he or she will identify a group of individuals who are HIV-seronegative but were at risk of the same exposure and similarly question them about their exposures. Ultimately, the investigator will calculate the odds of contracting the disease given the exposure, compared to the odds of contracting the disease without having had the exposure. This is known as the odds ratio. The odds ratio generally approximates the risk ratio.

Cohort studies and case-control studies both permit causal inferences because the design permits us to follow the participants over time and in both designs the exposure precedes the development of the disease. Retrospective cohort studies and case-control studies both have as a disadvantage, however, the need to reconstruct past history of exposure. This can be done through memory of the participant and reliance on documents such as medical records, union records, records of federal and other agencies, etc. However, these files may be difficult to locate or may have become nonexistent due to the passage of time.

Cross-Sectional Studies

Unlike cohort and case-control studies, which permit us to draw causal inferences, cross-sectional studies cannot provide the basis for such inferences because they are characterized by temporal ambiguity. That is, data relating to both the exposure and the outcome of interest are collected at the same time; we may not know whether the hypothesized outcome occurred after the exposure of interest or whether it actually preceded the exposure (Kelsey, Thompson, and Evans, 1986).

Suppose, for instance, we wished to examine the relationship between neurological impairment and HIV status. We could recruit a group of individuals and test them for both neurological impairment and for their HIV status. If we found an association between HIV seropositivity and decreased neurological functioning, we would not know whether the neurological impairment resulted from the HIV infection or existed prior to the infection. Cross-sectional studies may provide important clues for the formulation of hypotheses to be further tested with more rigorous study designs, but they cannot be used to draw conclusions about cause and effect.

Data Collection Strategies

A variety of data collection strategies can be used, regardless of which study design is employed. The strategies that will be discussed here include focus groups, individual interviews, and participant observation. Each of these strategies has methodological strengths and weaknesses, but our attention here is on the ethical issues that attend the use of each. The reader is referred to other sources in addition to those referenced in this discussion for a more detailed discussion of the methodological issues (e.g., Bernard, 1994; Wengraf, 2001).

Focus Groups

A focus group has been explained as "a situation in which a group moderator keeps a small and usually homogeneous group of about 6 to 12 people focused on the discussion of a research topic or issue" (Johnson and Turner, 2003). Focus groups are often conducted in order to obtain preliminary information about a particular issue, which can then be used as the basis for a more in-depth survey or individual interviews. They may also be helpful as a mechanism to test new instruments or at the end of a study to understand participants' reactions to the conduct of the study.

Focus groups usually last anywhere from one to three hours. They are usually conducted by a facilitator, with one or more other individuals responsible for the recording of the focus group discussion through notes, video, and/or tape recording. And, although focus groups may appear to be relatively informal, they often require extensive preliminary preparation in order to formulate appropriate questions that will yield the most in-depth information.

One of the primary difficulties with focus groups is that of maintaining confidentiality. Although the researcher may explain the need to maintain the confidentiality of the information that is shared, he or she cannot guarantee that information that is divulged in the context of the focus group will not be repeated outside of the group by another group participant. For this reason, it is important that the researcher stress the importance of confidentiality and that individuals need not disclose any more than what they can comfortably share.

Individual Interviews

Open-ended interviews help to expand knowledge about which little is known, identify new domains, break down domains into component factors and subfactors, and obtain information about the context in which the study is to occur (Schensul, Schensul, and LeCompte, 1999). The process of conducting open interviews is as follows.

The interviewer/researcher first develops general questions derived from a theoretical model. Probes can be used during the course of the interview to elicit more information. These include the repetition of the individual's statement in a questioning way, asking for more information, asking for clarification, asking for an opinion, and/or asking for clarification of terms used. The interviewer may also identify or return to topics to clarify information that is unclear or incomplete, try to define domains of culture by asking for lists of things, and/or ask the individual for a narrative of his or her experience. All of these techniques must be utilized in a manner that is designed to maintain the flow of the interview (Schensul, Schensul, and LeCompte, 1999).

Semi-structured interviews combine the flexibility of unstructured, open-ended interviews with the directionality of survey instruments. Questions are preformulated, but the answers are open-ended. This type of interview provides the basis for the development of surveys, helps to clarify the domains and factors in the study, assists with the development of preliminary hypotheses, and helps to operationalize factors into more precise variables.

Questions in a semi-structured interview may be organized temporally, according to their complexity, according to the domains or topics, by level of abstraction within domains, or according to the level of threat that they pose to the respondent. For instance, in the context of a semi-structured interview about sexual behavior, questions may be arranged in a chronological fashion, whereby the respondent narrates his or her experiences from the most distant sexual encounter to the most recent, or by their level of sensitivity, so that questions relating to participation in specified activities, such as anal intercourse or fetish practices, are reserved for later in the interview, after the interviewer has had a chance to develop rapport with the research participant and the participant feels more comfortable (Schensul, Schensul, and LeCompte, 1999).

Participant Observation

Participant observation is a strategy that is used in qualitative research. It has been defined as "a process of learning through exposure to or involvement in the routines or activities of the research participants" (Schensul, Schensul, and LeCompte, 1999: 91). This strategy provides an intuitive understanding of the ways things are organized and prioritized, how people relate to each other, and how boundaries are defined. It demonstrates or confirms patterns, e.g. in etiquette, status, hierarchies; endorses the presence of the researcher in the community; and allows the investigator to witness events that outsiders would not be permitted to attend.

The extent of the researcher's participation is on a continuum in terms of the types of activities and the level of actual participation in specific activity. Ultimately, the quality of observation depends on the observational, documentation, and interpretation skills of researcher. Individuals and communities may utilize any of numerous strategies to exclude the researcher, such as using language that is unfamiliar to investigator, "code switching," changing the subject, refusing to answer questions, positioning so that the researcher cannot hear, and/or not inviting the researcher to attend events (Schensul, Schensul, and LeCompte, 1999).

Observations include both events and settings. Various descriptive techniques can be utilized to explain the who, what, where, when, why, and for whom of a particular event or setting, including numerical counts, census taking, descriptive details, and/or ethnographic mapping. Field notes from such observations must be done on a regular basis. Accurate notes require that the individual conducting participant observation record the date, time, place, name of researcher at top of each page; define behaviors behaviorally; describe the appearance of individuals; describe the physical state of the environment; record personal reactions and interpretations separately; include exact quotes to the extent possible; use pseudonyms or coding to maximize confidentiality; describe activities in the sequence in which they occurred; and include relevant history related to events or individuals (Schensul, Schensul, and LeCompte, 1999).

Ethical Issues

Informed Consent

Capacity

The issue of capacity can arise in numerous ways in the context of even observational research. As an example, the question has been raised as to whether individuals who chronically abuse substances can ever have the capacity to consent (Woodhouse et al., 1995). However, many of these issues have been addressed in the context of specific chapters focusing on children and mentally ill individuals and, for this reason, this discussion will concentrate on elements other than capacity that also comprise informed consent.

Knowledge and Voluntariness

Ethical issues often arise in the context of observational studies conducted in conjunction with public health functions, such as surveillance, and epidemiological studies. In fact, one scholar called "the clash between the privacy rights of persons and the need for access to and disclosure of personal health-related information . . . the most frequent ethical dilemma to confront epidemiologists (Last, 1996: 57).

As an example, *blinded serosurveys* have been conducted under the rubric of public health surveillance to determine prevalence rates of HIV infection in specific communities or among specific populations. In such testing, the blood samples are unlinked to the persons from whom the blood is drawn. This can be accomplished by first asking the individual for their consent to have their blood tested for HIV prior to drawing the blood sample. Alternatively, the blood that is "left over" from blood drawn for a different purpose can be tested for HIV, without the individual's knowledge or consent.

Blinded serosurveys conducted without the knowledge of the individuals from whom the blood is taken pose serious ethical questions. First, the basic principle of respect for persons, encompassing the requirement of informed voluntary consent, is violated. One might argue that there is no harm to the individuals because their identity is not known. However, individuals may feel wronged. Additionally, the failure to advise individuals who test HIV-positive of their HIV test results is in direct opposition to the policy requiring that individuals participating in other types of studies be told of their HIV test results and, additionally, results in a missed opportunity to provide HIV risk reduction information to the individual.

It has been argued that practices such as blinded serosurveys do not constitute research and are therefore exempt from the need to inform the affected individuals and obtain their voluntary consent. The World Bank, for instance, has asserted that:

Surveillance is not research. Public health surveillance is essentially descriptive in nature. It describes the occurrence of injury or disease and its determinants in the population. It also leads to public health action . . . . If we confuse surveillance with research, we may be motivated to collect large amounts of detailed data on each case. The burden of this approach is too great for the resources available . . . . (World Bank, 2002).

The Centers for Disease Control and Prevention have similarly objected to characterizing such undertakings as research, but for other reasons:

The implications of calling public health surveillance research are broad and far reaching . . . If all surveillance activities were research, it might mean that each local health department would have to form institutional review boards (IRBs) (Centers for Disease Control and Prevention, 1996).

Instead, activities such as blinded serosurveys, it is argued, are to be characterized as public health practice when conducted by a state or governmental agency, and research when conducted by a nongovernmental individual or entity.

However, the demarcation between research and public health practice is often blurred, even when the activity is conducted through a governmental entity

(Fairchild and Bayer, 2004). It is also clear that individuals may be subjected to similar risks through public health surveillance as through participation in research, often with little or no direct benefit. Accordingly, at least some scholars have called for ethical review of public health surveillance activities at state and federal levels, regardless of whether such activities are to be considered public health practice, research, or somewhere between the two (Fairchild and Bayer, 2004).

The issue of knowledge and the need to re-inform participants may also arise due to a change in the larger context during the course of an investigation. This may occur in the context of all HIV-related research, but is particularly critical in situations involving natural history studies. For instance, at the beginning of the HIV epidemic, nothing was known about this new and baffling disease. Consequently, natural history studies constituted a critical step in the development of an understanding of the disease and its symptoms and risk factors for its transmission. However, as we learned more about the disease and developed strategies and medications to limit its transmission and reduce its impact, the continuation of such natural history studies would be ethically suspect. Indeed, responsible researchers conducting such studies made every effort to keep the participants in such studies apprised of research findings and new treatments.

Assessing Risks

Limitations on Confidentiality and Privacy

Confidentiality has been identified as one of the major ethical issues arising in the context of HIV-related research (Muthuswamy, 2005). Shifting ethical standards over the course of the HIV epidemic make it difficult for investigators to ensure the confidentiality of all data collected. As an example, in several longitudinal (cohort) studies initiated at the beginning of the epidemic, participants could choose whether or not to receive their HIV test results (Kaslow, Ostrow, Detels, Phair, Pols, and Rinaldo, 1987). At a later date, the National Institutes of Health decided that, ethically, all participants tested for HIV should receive their HIV test results. As a result, individuals who agreed to participate in the studies with one understanding found that the standards had shifted and their choice was no longer valid (Blanck, Bellack, Rosnow, Rotheram-Borus, and Schooler, 1992). Commentators have noted that the requirement of mandatory notification essentially nullifies a study participant's ability to withdraw from a study at any time, because he or she *has to* receive the test results (Avins and Lo, 1989). This possibility contravenes the ethical principle of respect for persons. Similarly, states that once did not require the reporting of the names of individuals who tested HIV-positive or partner notification later implemented such requirement(s) (Blanck et al., 1992).

The investigator's ability to assure confidentiality in the context of HIV-related research may also be limited due to a duty to warn, state-imposed reporting requirements, and legal attempts to access the data. Although these issues may arise during the course of any HIV-related research, they may be especially likely to arise in observational studies conducted over an extended period of time.

Duty to Warn. A "duty to warn" may exist as the result of a line of court cases that began in 1976 with the now famous case of *Tarasoff v. Regents of the University of California.*

The case involved a lawsuit by the Tarasoff family against the University of California and a psychologist at the Berkeley campus of the university for the death of their daughter Tatiana. Tatiana had refused the advances of another graduate student at Berkeley. The would-be suitor had revealed his intent to kill Tatiana during the course of counseling sessions with a psychologist at the school's counseling services. The psychologist and several colleagues sought to have this student involuntarily hospitalized for observation purposes, but he was released after a brief observation period, during which it was concluded that he was rational. He subsequently shot and killed Tatiana.

The majority of the court rejected the psychologist's claim that he could not have advised either the family or Tatiana of the threat because to do so would have breached the traditionally protected relationship between the therapist and the patient. Instead, the court held that when a patient "presents a serious danger . . . to another [person], [the therapist] incurs an obligation to use reasonable care to protect the intended victim against such danger." That obligation could be satisfied by warning the intended victim of the potential danger, by notifying authorities, or by taking "whatever other steps are reasonably necessary under the circumstances" (*Tarasoff v. Regents of the University of California,* 1976). The court specifically noted that the therapist-patient privilege was not absolute:

We recognize the public interest in supporting effective treatment of mental illness and in protecting the rights of patients to privacy and the consequent public importance of safe-guarding the confidential character of psychotherapeutic communication. Against this interest, however, we must weigh the public interest in safety from violent assault . . . . We conclude that the public policy favoring protection of the confidential character of patient-psychotherapist communications must yield to the extent to which disclosure is essential to avert danger to others. The protective privilege ends where the public peril begins.

Some later cases have followed the reasoning of the *Tarasoff* court. A New Jersey court ruled in *McIntosh v. Milano* (1979) that the doctor-patient privilege protecting confidentiality is not absolute, but is limited by the public interest of the patient. In reaching this conclusion, the court relied on the 1953 case of *Earle v. Kuklo,* in which the court had stated that "a physician has a duty to warn third persons against possible exposure to contagious or infectious diseases." A Michigan appeals court held in *Davis v. Lhim* (1983) that a therapist has an obligation to use reasonable care whenever there is a person who is foreseeably endangered by his or her patient. The danger would be deemed to be foreseeable if the therapist knew or should have known, based on a professional standard of care, of the potential harm.

Courts are divided, however, on whether the patient must make threats about a specific, intended victim to trigger the duty to warn. The court in *Thompson v. County of Alameda* (1980) found no duty to warn in the absence of an identifiable victim. Another court, though, held that the duty to warn exists even in the absence of specific threats concerning specific individuals, if the patient's previous history suggests that he or she would be likely to direct violence against a person (*Jablonski v. United States*, 1983).

Consider a similar scenario, but in the context of HIV. A patient discovers that he or she is HIV-infected. Angry, and refusing to accept the diagnosis and recommended changes in sexual behavior, such as use of a condom to protect himself and partners, the individual continues to engage in unprotected sex with his current sexual partner. This situation is most analogous to *Tarasoff*: a specifically identified person is placed at risk as a result of the patient's behavior. Less clear is the situation in which the individual continues to have unprotected intercourse with multiple partners, who may be unknown to him by name. The case study by Gore-Felton and DiMarco that follows chapter 10, Behavioral Intervention Studies, addresses this scenario in detail.

Despite the passage of time since the beginning of the HIV epidemic, there continues to be no legal consensus on the applicability of the *Tarasoff* doctrine to HIV-related situations. It has been argued that HIV is less likely to be contracted as the result of unprotected intercourse than death is to occur as the result of a shooting with a bullet. And, even if an individual is exposed to HIV, there is the possibility that he or she will be able to clear the virus from his or her system. And, unlike the situation in *Tarasoff*, involving the immediate death by a bullet, individuals infected with HIV may live for extended periods of time (Traver and Cooksey, 1988).

Rather than relying on court decisions to determine the applicability of California *Tarasoff* doctrine, many state legislatures have adopted legislation requiring that HIV-infected persons advise their prospective sexual and/or needle-sharing partners of their HIV status before engaging in any risky behavior with them; a failure to do so can result in criminal prosecution. As an example, the Kentucky statute defining wanton endangerment in the first degree and making it a felony has been applied in cases involving sexual intercourse of an HIV-positive person (Kentucky Revised Statutes Annotated, 2006; *Hancock v. Commonwealth*, 1998). Ohio's definition of felonious assault specifically includes sexual conduct by an HIV-infected person with another individual without first having disclosed his or her HIV seropositivity to the prospective sexual partner (Ohio Revised Code Annotated, 2006; *State v. Gonzalez*, 2003).

The duty to warn may be relevant in the research setting, even absent general legal consensus regarding the applicability of the *Tarasoff* doctrine to HIV-related situations, if for no other reason that researchers are often clinicians, as well, such as physicians, nurses, or social workers. However, its application is fraught with potential problems, as there are no clinically accepted standards for the evaluation of dangerousness (Lamb, Clark, Drumheler, Frizzell, and Surrye, 1989; Public Health Service, 1987) and psychotherapists often differ in their assessments of the extent to which the dangerousness of an HIV-infected individual and the identifiability of a victim should mandate a breach of confidentiality (Totten, Lamb, and Reeder, 1990).

Mandatory Reporting. All states require the reporting of newly diagnosed cases of AIDS to their state health departments. Many also require the reporting of newly diagnoses cases of HIV. Although some states permit anonymous testing, whereby only salient factors are reported with the test results, such as risk behaviors, a growing number of states require that individuals testing HIV-seropositive be reported by name.

Depending on the particular state, however, researchers may also be required to report instances of child sexual abuse, child abuse or neglect, elder abuse, or intimate partner violence that may be committed by or perpetrated on a research participant. Whether such an obligation exists often depends on the age and state of residence of the victim, the state's definition of the offense, the recency of the event, and the status of the reporter, that is, whether a researcher under that state's laws is a mandated reporter.

Subpoenas. A subpoena is an order from a court or administrative body to compel the appearance of a witness or the production of specified document or records. This discussion focuses on subpoenas issued to compel the production of records or documents associated with the research.

A subpoena can be issued by a court or administrative body at the state or federal level. The information sought may be believed to be important to the conduct of an investigation, a criminal prosecution, or a civil lawsuit. The issuance of subpoenas against researchers had become increasingly common (Brennan, 1990) and have been used as a mechanism to obtain data relating to identifiable research participants.

Certificates of confidentiality may potentially limit the extent to which research data may be obtained by subpoena. Certificates of confidentiality are issued by the appropriate institute of the National Institute of Health and other agencies of the United States Department of Health and Human Services. Authority for their issuance derives from the section 301(d) of the Public Health Service Act, which provides that:

The Secretary may authorize persons engaged in biomedical, behavioral, clinical, or other research (including research on mental health, including research on the use and effect of alcohol and other psychoactive drugs) to protect the privacy of individuals who are the subject of such research by withholding from all persons not connected with the conduct of such research the names or other identifying characteristics of such individuals. Persons so authorized to protect the privacy of such individuals may not be compelled in any Federal, State, or local civil, criminal, administrative, legislative, or other proceedings to identify such individuals.

Certificates are potentially available for research where the participants may be involved in litigation that relates to the exposure under study, such as occupational exposure to HIV; that collects genetic information; that collects data pertaining to participants' psychological well-being, their sexual attitudes, preferences, or practices or their substance use or other illegal activities or behaviors. Additional details relating to certificates are available from the various websites sponsored by the Office of Extramural Research of the National Institutes of Health (http://grants.nih.gov/grants/policy/coc/appl_extramural.htm;

http://grants.nih.gov/grants/policy/coc/back ground. htm;
http://grants.nih.gov/grants/policy/coc/faqs.htm).

Although the validity of these certificates was once upheld by a New York
court (*People v. Newman*, 1973), their validity is subject to question because, in
essence, they allow an agency of the federal government to limit the ability of the
states to investigate and prosecute possible criminal activity and the ability of the
courts and litigants in civil cases to obtain evidence that may be critical.

Stigmatization

The issue of stigmatization may also arise as group stigmatization, in the context of
population-based research, such as epidemiological studies of prevalence and incidence. For instance, a researcher may find that individuals of a particular ethnic
group or a particular area of a city or village have a higher prevalence of HIV than
other groups or communities, or that the prevalence of high-risk behaviors is
higher. The group or community involved may not wish to be specifically identified
in publications because of the possibility of such stigmatization. The history of HIV
research provides an example of how such stigmatization may result.

Initially, research relating to HIV/AIDS focused on the identification of routes
of transmission and risk factors for the disease. By 1982, within a year of identifying the first cases of what would come to be called AIDS, the Centers for
Disease Control and Prevention (CDC) had labeled Haitians a "risk group." This
emphasis on group membership as a risk factor, rather than relevant activities or
behaviors, ultimately resulted in the medical and social construction of "risk
groups," whose members were presumed to be at higher risk of contracting and
transmitting the infection by virtue of their membership in the specified group,
regardless of their individual behaviors (Schiller, Crystal, and Lewellen, 1994).
These four groups—Haitians, homosexuals, heroin addicts, and hemophiliacs—
came to be known as "the 4-H club."

Although the withholding of specific group identifying information, that is,
group confidentiality, may reduce the likelihood of such stigmatization, it may also
result in an inability to replicate the study findings and to develop interventions and
programs that may be necessary to address the identified problem. It is important
that the researcher(s) and the community involved in the research work together to
develop an approach that not only considers the need to disseminate the research
findings to benefit present and future members of the relevant population, but also
respects the needs and sensitivities of the affected communities and groups
(cf. Council for International Organizations of Medical Sciences, 2005).

Psychological Sequelae

Psychological risks may be a concern in studies of longer duration, during which the
research participants and the research staff establish a relationship. It is possible
that research participants may feel abandoned following the termination of the study,
particularly if the staff has been supportive during the intervening time through the
provision of referrals, a listening ear, etc. It is important to institute strategies during

the course of the study to reduce this likelihood. This can be done, for instance, through regular reminders to participants of the nature of the relationship.

Assessing Benefits

Frequently, there may be no direct benefit to the study participants of participating in an observational study. Benefits often associated with participation include social interaction with others; the ability to provide information that may be helpful to others at a later date; receiving up-to-date information about HIV prevention and transmission; receiving HIV testing and test results; and, depending upon the nature of the study, receiving referrals for needed services, such as substance abuse treatment and temporary shelter.

In some circumstances, what appears at first glance to be a benefit may, in fact, become an unforeseen risk. For instance, one research group conducting a demonstration project that targeted women who exchanged sex for drugs routinely provided HIV test results to its research participants to enable them to obtain appropriate medical attention (Siegal, Carlson, Falck, Reece, and Perlin, 1993). The investigators had not anticipated that a study participant might later be incarcerated and be unable to obtain the necessary treatment without first disclosing her HIV status to prison authorities, which could result in segregation from the general prison population and a threat to her safety. Although this situation arose in the context of an intervention study, it could easily occur, as well, in the context of an observational study in which HIV testing is a component.

Third Parties

Observational research, particularly research involving participant observation and genetic epidemiology, often requires the collection of data relating to third parties, such as family members or social network members (DeCamp and Sugarman, 2004; Woodhouse et al., 1995). As an example, researchers may wish to describe social network structures and their influence on HIV transmission and characterize the specific strain(s) of the virus that are transmitted. Such research necessarily requires that individuals divulge information about others with whom they have interacted, yet these third parties are not themselves enrolled as participants. Another example is provided by the strategy developed by the Centers for Disease Control for the conduct of a national survey to determine the prevalence of HIV infection (Hurley and Pinder, 1992). The survey initially was structured to obtain information from neighbors about the number, age, race, and sex of persons living in a household where no one was home. The survey strategy was later modified following the characterization of this plan by the Policy Advisory Panel as an undue invasion of privacy (Hurley and Pinder, 1992).

The question becomes: at what point has the privacy of these third parties and the confidentiality of data pertaining to them been violated for failure to obtain their informed consent to such observations and data collection? This is an important issue because these third party individuals, if identifiable, could potentially suffer many of the same risks as the study participants themselves, such as stigmatization, psychosocial harm, and family discord (DeCamp and Sugarman, 2004), but they will not have had the same opportunity as the study participants to assess and weigh these risks and any potential benefits for themselves.

Guidelines have been developed for the factors to be considered in determining when such third parties are potentially identifiable:

1. the quantity of information collected about the third party;
2. the nature of the information that is collected, including its sensitivity and the possibility that it might result in harm to the third party;
3. the ability of the investigators to record the information in a manner that safeguards the identity of the third party; and
4. the possibility that the classification of the third party as a study participant might have on the rights or welfare of the index (original) study participant, thereby requiring that the IRB protect the interests of both the original participant and the third party (National Human Research Protections Advisory Committee, 2002).

References

Avins, A.L. & Lo, B. (1989). To tell or not to tell: The ethical dilemmas of HIV test notification in epidemiologic research. *American Journal of Public Health, 79(11),* 1544–1548.

Bernard, H.R. (1994). Research Methods on Anthropology: Qualitative and Quantitative Approaches, 2nd ed. Walnut Creek, California: AltaMira Press.

Blanck, P.D., Bellack, A.S., Rosnow, R.L., Rotheram-Borus, M.J., & Schooler, N.R. (1992). Scientific rewards and conflicts of ethical choices in human subjects research. *American Psychologist, 47(7),* 959–965.

Centers for Disease Control and Prevention. (1996). Minutes of Meeting on Protection of Human Research Subjects in Public Health, Atlanta, Georgia. Cited in A.L. Fairchild & R. Bayer. (2004). Ethics and the conduct of public health surveillance. *Science, 303,* 631–632.

Council for International Organizations of Medical Sciences. (2005). *International Guidelines for Ethical Review of Epidemiological Studies.* Geneva, Switzerland: Author.

Davis v. Lhim. (1983). 124 Mich. App. 291, *aff'd on rem* 147 Mich. App. 8 (1985), *rev'd* on grounds of government immunity in *Canon v. Thumudo,* 430 Mich. 326 (1988).

DeCamp, M. & Sugarman, J. (2004). Ethics in population-based genetic research. *Accountability in Research, 11,* 1–26.

Earle v. Kuklo. (1953). 26 N.J. Super. 471 (App. Div.).

Fairchild, A.L. & Bayer, R. (2004). Ethics and the conduct of public health surveillance. *Science, 303,* 631–632.

Hancock v. Commonwealth. (1998). 998 S.W.2d 496 (Ky. Ct. App.).

Hurley, P. & Pinder, G. (1992). Ethics, social forces, and politics in AIDS-related research: Experience in planning and implementing a household HIV seroprevalence survey. *Milbank Quarterly, 70(4),* 605–628.

Jablonski v. United States. (1983). 712 F.2d 391 (9th Cir.).

Johnson, B. & Turner, L.A. (2003). Data collection strategies in mixed methods research. In A. Tashakkori & C. Teddle (Eds.). *Handbook of Mixed Methods in Social & Behavioral Research* (pp. 297–319). Thousand Oaks, California: Sage Publication.

Kaslow, R., Ostrow, D., Detels, R., Phair, J., Pols, B., & Rinaldo, C. (1987). The multi-center AIDS cohort study: Rationale, organization, and selected characteristics of the participants. *American Journal of Epidemiology, 126,* 310–318.

Kentucky Revised Statutes Annotated §508.060 (2006).

Lamb, D.H., Clark, C., Drumheller, P., Frizzell, K., & Surrey, L. (1989). Applying *Tarasoff* to AIDS-related psychotherapy issues. *Professional Psychology: Research and Practice, 20,* 37–43.

Last, J. (1996). Professional standards of conduct for epidemiologists. In S.L. Coughlin & T.L. Beauchamp (Eds.), *Ethics and Epidemiology* (pp.53–75). New York: Oxford University Press.

McIntosh v. Milano. (1979). 168 N.J. Super. 466.

Muthuswamy, V. (2005). Ethical issues in HIV/AIDS research. *Indian Journal of Medical Research, 121,* 601–610.

National Human Research Protections Advisory Committee. (2002). *Clarification of the Status of Third Parties When Referenced by Human Subjects in Research.* Available at http://ohrp.psophs.dhhs.gov/nhrpac/documents/thirs.pdf. Last accessed October, 2003.

Ohio Revised Code Annotated § 2903.11 (2006).

People v. Newman. (1973). 298 N.E.2d 651 (App. Div.).

Public Health Service, United States Department of Health and Human Services. (1987). *A Public Health Challenge: State Issues, Policies and Programs,* vol. 1, pp. 4–1 to 4–31.

Public Health Service Act § 301(d), 42 United States Code § 421(d).

Schensul, S.L., Schensul, J.J., & LeCompte, M.D. (1999). *Essential Ethnographic Methods: Observations, Interviews, and Questionnaires.* Walnut Creek, California: AltaMira Press.

Schiller, N.G., Crystal, S., Lewellen, D. (1994). Risky business: The cultural construction of AIDS risk groups. *Social Science & Medicine* 38: 1337–1346.

Siegal, H.A., Carlson, R.G., Falck, R., Reece, R.D., & Perlin, T. (1993). Conducting HIV outreach and research among incarcerated drug abusers: A case study of ethical concerns and dilemmas. *Journal of Substance Abuse Treatment, 10,* 71–75.

State v. Gonzalez. (2003). 154 Ohio App. 3d 9.

Tarasoff v. Regents of the University of California. (1976). 17 Cal. 3d 425.

Thompson v. County of Alameda. (1980). 27 Cal. 3d 741.

Totten, G., Lamb, G.H., & Reeder, G.D. (1990). Tarasoff and confidentiality in AIDS-related psychotherapy. *Professional Psychology: Research and Practice, 21,* 155–160.

Traver, L.B. & Cooksey, D.R. (1988). Defense argument. In J.A. Girardi, R.M. Keese, L.B. Traver, & D.R. Coksey. (1988). Psychotherapist responsibility in notifying individuals at risk for exposure to HIV. *Journal of Sex Research, 25,* 1–27.

Wengraf, T. (2001). *Qualitative Research Interviewing.* Thousand Oaks, California: Sage Publications Inc.

Woodhouse, D.E., Potterat, J.I., Rothenberg, R.B., Darrow, W.M., Klovdahl, A.S., & Muth, S.Q. (1995). Ethical and legal issues in social network analysis: The real and the ideal. *NIDA Research Monograph, 151,* 131–143.

World Bank Group. (2002). Public health surveillance toolkit. Available at http://www.survtoolkit.worldbank.org. Last accessed October 2003.

Case Study Three
Challenges and Strategies
for Personal Safety in Fieldwork

Nancy Méndez and Ingrid Vargas

I drove up to Rita's house, a Puerto Rican woman with a long history of heroin use, legal problems and severe mental illness who had recently been recruited into our research program. I only carried my cell phone, participant stipend, identification and study instruments. Rita had recently moved into a newly rented house. The location was very isolated. The small house sat near abandoned railroad tracks and the house looked uninhabited. The paint was well worn, many of the windows were broken and the screen door hung by one hinge. I walked into Rita's home for our first interview. The living room and kitchen were cluttered with garbage. Rita was disheveled. Her hair was uncombed, her clothes were wrinkled. The first 20 minutes into the interview she seemed preoccupied and nervous. Suddenly her boyfriend stormed in through the back door and began a heated argument with Rita about missing drugs. Another man entered behind Rita's boyfriend who had just realized that I was sitting in the living room. Rita introduced me as her social worker. I quickly terminated the interview and left the home.

Introduction

This excerpt from research transcripts highlights the dangers faced by many of our severely mentally ill participants who struggle with drug use. The potential difficulties faced by ethnographers in the field are underscored in the passage. Using examples drawn from transcripts taken during a five year research project, we discuss safety concerns that arise when conducting ethnographic research with women who suffer from severe mental illness and substance use and examine strategies to help ethnographers minimize risks in the field. Although our focus here is on HIV-related research, these strategies may be helpful to researchers engaged in fieldwork that targets other diseases.

Background

The number of women with HIV infection and AIDS has increased steadily worldwide. By the end of 2005, 17.5 million women worldwide were infected with HIV (National Institute of Allergy and Infectious Diseases, 2006). Hispanics have been disproportionately impacted by HIV/AIDS. According to the Centers for Disease Control and Prevention (2005), in 2004 Hispanics accounted for 20% of all new AIDS cases diagnosed that year. Furthermore, there is growing concern that severe mentally ill adults living in the community have a high risk for HIV infection. Numerous articles have reported variable rates of HIV infection among persons with SMI, ranging from 3.1% to 22.9%, compared to an estimated prevalence rate of 0.4% in the general population (Cournos, 1997)

This observational study was designed to improve our understanding of the HIV risk of Puerto Rican and Mexican severely mentally ill (SMI) women ages 18 to 50 and to describe the context in which these behaviors occurred. For the purposes of this study, we defined severely mentally ill as schizophrenia, bipolar disorder, or major depression. Secondary objectives included the acquisition of additional insights that could be incorporated into the informed consent process and the content and delivery of a future HIV intervention, examination of the feasibility and acceptability of various modes of delivery of HIV prevention for this population, and field testing of an HIV peer education model for SMI. The study was guided by reference to social cognitive theory, the theory of gender and power, and the leadership-focused model.

The study was conducted in four phases. Phase 1 included focus groups with Latina SMIs, family members of Latina SMIs, and key informants that aimed to refine strategies to optimize the informed consent process, recruitment, and retention of Latina SMIs during Phase 2. Phase 2 consisted of ethnographic interviews, the administration of standardized instruments, up to 100 hours of participant observation per participant, and key informant interviews. Interviews and shadowing in the second phase took place in a variety of locations including participant's homes, nightclubs, clubs, churches, stores, and libraries. The third phase examined through focus groups and interviews the acceptability and feasibility of various potential prevention interventions for severely mentally ill Latinas and will field-test an HIV peer education model for SMI. This research was deemed critical to inform the later development of an HIV prevention intervention trial to test the efficacy of an intervention that is culture and gender sensitive to this high risk population.

Training

Interviews were conducted by one of several trained bilingual female ethnographers in English and/or Spanish. Ethnographers received training on questionnaire administration, and on transcription and the conduct of interviews, both designed to maximize privacy and confidentiality. The research team set guidelines for the protection of the study participants in accordance with the ethical principles found in the *Belmont Report*, the Nuremberg Code, the Helsinki Declarations, and the guidelines promulgated by the Council for International Organizations of Medical Sciences (1991, 2002). Our study procedures were approved by the IRB.

The ethnographers sought to establish relationships with the research participants while simultaneously minimizing the amount of involvement and impact on participants' lives. Ethnographers learned how to gain the participants trust in order to obtain very personal sensitive information while, to the extent possible, avoiding answering personal questions (Broadhead, 1990). Acquiring this skill was critical to the success of the ethnographer. Researchers suggest that ethnographers should make a conscious decision to go along with whatever role a participant may ascribe to them, be it sister, friend, or parent in order to relate on a level that is most comfortable for the participant (Boynton, 2002).

Ethical Issues and Resolutions

Balancing Risks and Benefits

As indicated in the preceding chapter, the ethical principles of beneficence and nonmaleficence require that the risks and benefits of study participation be assessed and the risks of participation be minimized. What is rarely recognized, however, is that a failure to attend to the risks to study staff may inadvertently increase the risks faced by individuals participating in the study.

As an example, consider a situation in which a study team member is to shadow a study participant while she is "on the stroll," looking for a "date" for sex. If the study team member appears uncomfortable or panics, the team member's response could inadvertently trigger an emotional response in others who are present, resulting in violence. Consequently, it is critical that study team safety be considered, not only to safeguard the study team members themselves, but also in the context of identifying procedures to reduce participant risk.

Safety Issues

In order to address safety issues that might arise during the course of shadowing and while conducting interviews, the research team created safety protocols for fieldwork. These protocols evolved further as different situations arose. This evolution is reflected in the strategies developed by team members while shadowing Rosa (fictitious name).

The research team quickly determined that women who were currently using illicit drugs posed greater challenges. Rosa (fictitious name) was one of the participants with a dual diagnosis of mental illness and drug abuse (heroin/alcohol). She had been arrested several times for solicitation and drug possession. She lived alone in an apartment located on the second floor of a building that was located in a neighborhood known for high drug activity and prostitution, as was the apartment building itself. The first few interviews with Rosa were without incident. She was not using drugs and she told the ethnographers about warning signs they could look for if she relapsed. *"You see how I have all this jewelry on. I have my house furnished. If I start using again this all end up at the pawn shop."* By the third visit we indeed observed changes in Rosa's physical appearance as well as that of her apartment. Rosa did eventually relapse and the ethnographers subsequently encountered instances of prostitution, violence, and substance abuse, and were confronted by threats of theft, threatening neighbors, and sexual advances by Rosa's family members and friends.

It had been standard practice for the ethnographers to carry cell phones and identification at all times and to leave their purses and brief cases at the office. After Rosa's relapse and a few such difficult encounters, the research team decided to have two ethnographers shadow Rosa at all times. During this time Rosa's drug use increased as did her openness about her involvement in prostitution and theft.

Rosa said she likes the viejitos (older ones). I asked why and she said because they have money and they are always flattered. She said she has slept with a guy for $200. I asked for how long and she said five minutes. She started laughing and said she steals their wallets while fucking them. We asked how and she said while she is rubbing their dick she turns off the lights and takes their money with her other hand. I asked her if she has done that a lot and she slapped her hands (laughing) and said plenty times. She said there's no man that uses her.

We briefly lost contact with Rosa. A month later the ethnographer was able to track her down. *"I asked her what had happened and she said the cops saw her walking the street and arrested her for soliciting."* A recurring issue when interviewing Rosa was sexual advances from her "dates." Rosa was well-known in the neighborhood for trading sex for money and drugs. Men who approached Rosa would sometimes make advances towards the female ethnographers as well. It was very important that we did not demonstrate noticeable awkwardness. The ethnographers' state of mind is extremely important. If the ethnographer shows fear, the participant may become uncomfortable and it could turn both the participant and the researcher into potential targets. (Kotarba, 1990; Williams, 1992).

A man walked down the side walk and approached us. I asked Rosa if she knew him and she said he lived in the apartments next door. He walked up to us and asked us which one of us he can have. I looked at him and asked him what he meant. He said we were all beautiful and he would not mind having us all. He introduced himself as Jamal. I told him I was married and he said not to worry about it because it wasn't my husband he was trying to get with. Rosa smiled and told us he didn't care if we had men at home.

Rosa began to report assaults by her neighbors and frequent drug busts by the police. As the situation in her apartment building deteriorated, the research team decided it would be in the best interest of the participant and the ethnographers to meet Rosa in public settings. We asked Rosa if it would be acceptable to meet at restaurants. She agreed as long as we picked her up from the apartment building.

This visit was hard to document because Rosa speech was slurred. Her eyes were red and I smell the alcohol on her. She moved around a lot and repeat things a lot. During this visit Rosa told us she was now smoking crack everyday. It was after this visit that we decided that it might be unsafe for us to be in the building with Rosa. Rosa disclosed that her neighbors were using drugs and some were selling. At previous times when I had been there, Jasmine (Rosa's neighbor)(fictitious name) walked up to Rosa and asked for money. Rosa is now asking us for money. We also decided it might be a good idea to shadow Rosa outside the building and take her to places like Mickey D's [local restaurant].

We continued to interview Rosa for one more year without incident. We followed all safety protocols and continued to reporting any changes in the participant's situation.

Recommendations

The protocols developed by the research team to address researcher safety in the field were formulated in reliance on the ethical principles that guide research involving human participants. Feeling secure as ethnographers helped our participants to feel more comfortable, less stressed, and potentially decreased their risk as potential targets (Boynton, 2002). The safety protocols served a dual purpose: keeping the ethnographer safe and protecting the participant. The participant should not be signaled out by their peers because they brought in an outsider.

Specific Recommendations Include:

- Carry mobile phones at all times.
- Leave purses and other valuable items in the office during field visits.
- Wearing appropriate clothing for the setting in order to avoid undue attention from onlookers.
- Always carry an emergency contact card with the Principal Investigator's contact information.
- When giving stipends, have only the exact dollar amount.
- Become familiar with the neighborhood and its residents.
- Be aware of space and surroundings during home visits. Immediately identify exits in the room during all home visits.
- Partner with another researcher/ethnographer on the team to conduct visits with participants who are aggressive and/or volatile.
- Always inform the research team of your whereabouts. A posted wall calendar with researchers'/ethnographers' schedule is encouraged.
- Have regular debriefing sessions with the study team regarding fieldwork experiences.

• Have the Project Coordinator conduct frequent checks-ins with ethnographers to discuss potentially volatile situations or debrief field visits that are particularly emotionally laden.

Conclusion

Ethnographic research is essential in gathering information that cannot easily be gathered in a clinical setting. In conducting such research, ethnographers often work on their own and engage in one -to-one relationships with participants. It is essential that the ethnographer receive proper training on how to engage study participants without overstepping boundaries. A critical aspect of safety in the field is anticipating what dangers might be present to researchers so that they can ameliorate or effectively manage them. Not all danger can be anticipated, but many situations can be avoided or controlled by careful reflection prior to data collection. (Sluka, 1990).

The issue of personal safety is rarely addressed as a methodological or an ethical issue. There is relatively little discussion on how to minimize risk that ethnographers may face in the field (Vanderstaay, 2005; Williams 1992), even though a decrease in the risk to the ethnographers may also decrease risk to the participants.

Acknowledgements. This work was supported through a grant from the National Institute of Mental Health R01 MH-63016.

References

Boynton, P.M. (2002). Life on the streets: The experiences of community researchers in a study of prostitution. *Journal of Community & Applied Social Psychology, 12,* 1–12.

Broadhead, R.S. & Fox, K.J. (1990). Takin' it to the streets: AIDS outreach as ethnography. *Journal of Contemporary Ethnography, 19(3),* 322–348.

Centers for Disease Control and Prevention. (2005). *HIV/AIDS Surveillance Report, 2004.* Atlanta, Georgia: Author.

Council for International Organizations of Medical Sciences. (2002). *International Ethical Guidelines for Biomedical Research Involving Human Subjects.* Geneva, Switzerland: Author.

Council for International Organizations of Medical Sciences. (1991). *International Guidelines for Ethical Review of Epidemiological Studies.* Geneva, Switzerland: Author.

Cournos, F. & McKinnon, K. (1997). HIV seroprevalence among people with severe mental illness in the United States: A critical review. *Clinical Psychology Review, 17,* 259–269.

Kotarba, J. (1990). Ethnography and AIDS: Returning to the streets. *Journal of Contemporary Ethnography, 19,* 259–271.

National Institute of Allergy and Infectious Diseases. (2006). HIV Infection in Women. Last updated May 18, 2006; Last accessed January 22, 2007; Available at http://www.niaid.nih.gov/factsheets/womenhiv.htm.

Sluka, J.A. (1990). Participant observation in violent social context. *Human Organization, 49(2),* 114–126.

Vanderstaay, S.L. (2005). One hundred dollars and a dead man: Ethical decision making in ethnographic fieldwork. *Journal of Contemporary Ethnography, 34(4),* 371–409.

Williams, T., Dunlap, E., Johnson B.D., & Hamid, A. (1992). Personal safety in dangerous places. *Journal of Contemporary Ethnography, 2(3),* 343–374.

Chapter 10
Behavioral Intervention Studies

The Design of Behavioral Intervention Studies

Like clinical trials, which are designed to examine the toxicology and efficacy of an experimental drug or device (the intervention), behavioral intervention studies examine the efficacy of an experimental behavioral therapy or treatment. Accordingly, many of the ethical and design issues that are raised in the context of clinical trials are also notable in the context of behavioral intervention studies. And, like clinical trials, behavioral intervention studies are structured by phase.

Three stages of behavioral treatment research have been identified. Stage I consists of an iterative process in which the investigator identifies clinical, behavioral, affective, and cognitive scientific findings that are relevant to treatment. New behavioral treatments are formulated or existing treatments are modified based upon these findings. Then, principles and techniques of the therapies must be operationalized and standardized in manuals. Finally, these therapies must be pilot tested and refined (National Institute on Drug Abuse, 2006).

Stage II of behavioral intervention studies are designed to assess the efficacy of the intervention developed in Stage I, to examine the components of therapies, to study the mechanism of action of efficacious therapies, to examine the dose-response of therapies, to examine individual differences in responding to therapies, and/or to replicate at additional sites efficacy studies with treatments that have been found to have positive effects.

Stage III of behavioral intervention studies focus on one or more of several issues:

- how an efficacious treatment can be utilized in community settings;
- the degree to which a behavioral intervention maintains its "potency" when it is translated to a community setting; and
- how therapists and counselors in a community setting can be trained to administer the intervention (National Institute on Drug Abuse, 2006).

Ethical Issues

Behavioral intervention research may involve individuals who can be considered to be vulnerable, such as mentally ill persons, children, etc. (Devries, DeBruin, and Goodgame, 2004). To the extent that it does, the ethical issues that arise in the context of behavioral intervention studies overlap with those that relate to cognitively impaired research participants. And, because the design of behavioral intervention studies often mirrors that of clinical trials, many of the ethical issues associated with study design are similar to those we find in conjunction with clinical trials.

Informed Consent

It is presumed at the commencement of research studies that a prospective participant has capacity to consent unless there is *a priori* reason to believe that he or she does not or that the capacity to consent may be limited in some way. Decisionmaking ability in the context of participation in research requires that the individual be able to understand basic study information, including the procedures to be performed, the risks associated with participation, the potential benefits that may inure to him- or herself or others, alternatives to study participation, the difference between research interventions and established therapy, and their ability to refuse to participate without suffering a penalty (Dresser, 2001).

Socioeconomic disadvantage, in particular, is believed to be "a critical concern in the context of behavioral health research" (De Vries, DeBruin, and Goodgame, 2004). This stems from existing inequitiies and lack of access to income, housing, employment, and health care (National Bioethics Advisory Commission, 2001).

The controversy concerning the enrollment of heroin-dependent or -addicted persons into a behavioral intervention trial illustrates a number of ethical issues related to informed consent that might arise, including issues relating to socioeconomic disadvantage. One writer has asserted that heroin addicts can never be competent to consent to enrollment in trials to test the efficacy of prescription heroin as a behavioral intervention to reduce HIV risk because they are obsessed with the drug, they lack a stable set of values because of their addiction, whatever values they espouse are no longer truly theirs due to the impact of their addiction to heroin and, consequently, they cannot be accountable for any decision (Charland, 2002).

This perspective, though, has met with harsh criticism for several reasons. First, it presumes that all heroin addicts lack capacity to consent to participate despite the general presumption at the commencement of research studies that a prospective adult participant has capacity to consent, unless there is some reason to believe that he or she does not have capacity or that capacity is limited in some way (National Bioethics Advisory Commission, 1998). This position essentially equates an inability or unwillingness to say "no" to a lack of capacity (Ling, 2002). This position also rests on a gross exaggeration of the impact of addiction; the fact of a diagnosis of addiction or drug dependence is relevant to, but not determinative of, the issue of capacity (Hugh et al., 1994). Finally, a determination that heroin-dependent persons could not be held responsible for their decisions and their conduct in the research context due to their drug dependence raises additional issues regarding their capacity in the clinical care and criminal law contexts.

In contrast to this unequivocal view of heroin addicts as lacking a stable set of values, another scholar has suggested that addicts are cognizant of the choices available to them—participation in an unproven heroin prescription treatment versus life on the streets supported through begging and criminal activity —and that they are able to assess the extent to which each such choice is consistent with their values to arrive at a decision (Perring, 2002). The use of needle exchange programs, for instance, has established that injecting heroin-dependent persons are, despite their addiction, able to weigh the risks and benefits of using such a service in order to reduce potential health threats.

Nevertheless, heroin-dependent individuals are to be considered vulnerable persons within the context of research. Vulnerable participants are those individuals with "insufficient power, prowess, intelligence, resources, strength or other needed attributes to protect their own interests through negotiations for informed consent" (Levine, 1988). The ability of heroin addicts to protect their own interests may be temporarily diminished if they are undergoing the acute effects of heroin use or of withdrawal, but the use of the drug has not been found to affect attention or memory (Lundqvist, 2005). Indeed, the capacity to provide informed consent may be understood as fluctuating, a phenomenon that has been recognized in considering the enrollment of mentally ill persons in research (National Bioethics Advisory Commission, 1998). As a group, heroin-addicted persons may be disempowered due to poverty, imprisonment, and/or stigmatization. This status does not, however, negate their ability to participate. Rather, it requires that investigators develop protections to maximize the likelihood that prospective participants have capacity to consent at the time that they are solicited for their participation. A refusal to enroll any heroin-dependent person in such a trial based on their membership in the class of heroin-dependent persons, absent an individualized assessment of their capacity to consent, would contravene the ethical principle of justice.

The issue of voluntariness may also arise in the context of informed consent to participate in a behavioral intervention study. As an example, some might argue that individuals who are addicted to a drug are incapable of giving consent voluntarily to participate in an intervention, such as an intervention trial to assess the efficacy and/or feasibility of prescription heroin or needle exchange to reduce HIV risk, specifically because of their addiction and the hold that their addiction has over their behaviors. Others might assert, however, that the choice facing heroin addicts is not whether to obtain heroin, but from whom (the dealer or the clinical trial) and at what cost (life on the streets, privacy, and personal freedom, risk of disease versus supervision and loss of privacy).

Privacy and Confidentiality

Researchers may be obligated by state law to report various circumstances that may be reported to them by participants, such as child abuse, sexual abuse, partner abuse, elder abuse, and/or specified infectious diseases. They may also be obligated to warn third parties if they believe that the third party is in immediate danger of being harmed by the research participant. These issues are addressed in great detail in chapter 9, which addresses observational studies, and will not be repeated here. The case study by Gore-Felton and DiMarco, which follows this chapter, illustrates well the issues that may face researchers in the context of conducting a behavioral intervention trial.

Internet-based research, is increasingly utilized in behavioral intervention studies. The CIOMS' *Special Ethical Considerations for Epidemiological Research* (2006) recognizes in commentary to Guideline 23 the diverse strategies for Internet utilization in the context of research:

First, while *collecting data*, researchers may use the Internet to actually perform the research itself (online research); visitors to sites may be enrolled as respondents and questionnaires may be made accessible through the Internet. In open Internet locations, investigators may observe, as a source of data, what others are saying and doing without necessarily interacting with other visitors to the site in question. Second, in *building databases*, researchers may send files containing the results of their research to other researchers. This is the case, for instance, in multi-centre trials. Finally, after *completion of the study*, researchers may want to make some results available through the Internet.

These varied uses of the Internet raise unique ethical issues related to privacy and confidentiality. Strategies to ensure online confidentiality include limiting and/or encrypting e-mail communications, advising prospective participants of the security limitations of the system (Childress and Asamen, 1998), and limiting access to any back-up copies of the data (Council for International Organizations of Medical Sciences, 2006).

Standard of Care

The controversy that exists in the context of clinical trials with respect to the standard of care to be provided to participants randomized to the comparison treatment also exists in the context of behavioral intervention research. To illustrate this issue, assume that a behavioral intervention trial wishes to assess the efficacy of a behavioral intervention to reduce the incidence of HIV infection within a specific community. The experimental group is to receive a full-blown intervention. What should the comparison group receive? Pamphlets describing HIV transmission and basic prevention strategies? A time-matched intervention to teach employability skills? Wait listing for the experimental intervention? Nothing, because that is what the community in question currently receives? A failure to provide an appropriate comparison "treatment" in the context of such behavioral research means that investigators will watch as individuals become HIV-infected. Clearly, utilization of a comparison intervention that may reduce risky behavior and the incidence of infection renders the analysis more complex. However, failure to do so may well be violative of the principles of beneficence and nonmaleficence.

Issues also relate to the choice of the intervention to be provided to the experimental group. Must this intervention be "evidence-based"? Whatever the decision, it must be remembered that "[w]hether it is a drug to behavioural intervention, the ethical argument for randomized trials rests on 'equipoise'—that is, genuine uncertainty that the intervention will actually result in more good than harm" (Stephenson, 1999).

The background standard of care available in a community may also be relevant to the conduct of behavioral intervention research. As an example, consider the issues involved in conducting an intervention trial deigned to reduce HIV transmission. As a component of participation, participants are provided with free periodic HIV and STI screening and are provided appropriate treatment for their infections. If such care is not available and/or easily accessible within the community from which the participants are drawn, a dual standard of care within the community will result (Shapiro and Benatar, 2005). This higher standard of care associated with participation in the trial may be perceived as coercive, may not be sustainable following the conclusion of the research, and may diminish the level of care available in the larger community due to the employment of community health providers as part of the research team. Although it has been argued that the principles of justice and beneficence require demand that researchers work to improve the broader standard of care available in the community, and identify mechanisms to sustain the improvement (Shapiro and Benatar, 2005), such goals may be difficult to achieve in view of limited resources, funding restrictions, and political considerations.

Ethical Issues in a Community Setting

As indicated above, the implementation of behavioral intervention studies often requires the active involvement of community, for instance, in evaluating the acceptability and/or effectiveness of the intervention in a community setting. Accordingly, in considering both the design and the ethical issues that are involved in a community-focused behavioral intervention study, the researcher must consider not only the individual participants, but also the community as a whole. These ethical issues include:

• Obtaining informed consent from the community and the individual participants
• The acceptability of the intervention for the various categories of stakeholders such as consumers, family members, community members, insurers, etc.
• Ownership and control of the research and the research results
• The relevance of the outcomes to the goals of the community and the individual participants
• Potential mechanisms to ensure the sustainability of the intervention following the completion of the research (Giesbrecht and Ferris, 1993; Hohmann and Shear, 2002).

Each of these is considered in more detail below as they relate to the four basic ethical principles of respect for persons, beneficence, nonmaleficence, and distributive justice.

Obtaining Informed Consent

Recall that the principle of respect for persons encompasses the two components of autonomy and protections for vulnerable persons. In turn, autonomy is reflected in the requirement of informed consent. In a community context, the operationalization of this requirement may mean that the informed consent of the community will be required in addition to the informed consent of each individual participant. This potentiality is noted in the various guidelines governing research that have been promulgated by the Council of International Organizations for Medical Sciences (2002, 2006). A draft of CIOMS' *Special Ethical Considerations for Epidemiological Research*, provides in commentary to Guideline 4:

Community review of and permission for studies. Investigators carrying out biomedical research often include a process of consultation with representatives of the community in which it is proposed to conduct the study, particularly when the research originates outside that community or even the country in which the community is located. Such consultation can taken the form of a "dialogue" with the community about the proposed study and its potential implications, or a more structured consultation that would document the concerns of a socially identifiable group. In some cases, formal approval may be legally required; for example, under US law, a Native American tribal council must formally approve any research conducted within tribal jurisdiction. In industry-based occupational epidemiology, the agreement and collaboration of employers and employees is a necessary requisite to the conduct of a study. Epidemiologists should follow in general the same approach when developing field investigations, especially when research findings may be presented or interpreted in ways that directly relate to a community or other identifiable group of people or in which the collectivity itself is the unit of analysis. Those consulted should be in a position to speak on behalf of the community or to reflect its views; researchers should have adequate time and resources to discern how the study population is organized socially and politically and which groups can best speak with authority for the population. Care should, of course, be taken to ensure that those consulted include all relevant groups and do not exclude, for instance, women or members of minority groups.

Similar issues arise in the context of a cluster randomized trial. For instance, school districts might be randomized to receive a particular HIV prevention intervention. Potentially, some of the children within a particular district might receive the intervention and may or may not be afforded an opportunity to decline participation. In such instances, it is important that their parents receive information about the study. The scope of authority of the individual(s) providing consent for the study to proceed must encompass the authority to give consent to proceed with such research (Council for International Organizations of the Medical Sciences, 2006).

Ownership and Control of the Research and the Research Results

It is possible that the dissemination of research findings involving a particular community or group may bring about or exacerbate stigma, bigotry, and discrimination towards that group. This is particularly true of research involving HIV prevention, where even participation in a study may be perceived as evidence that an individual is seropositive, despite the inclusion of HIV-positive and -negative individuals in the research (MacQueen, Shapiro, Karim, and Sugarman, 2004).

Relevance of the Outcomes

Four different scenarios may occur with respect to the relevance of the proposed research. If the research is of high priority to both the researcher and the community, or if both the researcher and the community agree that the proposed research is of low priority, there will be little controversy or conflict regarding the need to carry out the research (Giesbrecht and Ferris, 1993). However, if the researcher and the community disagree with respect to the relevance of the research, its implementation may be controversial and a source of conflict.

Ethical guidelines and the ethical principles of justice and beneficence underscore the need to develop research priorities in conjunction with the community that will be participating in the research. For instance, Guideline 10 of CIOMS' *International Ethical Guidelines for Biomedical Research Involving Human Subjects* specifies that

Before undertaking research in a population or community with limited resources, the sponsor and the investigator must make every effort to ensure that:

- the research is responsive to the health needs and the priorities of the population or community in which it is to be carried out; and
- any intervention or product developed, or knowledge generated, will be made reasonably available for the benefit of that population or community.

This guideline is equally relevant to the conduct of behavioral research (CIOMS, 2006).

Ensuring Sustainability

Guideline 20 of CIOMS' *International Ethical Guidelines for Biomedical Research Involving Human Subjects* focuses on the sustainability of research expertise, infrastructure, and products developed in the context of biomedical research in other countries:

Many countries lack the capacity to assess or ensure the scientific quality or ethical acceptability of biomedical research proposed or carried out in their jurisdictions. In externally sponsored collaborative research, sponsors and investigators have an ethical obligation to ensure that biomedical research projects for which they are responsible in such countries contribute effectively to national or local capacity to design and conduct biomedical research, and to provide scientific and ethical review and monitoring of such research.

Capacity-building may include, but is not limited to, the following activities:

• establishing and strengthening independent and competent ethical review processes/ committees
• strengthening research capacity
• developing technologies appropriate to health-care and biomedical research
• training of research and health-care staff
• educating the community from which research subjects will be drawn.

These provisions are equally relevant to the conduct of behavioral research (CIOMS, 2006), whether conducted with a local community or a community in a different country.

References

Charland, L.C. (2002). Cynthia's dilemma: Consenting to heroin prescription. *American Journal of Bioethics, 2,* 37–47.

Childress, C.A. & Asamen, J.K. (1998). The emerging relationship of psychology and the Internet: Proposed guidelines for conducting Internet intervention research. *Ethics & Behavior, 8,* 19–35.

Council of International Organizations for Medical Sciences. (2002). *International Ethical Guidelines for Biomedical Research Involving Human Subjects.* Geneva, Switzerland: Author.

Council for International Organizations of Medical Sciences. (2005). *Special Considerations for Epidemiological Research.* Geneva, Switzerland: Author.

De Vries, R., DeBruin, D.A., & Goodgame, A. (2004). Ethics review of social, behavioral, and economic research: Where should we go from here? *Ethics & Behavior, 14(4),* 351–368.

Dresser, R. (2001). Dementia research: Ethics and policy for the twenty-first century. *Georgia Law Review, 45,* 661–690.

Giesbrecht, N. & Ferris, J. (1993). Community-based research initiatives in prevention. *Addiction, 88(Supplement),* 83S-93S.

High, D.M., Whitehouse, P.J., Post, S.G., & Berg, L. (1994). Guidelines for addressing ethical and legal issues in Alzheimer research: A position statement. *Alzheimer's Disease & Associated Disorders, 8,* 66–74.

Hohmann, A.H. & Shear, M.K. (2002). Community-based intervention research: Coping with the "noise" of real life in study design. *American Journal of Psychiatry, 159(2),* 201–207.

Levine, R.J. (1988). *Ethics and Regulation of Clinical Research.* New Haven, Connecticut: Yale University Press.

Ling, W. (2002). Cynthia's dilemma. *American Journal of Bioethics, 2,* 55–56.

Lundqvist, T. (2005). Cognitive consequences of cannabis use: Comparison with abuse of stimulants and heroin with regard to attention, memory and executive functions, *Pharmacological and Biochemical Behavior, 81,* 319–330.

MacQueen, K.M., Shapiro, K., Karim, Q.A., & Sugarman, J. (2004). Ethical challenges in international HIV prevention research. *Accountability in Research, 11,* 49–61.

National Institute on Drug Abuse. (2006). Behavioral Therapies Development Program. Available at http://grants.nih.gov/grants/guide.Last accessed August 29, 2006.

National Bioethics Advisory Commission. (2001). *Ethical and Policy Issues Involving Human Participants*, vol. 1. Rockville, Maryland: U.S. Government Printing Office.

Perring, C. (2002). Resisting the temptations of addiction rhetoric. *American Journal of Bioethics*, 2, 51–52.

Shapiro, K. & Benatar, S.R. (2005). HIV prevention research and global inequality: Steps towards improved standards of care. *Journal of Medical Ethics, 31,* 39–47.

Stephenson, J.M. (1999). Evaluation of behavioural interventions in HIV/STI prevention. *Sexually Transmitted Infections, 75,* 69–71.

Case Study Four
Protecting the Rights of Human Subjects Who Participate in HIV/AIDS Research: Balancing the Ethics of Beneficence and Justice

Cheryl Gore-Felton, Ph.D. & Michael DiMarco, Psy.D.

Introduction

At the end of 2005, there were an estimated 33 million to 46 million people living with HIV (PLH; UNAIDS, 2006). Moreover, worldwide surveillance data indicate that AIDS has been diagnosed in virtually every country, and HIV infection and illness have increased at alarming rates in developing countries in sub-Saharan Africa and Asia (UNAIDS, 2006). In the United States, the estimated number of deaths among persons with HIV decreased by 25% during 1995–1996 (CDC, 2003), by 46.4% in 1997 (Holmes, 1998), by 21% from 1997–1998 (Martin, Smith, Mathews, & Ventura, 1999), and remained relatively stable from 1999 – 2003 (CDC, 2004). The number of new HIV infections in the U.S. is estimated to be 40,000 per year (CDC, 2003).

The HIV mortality rate sharply declined between 1995 and 2002 (CDC, 2005). The introduction of highly active antiretroviral therapy (HAART) is credited with this decline and has not only extended the life expectancy but has improved the physical health and quality of life of PLH (Emini, Schleif, Deutsch, & Condra, 1996; Rabkin & Ferrando, 1997; Vittinghoff et al., 1999). Although HAART affords PLH the opportunity for longer, healthier lives, there are increased challenges in reducing transmission risk behavior. For instance, there is evidence that among a diverse sample of HIV-positive adults the likelihood of contracting a post-HIV STI increased the longer individuals knew their HIV diagnosis (Gore-Felton et al., 2003). Additionally, an epidemiological study among over 16,000 MSM found that as HAART use increased over time, so did the reports of multiple sexual partners and unprotected anal intercourse (Katz et al., 2002). It is important to note that following notification of their HIV-positive serostatus, most PLH make and maintain changes in their sexual and injection drug use practices to avoid transmitting HIV to others (Kalichman, Rompa, & Cage., 2000; Rotheram-Borus et al., 2001; Crepaz & Marks, 2002). A review of the research on risk behavior among PLH suggests that high-risk behaviors are more likely with other infected persons, but significant rates of risk behaviors are observed with HIV-negative partners and partners of unknown serostatus (Kalichman, 2000). As a result, it is critical to develop interventions that can assist

HIV-positive persons in reducing high-risk sexual risk behavior and provide assistance in maintaining good physical and psychological health.

While implementing and conducting prevention research among HIV-positive individuals, the authors have been confronted with complex ethical dilemmas that required consideration of the ethical conduct of research with human subjects and state law. While the dilemmas have presented themselves in many research participants over the years, we will use one case study to discuss the following two major ethical concerns that are likely to confront researchers conducting HIV prevention intervention research among persons living with HIV and AIDS:

1. Beneficence. Two general rules are applied which are a) do not harm and b) maximize possible benefits and minimize possible harms. In applying these edicts, when is it necessary to break confidentiality and privacy?
2. Justice. Whose views determine the intervention content and design?

The Research Study

In the case study that follows, we describe feelings, situations, and contexts that typically confront HIV researchers. We changed circumstances and characteristics to protect the privacy and confidentiality of research participants. Thus, any resemblance to an individual living or dead is unintended and is merely chance.

Mr. B

Mr. B is a 40 year old Caucasian male who enrolled in a randomized behavioral HIV prevention intervention trial that was being conducted in four U.S. cities for persons with HIV (PLH). The intervention focused on developing coping skills to deal with stress, reducing transmission risk behavior and increasing medication adherence. (For complete description of intervention, see Gore-Felton et al., 2005).

Mr. B had been diagnosed with HIV for the past 13 years. He unknowingly contracted HIV through sexual activity with another man. He lives alone and is a self-employed businessman. He has a four-year college degree and has taken some graduate courses. He admits he is more comfortable being by himself and doesn't have many friends but he has been able to maintain friendships with a few individuals he describes as "close friends," These friends do not know he is HIV positive. He does not have children. His family knows he is gay, but do not know he is HIV positive. He self-identifies as gay. He has never been in a long-term or steady relationship. Although he stated on a few occasions that it would be nice to meet someone and settle down, he really does not believe in the longevity of relationships.

Mr. B is of average height and weight and reported that his health is good. His HIV care is managed by a physician who specializes in infectious diseases. At the time of study enrollment, Mr. B was stable on HAART and his CD4 count was in the 500 to 600 range and his viral load was undetectable. He denied a previous history of psychiatric or drug and alcohol treatment. He did say that he abused cocaine in his late teens and early 20's.

Mr. B has casual male sex partners that he meets through the internet or anonymously in sex clubs and public gyms. For men he meets through the internet, he arranges a meeting at a coffee shop or bar in order to decide whether he wants to "hook up" for sex. In regards to sex in sex clubs and gyms he does not have conversation. He says he uses condoms with partners that initiate use; however, if the partner does not bring up the use of condoms, then condoms are not used. In his opinion, if men don't insist on using a condom he assumes that they are already positive or they don't care about getting HIV. He maintains one exception to the rule when it comes to having sex with men who are younger than he is (in their 20's), where he does insist on using condoms. Although he does not disclose his HIV status to partners, he does make sure that younger men are protected as he perceives that younger men are not as aware of the risks as older generations are given the history of HIV/AIDS in the 1980s and early 1990s. He said that "older men," including men around his own age, should know the risks.

Mr. B harbors anger toward men who have sex with men who don't claim a gay identity. He believes that bisexuality is a "cover up" for homosexuality and that married men who have sex with men want the "best of both worlds." He said that if he encounters a man at the gym with a wedding ring on, looking for sex, he assumes that the man is married. He said that he gets so angry with those men that he purposefully pursues and flirts in order to have unprotected sex— secretly hoping that he will transmit the virus. Mr. B engages only in insertive sex (i.e., Mr. B is the top) with these men. This sexual behavior presents the greatest probability of transmitting the virus and Mr. B is aware of this. In fact, he actively encourages these men to have receptive anal sex with him and rationalizes that if these men did get infected they would finally have to deal with their homosexuality. Mr. B is angry that he has HIV and recollects how he was infected by someone older than he and no one ever cared about protecting him from HIV, so why should he be any different.

The Ethical Dilemma and Analysis

What are the ethical and legal duties of clinical researcher who learn that an HIV-infected client is endangering others by knowingly exposing them to the virus? Does the duty to protect apply in the research setting?

The case illustrated above poses an ethical dilemma for clinical researchers in two distinct areas, beneficence and justice. Sound scientific methodology to decrease HIV-related risk among PLH demands that individuals who are at risk be recruited into the study. The difficulty many clinical researchers face is that there will be research participants like Mr. B who have not disclosed and do not plan to disclose, their HIV status to sex partners and continue to engage in unsafe sexual practices. One could argue that in order to protect the public, Mr. B's confidentiality should be broken and he should be reported to authorities.

However, if we apply beneficence to this case, it can be argued that if there was a mandate to report Mr. B then our study is designed to harm because we are specifically targeting individuals who report risk behavior and we will have broken the tenet, "do not harm." Moreover, it can be argued that in order to maximize the benefit to society in reducing HIV transmission, Mr. B's confidentiality should not be broken so that he can continue to be exposed to the intervention which is designed specifically to protect the public by reducing HIV-related risk behavior, thereby, preventing new infections.

In the case of Mr. B, some have argued that it is unethical to design a study that recruits persons with HIV who are knowingly engaging in behavior that poses a transmission risk. This question directly addresses justice and whether or not PLH have a right to be included in research trials that may not only benefit them directly, but their community. Given that there is strong evidence for the efficacy of HIV prevention methods that target risk behavior, it seems that it would violate the ethical principle of beneficence to withhold prevention research from individuals who pose the greatest threat to public health.

What if Mr. B had identified a specific person that he was going to try and infect? For more than a decade, researchers have debated the applicability of the concept of a "duty to warn" based on the Tarasoff doctrine (1976) to ethical dilemmas involving HIV-infected individuals (VandeCreek & Knapp, 1993; DiMarco & Zoline, 2004). It has been argued that the duty to protect was formulated in and only applies to clinical settings and therefore it does not apply to the research setting. This argument is too simplistic because research and clinical functions are often integrally related. Clinical researchers have to uphold the state laws as well as the ethical standards imposed on them by their profession. Thus, it may be that in the circumstance that a research participant is purposely trying to infect an identified individual that the research would need to break confidentiality and notify the appropriate authorities. In this case, the research participant should have been aware of this consequence because it should have been articulated in the informed consent form which the participant had to sign and acknowledge his or her understanding prior to being enrolled in the study.

Balancing the rights of research participants and those of the larger community can be difficult. When developing secondary prevention interventions aimed at recruiting PLH who are at risk for transmitting the virus, it is imperative that researchers develop the intervention content and design (justice) so that it maximizes benefit to the participant and the public (beneficence). After over twenty years of HIV prevention research, we know how to behaviorally reduce the risk of HIV transmission. Accordingly, it seems that ethical research among PLH that is aimed at reducing HIV-related risk behavior should expose all of the participants to HIV prevention educational material that specifically details how to reduce sexual- and substance-related HIV risk behaviors. Indeed, we would argue that this should be the standard-of-care or control condition in any prevention-focused, randomized clinical trial for PLH.

Researchers need to have an adverse event protocol in place that dictates when and to whom confidentiality should be broken. Researchers are encouraged to work with their institutional review boards (IRB) to determine whether their protocol is consistent with state mandates and professional ethics. In the spirit of beneficence and justice, researchers conducting clinical intervention trials must have protocols in place to train intervention facilitators on how to manage research participants who are recklessly or intentionally exposing partners to HIV. Notably, psychotherapy studies have found that prejudice and negative biases, including homophobia, can affect professional ethical decisions when judgment related to HIV transmission risk is required (e.g. Kozlowski, Rupert & Crawford, 1998; McGuire, Nieri, Abbott, Sheridan, and Fisher, 1995; Stewart & Reppucci, 1994; Totten, Lamb, & Reeder, 1990). In clinical studies using hypothetical vignettes, clinicians who had less clinical experience working with HIV-positive clients were more likely to breach confidentiality than clinicians with more experience (Totten, Lamb, & Reeder, 1990; DiMarco & Zoline, 2004). The best antidote against premature breaching behavior and unjust withholding of secondary prevention interventions intended to lower risk behavior is education. Therefore, providing intervention facilitators with the opportunity to confront their own biases and develop a working knowledge base on HIV-related risk behavior, research ethics, and issues associated with confidentiality and privacy will systematically control for possible counter-transference reactions that influence duty to warn decisions. In our experience, training that includes didactics and the opportunity to role play sensitive situations in a safe environment is highly effective at achieving this aim.

Prior to beginning the clinical research trial that Mr. B was enrolled in, the study investigators sought IRB approval. During the initial approval process, the IRB raised the concern that HIV positive individuals may report current risk during the trial, which would require the investigators to report it. The researchers argued that the study investigators should not have to break confidentiality when a research subject reported risk behavior, even deliberate risk behavior because the intervention content was designed to reduce and even prevent such behavior. Therefore, the public would be harmed more by not having individuals like Mr. B enrolled in the study. The IRB agreed. So, when the investigators were confronted with the circumstance of Mr. B they did not have to break confidentiality and Mr. B was able to remain in the study.

Mr. B's involvement in the study enabled him to learn about his disease and how his behavior was not only posing a risk to others but himself in the form of subsequent sexually transmitted infections as well as the possibility of being reinfected with a viral strain that was resistant to his current medication regimen, both of which could speed the progression of his HIV disease. After being in the study, Mr. B stated he was going to practice safer sex but stated that he remained angry at men living a "double life" or "being on the down low" and reaping the benefits of "both worlds" without making any sacrifices.

Some individuals who are exposed to a psychosocial risk reduction intervention will warrant a referral for ongoing clinical and social services. Thus, researchers have an ethical responsibility to provide appropriate referrals as well as conduct follow up assessments to ensure uptake of needed services.

Clearly, HIV prevention interventions are not the answer for all of the social and psychological issues facing PLH, however, they can be very effective at reducing morbidity and mortality. Until there is a vaccine or cure for HIV, behavioral interventions remain the best method of reducing new infections particularly among populations vulnerable to infection.

References

Centers for Disease Control and Prevention. (2003). *HIV/AIDS Surveillance Report. Cases of HIV infection and AIDS in the United States, 2002* Retrieved December 23, 2003 from http://www.cdc.gov/hiv/stats/hasr1402.htm.

Center for Disease Control and Prevention (2004). *HIV/AIDS Surveillance Report, Cases of HIV Infection and AIDS in the United States, 2003, 15*, 1–46. Retrieved December 12, 2004 from http://www.cdc.gov/hiv/stats/2003SurveillanceReport.pdf.

Center for Disease Control and Prevention (2005). *Mortality L285 Slide Series (through 2002)*. Retrieved July 28, 2006 from http://www.cdc.gov/hiv/graphics/mortalit.htm.

Crepaz, N. & Marks, G. (2002). Towards an understanding of sexual risk behavior in people living with HIV: a review of social, psychological, and medical findings. *AIDS, 16*(2), 135–149.

DiMarco, M., & Zoline, S.S. (2004). Duty to warn in the context of HIV/AIDS-related psychotherapy: decision making among psychologists. *Counseling and Clinical Psychology Journal, 1*(2), 68–85.

Emini, E.A., Schleif, W.A., Deutsch, P., & Condra, J.H. (1996). In vivo selection of HIV-1 variants with reduced susceptibility to the protease inhibitor L-735,524 and related compounds. *Advances in Experimental Medicine and Biology, 394*, 327–331.

Gore-Felton, C., Rotheram-Borus, MJ., Kelly, J.A., Weinhardt, L.S., Catz, S.L., Chesney, M.A., et al. (2005). The Healthy Living Project: Individually-Tailored Multidimensional Intervention for HIV-Infected Persons. *AIDS Education and Prevention, 17* (1 Suppl A) 21–39.

Gore-Felton, C., Vosvick, M., Bendel, T., Koopman, C., Das, B., Israelski, D., et al. (2003). Correlates of Sexually Transmitted Disease Infection among Adults Living with HIV. *International Journal of STDS & AIDS, 14*, 539–546.

Holmes, S.A. (1998, October 8). AIDS deaths in U.S. drop by nearly half as infections go on. *New York Times*, p. A1.

Kalichman, S.C. (2000). HIV transmission risk behaviors of men and women living with HIV/AIDS: Prevalence, predictors, and emerging clinical interventions. *Clinical Psychology: Science and Practice, 7*(1), 32–47.

Kalichman, S.C., Rompa, D., & Cage, M. (2000). Sexually transmitted infections among HIV seropositive men and women. *Sexually Transmitted Infecions, 76*(5), 350–354.

Katz, M.H., Schwarcz, S.K., Kellogg, T.A., Klausner, J.D., Dilley, J.W., Gibson, S. et al. (2002). Impact of highly active antiretroviral treatment on HIV seroincidence among men who have sex with men: San Francisco. *American Journal of Public Health, 92*, 388–394.

Kozlowski, N.F. Rupert, P.A., Crawford, I. (1988). Psychotherapy with HIV-infected clients: Factors influencing notification of third parties. *Psychotherapy, 35, 1*, 105–115.

Martin, J.A., Smith B.L., Mathews, T.J., & Ventura, S.J. (1999). Births and deaths: preliminary data for 1998. *National Vital Statistics Reports, 47*(25), 1–8.

McGuire, J., Nierie, D. Abbott, D., Sheridan, K, & Fisher, R. (1995). Do Tarasoff principles apply in AIDS-related psychotherapy?: Ethical and decision making and the role of therapist homophobia and perceived dangerousness. *Professional Psychology: Research and Practice, 6,* 608–611.

Rabkin, J.G. & Ferrando, S. (1997). A 'second life' agenda. Psychiatric research issues raised by protease inhibitor treatments for people with the human immunodeficiency virus or the acquired immunodeficiency syndrome. *Archives of General Psychiatry, 54*(11), 1049–1053.

Stewart, T.M. & Reppucci, N.D. (1994). AIDS and murder: Decisions regarding the maintenance of confidentiality versus the duty to protect. *Law and Human Behavior, 18,* 107–120.

Tarasoff v. Regents of the University of California (1976). *West's California Report, 131,* 14–42.

Totten, G., Lamb, D.H., & Reeder, G.D. (1990). Tarasoff and confidentiality in AIDS-related psychotherapy. *Professional Psychology: Research and Practice, 21* (3), 155–160.

UNAIDS/WHO Joint United Nations Programme on HIV/AIDS. 2006 Report on the Global AIDS Epidemic. Retrieved July 25, 2006 from http://www.unaids.org/en/HIV_data/2006GlobalReport/default.asp.

VandeCreek, L. & Knapp, S. (1993). *Tarasoff and beyond: Legal and clinical considerations in treatment of life-endangering patients.* Sarasota, FL: Professional Resource Exchange.

Vittinghoff, E., Scheer, S., O'Malley, P., Colfax, G., Holmberg, S.D., & Buchbinder, S.P. (1999). Combination antiretroviral therapy and recent declines in AIDS incidence and mortality. *Journal of Infectious Diseases, 179*(3), 717–720.

Chapter 11
Ethical Issues in Multicenter/ Multisite Studies

Introduction

Multicenter studies, which are relatively common in the context of HIV research, may be advantageous for various reasons. Because multicenter studies are conducted at multiple sites, the external validity and, consequently, the generalizability of the findings may be enhanced (Weinberger et al., 2001). Multicenter studies also enhance our ability to investigate diseases or exposures of interest that are of low incidence because they permit enrollment of a larger number of study participants than could be achieved through reliance on one site alone. For instance, investigators may wish to examine specific sequelae of HIV infection that may be of relatively low incidence among HIV-infected individuals. A multicenter study increases the likelihood that a sample size will be achieved that is sufficiently large to assure statistical power. Too, multicenter studies permit enrollment to occur at a faster rate, potentially reducing the costs and logistical difficulties that may be associated with a lengthier recruitment period.

However, multicenter/multisite studies also give rise to numerous ethical challenges because they are often conducted across diverse locales, cultures, and political boundaries. The operationalization of informed consent may be particularly difficult, due to varying definitions of autonomy and difficulties associated with reliance on interpreters. Additional issues may be confronted due to differing applications of the concept of vulnerability across sites, resulting in differing standards for the protection of the vulnerable persons; varying confidentiality protections across sites due to differences in legal provisions that prevail or concerns that arise at each site; and inconsistencies in the demands of the various local ethics review committees at the participating sites. Unfortunately, relatively few of these issues have been addressed as they apply to multicenter studies in the specific context of HIV-related research. Accordingly, the discussion of each of these topics that follows draws on literature from outside of the HIV context. The case study by Lounbury and colleagues that follows this chapter highlights many of the ethical concerns that may arise in conducting multi-site HIV-related research.

Informed Consent

The informed consent of each individual is a prerequisite to their enrollment in research. This requirement derives from the principle of respect for persons, first enunciated in the Nuremberg Code. The consent must reflect the presence of four elements: adequate information, understanding of that information, the capacity to consent, and the voluntary nature of that consent. Accordingly, the information must be communicated in a manner and language that are appropriate to the prospective participant.

Defining Personhood

The concept of autonomy may differ across locales, rendering it more difficult to decide who must be involved in the informed consent process and whose consent to participate must be sought. Unlike the United States' concept of personhood, which tends to view individuals as completely autonomous decisionmaking agents, other societies may define persons in the context of their relationships with others and as a part of a larger, related network. In these contexts, the investigator may be required to obtain the consent of local leaders or family elder in addition to that of the individual. Barry (1988: 1083) noted in his discussion of AIDS research in Africa that "Personhood is defined by one's tribe, village, or social group." Similarly, Loue and colleagues (1996: 49) observed that civil law in Uganda provides

that an eighteen-year-old male living at home has a legal right to make his own decisions. Customary law, however, dictates that the son obtain his father's consent prior to entering any obligation. Women . . .often refuse to make a decision regarding their own participation or their child's participation absent the consent of their partner.

It is critical, then, that investigators be cognizant of and integrate into the informed consent process provisions and procedures that adequately integrate such concepts of personhood and autonomy.

Providing Information and Ensuring Understanding

Multicenter studies conducted across different cultures and language groups may be difficult because the prospective participants may speak a language that is different from that of the investigative team, or their ability to communicate in the language of the investigators may be limited. These problems may be ameliorated, to an extent, through the use of interpreters, who the investigators may rely on both to communicate the information related to participation and to obtain consent to participate. However, difficulties may continue to exist due to the inability to translate equivalent expressions from one language to another, the omission or erroneous substitution of terms that may result from attempts to paraphrase material, and variations in the prospective participants' understanding of terms used by the interpreters.

Voluntariness

The voluntary nature of the consent that is obtained may be questionable in situations in which there exists a differential in the social status and educational level between the interpreter and the prospective participants from whom he/she is to obtain consent (Marshall, 1992). Individuals of lower social standing or education may be less likely to ask questions of an interpreter who is seen as more powerful.

Conversely, a higher level of education may be thought to obviate the likelihood of coercion and, therefore, the need to obtain informed consent to participate in research. Treloar and Graham (2003) reported from their observations that in some, but not all, of the countries participating in cross-national multidisciplinary qualitative research, a number of the medical scientists and the corresponding review committees of their institutions believed that it was unnecessary to obtain the informed consent of health care professionals who were interviewed as part of the study protocol, because such individuals were not "patients" and were highly educated and, therefore, were autonomous agents who were not susceptible to any form of coercion. Other investigators involved in the study, however, maintained a broader perspective and believed that informed consent was ethically required.

The appropriateness of any incentive that is offered must be assessed within each of the contexts in which the study is to be conducted. The "Common Rule," consisting of the regulations and accompanying guidance formulated by U.S. agencies for the conduct of research, states that

An investigator shall seek consent only under circumstances that provide the prospective subject or representative sufficient opportunity to consider whether or not to participate and that minimize the possibility of coercion or undue inducement (45 C.F.R. 46, 2006; Council for International Organizations of Medical Sciences, 2002, 2005).

Inducements are believed to be harmful because they may impair individuals' judgment, causing them to engage in activities that may actually be harmful to them. Guidelines developed by the Council of Organizations for Medical Sciences advise:

Payments in money or in kind to research subjects should not be so large as to persuade them to take undue risks or volunteer against their better judgment. Payments or rewards that undermine a person's capacity to exercise free choice invalidate consent (CIOMS, 2002).

Emanuel (2004) has identified four key characteristics of undue inducements that are reflected in the various advisories: (1) incentives offer a positive good; (2) the incentive offered is somehow irresistible; (3) the incentive produces a bad judgment; and, (4) the resulting action "entails a substantial risk of serious harm that contravenes a person's interest" (Emanuel, 2004, p. 101). The application of these criteria could potentially result in differing determinations about the use of a particular incentive across various sites yet, in order to ensure the validity of the research protocol, there must be comparability across participating sites.

Vulnerable Persons

The principle of respect for persons requires not only that persons who are capable of deliberation about their personal choices be treated with respect for their capacity for self-determination, by requiring that they provide informed consent as a prerequisite to participation in a study, but also that persons with impaired or diminished autonomy be afforded additional protections against potential harm or abuse. These precepts are clearly enunciated in the *International Ethical Guidelines for Biomedical Research Involving Human Subjects* (Council for International Organizations of Medical Sciences, 2002) and the *International Guidelines for Ethical Review of Epidemiological Studies* (Council for International Organizations of Medical Sciences, 2005).

The concept of vulnerability has been explained as referring to individuals who have "insufficient power, prowess, intelligence, resources, strength or other needed attributes to protect their own interests through negotiations for informed consent" (Levine, 1988: 72). However, jurisdictions may vary with respect to the classes of individuals encompassed within this definition and, consequently, the protections required for prospective participants may vary across sites. As an example, U.S. regulations delineate only pregnant women, children, and prisoners as being in need of special protection, (45 Code of Federal Regulations) whereas Uganda enumerates a significantly greater number of groups in its ethical guidelines for researchers, including pregnant women, children, refugees, prisoners, soldiers on command, and those suffering from mental illness and/or behavioral disorders (National Consensus Conference on Bioethics and Health Research in Uganda, 1997). An even greater number of groups are listed in international documents as being potentially vulnerable, including pregnant women, institutionalized persons, children, those with diminished capacity for understanding, refugees, patients in emergency rooms, homeless persons, and members of some ethnic and racial groups, among others (Council for International Organizations of Medical Sciences, 2002, 2005). Many of these categories are relevant to HIV research because of the high prevalence of HIV in minority communities and in homeless and mentally ill populations.

As a result of such differences, review boards in diverse jurisdictions may impose mechanisms for the protection of vulnerable individuals that will not be required by other review boards. Investigators may always provide enhanced protection to all participants, regardless of the site at which they are located. However, depending upon the nature of the protection afforded, the autonomy of prospective participants may be compromised. As an example, in jurisdictions that consider all refugees to be vulnerable in the context of research, the investigators may wish–or may be required to–appoint a participant advocate to provide information to the participants and address their concerns. However, it could also be argued that the provision of an advocate serves to disempower the participants themselves and diminish their ability to act autonomously.

Confidentiality Protections

The mechanisms that are potentially available to protect the confidentiality of the data and the privacy of the participants may vary across sites, as a function of differences in protections afforded by local law, available technology, and concepts of privacy and confidentiality. Consequently, the potential risks of participation in a given study may also vary across sites. This may have implications for recruitment and enrollment of participants, as greater levels of confidentiality and privacy may lessen the barriers to participation. Methodologically, it may be impossible to determine the impact of these differences.

As an example, under U.S. regulations, a certificate of confidentiality is potentially available to protect the identity and identifying characteristics of individuals participating in studies in which highly personal and potentially damaging information is gathered. This includes such things as drug and alcohol use and sexual behavior. A certificate of confidentiality protects such data from being accessed by attorneys, courts, and law enforcement officials for use in civil, criminal, and administrative proceedings. However, this mechanism is available to protect only data collected in the U.S.; it does not apply to data collected from sites outside of the U.S. Accordingly, a multicenter study which includes sites inside and outside of the U.S. might provide differing levels of protection for the data across the participating sites.

This difference may have methodological, as well as ethical, implications. For instance, it may be impossible to know to what extent the provision of this additional level of confidentiality protection may have on the recruitment and retention of study participants. Although one can attempt to estimate this effect by asking those at the sites offering this additional protection whether they would have chosen to participate absent the certificate of confidentiality, the response to this question may itself reflect bias depending upon individuals' subjective experience with participation and personal events that occur during the course of their study participation.

Ethics Review Committees

Numerous studies have reported delays in the initiation of multisite studies due to variations in the requirements of local ethics review committees across participating sites and delays in the processing of reviews. Although some of these delays may result due to differing perspectives of what may be required ethically in order to safeguard study participants, it appears that they may also be due to administrative obstacles. At least one research group concluded that the multiple reviews that are often necessary result in inefficiency, duplication of effort, overemphasis on some monitoring aspects of the process and underemphasis on others, as well as confusion relating to responsibilities for the safety of the participants (Califf et al., 2003). These problems may be compounded when changes in the protocol become necessary. In order to ensure the validity of multicenter studies, any change in the protocol must be made at every collaborating center or institution or explicit comparability procedures must be implemented. These changes will necessitate review and approval by each of the relevant review committees.

As an example, Silverman and colleagues (2001) reported considerable variability across 16 local ethics review committees in their review of survey and informed consent forms pertaining to a multicenter trial that compared lower and traditional tidal volume ventilation in patients with acute lung injury. One of the institutional review committees waived the requirement for informed consent, while five permitted the use of telephone consent and three permitted the enrollment of prisoners into the study. The reading levels of the approved forms ranged from grade 8.2 to grade 13.4, with a mean reading level of 11.6. Thirteen of the approved forms lacked some of the elements of information that are required by U.S. law.

McWilliams and colleagues (2003) similarly encountered considerable variability in interactions with local review committees associated with the 42 sites participating in a multicenter genetic epidemiological study. Among the 31 sites that responded, it was found that 15 of the review committees required at least two informed consent forms and 10 did not require any form of assent from children. Seven of the review committees furnished expedited review, while the remaining 24 required a full review of the protocol.

Burman and colleagues (2003) found in their examination of the reviews afforded by ethics review committees at 25 different sites participating in multicenter study that a median of 46.5 changes were required on each consent form. More surprisingly, the changes mandated by the local review committees often resulted in an increase, rather than a decrease, in the reading level required for comprehension of the informed consent document, potentially reducing the likelihood that participants would be able to understand the form.

In one study of birth weight and child development, 118 of the 145 committees to which investigators applied for approval for the study required completion of different application forms (Middle, Johnson, Petty, Sims, and Macfarlane, 1995). Although almost three-quarters of the committees approved the protocol with no objections, a number of them expressed reservations relating to confidentiality, the wording of the information sheets, and the questionnaire that was to be utilized in the study.

Hewson, Weston, and Hannah (2002) reported on their experience obtaining approval from local ethics committees for the Term Breech Trial (TBT), which was a multicenter, international randomized trial that compared cesarean section with planned vaginal birth for specified pregnancies that presented with a breech presentation at birth. The trial involved 2088 women recruited through 121 centers in 26 different countries. The length of time needed to obtain approval from the various ethics review committees ranged from 3 months to 18 months; once the ethics approval had been received, the average time to recruitment was 2.6 months. Ethics concerns arose at several of the sites. At two of the Asia-based sites, the investigators felt that it would be unethical to tell the prospective participants that the doctor did not know which treatment was better because such a statement would arouse anxiety. At some sites, the informed consent process was revised to incorporate procedures for oral consent because of the relatively high illiteracy rates. Similar issues relating to the form of the consent process (oral, written, witnessed) were encountered by van Raak and colleagues (2002) in conducting an international multicenter trial to evaluate the neuroprotective effect of diazepam in acute stroke.

A number of suggestions have been made to alleviate or eliminate altogether the problems encountered in dealing with multiple review committees. Reliance on national coordinators to facilitate the review process has been found to be helpful (van Raak, Hilton, Kessels, and Lodder, 2002). Where permitted by relevant legislation, reliance on one centralized committee for approval for all sites within a country will also expedite the process (Gold and Dewa, 2005; van Raak, Hilton, Kessels, and Lodder, 2002). The development and use of standardized documents and procedures for their use, electronic access to documentation, and focused training for ethics review committee members may also be critical to improve the process (Gold and Dewa, 2005).

Rights of Participants to Treatment and Compensation

The *International Ethical Guidelines for Biomedical Research Involving Human Subjects* (2002) requires in Guideline 19 that investigators

ensure that research subjects who suffer injury as a result of their participation are entitled to free medical treatment for such injury and to such financial or other assistance as would compensate them equitably for any resultant impairment, disability or handicap. In the case of death as a result of their participation, their dependants are entitled to compensation. Subjects must not be asked to waive the right to compensation.

The first component refers to the provision of free medical treatment and compensation for accidental injury resulting from procedures or interventions that were performed solely to accomplish the purposes of the research. The second component pertains to the provision of material compensation for death or disability that is the direct result of participation in a study. Commentary to the guideline further provides that before

the research begins, the sponsor, whether a pharmaceutical company or other organization or institution, or a government (where government insurance is not precluded by law), should agree to provide compensation for any physical injury for which subjects are entitled to compensation, or come to an agreement with the investigator concerning the circumstances in which the investigator must rely on his or her own insurance coverage (for example, for negligence or failure of the investigator to follow the protocol, or where government insurance coverage is limited to negligence). In certain circumstances it may be advisable to follow both courses. Sponsors should seek adequate insurance against risks to cover compensation, independent of proof of fault.

These provisions stem from the ethical precepts of nonmaleficence, to avoid harm to research participants. Their operationalization may, however, be significantly more complex in the context of multicenter studies conducted in locales that may have widely differing legal systems and mechanisms for the compensation of injured parties. Consequently, although the same ethical precepts may apply, the extent to which they can be implemented may depend to a large extent on the legal and social context in which the injury or disability occurs.

Obligations to the Community

Guideline 20 of the *International Ethical Guidelines for Biomedical Research Involving Human Subjects* (2002) notes that

[m]any countries lack the capacity to assess or ensure the scientific quality or ethical acceptability of biomedical research proposed or carried out in their jurisdictions. In externally sponsored collaborative research, sponsors and investigators have an ethical obligation to ensure that biomedical research projects for which they are responsible in such countries contribute effectively to national or local capacity to design and conduct biomedical research, and to provide scientific and ethical review and monitoring of such research.

Capacity-*building* may include, but is not limited to, the following activities:

• establishing and strengthening independent and competent ethical review processes/ committees
• strengthening research capacity
• developing technologies appropriate to healthcare and biomedical research
• training of research and healthcare staff
• educating the community from which research subjects will be drawn.

Commentary to the Guideline suggests that the amount of capacity building should be in proportion to the magnitude of the research project.

Again, the operationalization of this proposition is rendered significantly more complex when the research is conducted at multiple centers. First, the needs of each collaborating locale may be quite different. Second, even where the needs are similar, the costs associated with the development of the same infrastructural component may be wildly different due to vast differences in the costs of labor or materials. If, then, differing sites are participating to the same degree in the research but the costs of infrastructural development and technology development are significantly greater at one site, the question becomes whether funding should be contributed in the same amount at each site or whether the resulting level of technology or infrastructure should control, regardless of the difference in cost. The question is one of equity and fairness across collaborators.

References

Barry, M. (1988). Ethical considerations of human investigation in developing countries: The AIDS dilemma. *New England Journal of Medicine, 319*, 1081–1083.

Burman, W.P., Breese, S., Weiss, N., Bock, J., Bernardo, A., & Tuberculosis Trial Consortium. (2003). The effects of a local review on informed consent documents from a multicenter clinical trials consortium. *Controlled Clinical Trials, 24*, 245–255.

Califf, R.M., Morse, M.A., Wittes, J., Goodman, S.N., Nelson, D.K., DeMets, D.L., Iafrate, R.P., Sugarman, J. (2003). Toward protecting the safety of participants in clinical trials. *Controlled Clinical Trials, 24*, 256–271.

Council for International Organizations of Medical Sciences. (2002). *International Ethical Guidelines for Biomedical Research Involving Human Subjects*. Geneva, Switzerland: Author.

Council for International Organizations of Medical Sciences. (2005). *International Guidelines for Ethical Review of Epidemiological Studies*. Geneva, Switzerland: Author.

Emanuel, E.J. (2004). Ending concerns about undue inducement. *Journal of Law, Medicine, & Ethics, 32*, 100–105.

Gold, J.L., & Dewa, C.S. (2005). Institutional review boards and multisite studies in health services research: Is there a better way? *HSR: Health Services Research, 40(1)*, 291–307.

Hewson, S.A., Weston, J., & Hannah, M.E. (2002). Crossing international boundaries: Implications for the Term Breech Trial data Coordinating Centre. *Controlled Clinical Trials, 23*, 67–73.

Levine, R.J. (1988). *Ethics and Regulation of Clinical Research*. New Haven, Connecticut: Yale University Press.

Loue, S., Okello, D., & Kawuma, M. (1996). Research bioethics in the Ugandan context: A program summary. *Journal of Law, Medicine, & Ethics, 24*, 47–53.

Marshall, P.A. (1992). Research ethics in applied anthropology. *IRB: A Review of Human Subjects Research, 14*, 1–5.

McWilliams, R.J., Hoover-Fong, A., Hamosh, S., Beck, S., Beaty, T., & Cutting, G. (2003). Problematic variation in local institutional review of a multicenter genetic epidemiology study. *Journal of the American Medical Association, 290(3)*, 360–366.

Middle, C., Johnson, A., Petty, T., Sims, L., Macfarlane, A. (1995). Ethics approval for a national postal survey: Recent experience. *British Medical Journal, 311*, 659–660.

National Consensus Conference on Bioethics and Health Research in Uganda. (1997). *Uganda's Guidelines for the Conduct of Health Research Involving Human Subjects in Uganda*. Kampala, Uganda: Author.

Silverman, H., Hull, S.C., & Sugarman, J. (2001). Variability among institutional review boards' decisions within the context of a multicenter trial. *Critical Care Medicine, 29(2)*, 235–241.

Treloar, C. & Graham, I.D. (2003). Multidisciplinary cross-national studies: A commentary on issues of collaboration, methodology, analysis, and publication. *Qualitative Health Research, 13(7)*, 924–932.

Van Raak, L., Hilton, A., Kessles, F., & Lodder, J. (2002). Implementing the AGASIS trial, an international multicenter acute intervention trial in stroke. *Controlled Clinical Trials, 23*, 74–79.

Weinberger, M., Oddone, E.Z., Henderson, W.G., Smith, D.M., Huey, J., Giobbie-Hurder, A., et al. (2001). Multisite randomized controlled trials in health services research: Scientific challenges and operational issues. *Medical Care, 39*, 627–634.

45 Code of Federal Regulations 46 (2006).

Case Study Five
Managing Multisite
Collaborative Research with
Community-Based Agencies

David Lounbury, Ph.D., Paulette Murphy, Psy.D.,
Janice Robinson, and Bruce Rapkin, Ph.D.

In this chapter we present our experiences in a collaborative community research project that employed a process of negotiated partnership between 22 HIV care service organizations and a major academic medical center in New York City ['Family Access to Care Study' (FACS); NIMH Grant 5R01-MH06304; Bruce D. Rapkin, PI]. The major purpose of the research was to study the feasibility of building new partnerships between academic researchers and community-based providers that address the needs and concerns of families affected by HIV/AIDS.

It is now widely understood that families affected by HIV/AIDS face a broad range of problems and often require multiple services (Blackwell, Gruber, & vonAlmen, 1997; Bor & Du Plessis, 1997; Diamant, Wold, Spritzer, & Gelberg, 2000; Rotheram-Borus et al., 2002). In turn, this understanding has fostered research to test and develop new programs and resources to help families cope more effectively (Bauman, Camacho, & Silver, 2002; Biggar et al., 2000; Crosby, DiClemente, Wingood, & Harrington, 2002; Kmita, Baranska, & Niemiec, 2002; Mellins, Ehrhardt, Rapkin, & Havens, 2000; O'Hara & D'Orlando, 1996; Rotheram-Borus, Lee, Gwadz, & Draimin, 2001; Schaaf, Sherwen, & Youngblood, 1997; Vandehey & Shuff, 2001; Wood & Tobias, 2004). The NIMH has taken a key leadership role in launching this area of inquiry, including three major funding initiatives, formation of an active, multidisciplinary consortium of researchers, and sponsorship of 15 widely-attended annual conferences (Pequegnat et al., 2002; Trickett & Pequegnat, 2005). Throughout the chapter the term "family" should be taken to mean a "network of mutual commitment." This definition, which has been adopted by the NIMH consortium, is intended to accommodate the wide diversity evident among families (Mellins, Ehrhardt, Newman, & Conard, 1996).

Our group at Memorial Sloan-Kettering Cancer Center (MSKCC), a large and well-established clinical care and research facility located on Manhattan's Upper East Side, has been involved in the NIMH consortium from its first meeting in 1993. Research products of the consortium have been described in an edited volume presenting family-based interventions being tested in NIMH-sponsored Phase III trials (Pequegnat et al., 2002). This volume, along with the abstracts from the annual NIMH conferences, demonstrate considerable strides in framing programs to address the unique needs of families affected by HIV.

Although many of these programs are still being tested, findings concerning intervention efficacy have started to become available. It is reasonable to expect that evidence to warrant dissemination of a number of these programs will be forthcoming. Moreover, there is need to organize practice-based networks of community agencies that can efficiently and rigorously begin to assess the utility and fit of these programs to their clienteles' needs and interests (Macaulay & Nutting, 2006; Westfall, VanVorst, Main, & Herbert, 2006). In sum, there is a great deal of behavioral intervention technology pertaining to the specific needs of families affected by HIV/AIDS in the pipeline.

Conducting a study of dissemination of this work to determine whether and how it can benefit a much broader audience is an important, logical next step. In our study, we viewed each partnership formed as a unique opportunity to examine family needs in HIV care and to study the process of 'technology transfer'. Questions concerning the feasibility of technology transfer are particularly important with respect to HIV/AIDS and families (Kelly, Sogolow, & Neumann, 2000). HIV/AIDS-related medical and social services are largely organized around individuals. There are few institutional or financial incentives to involve families in care or to offer them support or services (Sherer et al., 2002; Walkup, Satriano, Barry, Sadler, & Cournos, 2002; Winiarski, Beckett, & Salcedo, 2005). Indeed, there are many potential disincentives, including pressures to maximize caseloads, limited clinic space and time, difficulty engaging families in care, and reticence to address needs that are not "AIDS-specific." In light of these disincentives, it is reasonable to expect wide variation among providers in terms of programs that they would be willing or able to undertake. In our study, the technology to be transferred came in the form of a variety of new family-focused behavioral and mental health interventions.

Technology transfer encompasses a wide variety of approaches, ranging from top-down implementation of a standardized program to interventions that draw upon grassroots organizing around unique and complex social issues in a given community (Brown, 2000; Green, 2001; Kerner, Rimer, & Emmons, 2005; Lamb, Greenlick, & McCarty, 1998; Lasker & Weiss, 2003; Merzel & D'Afflitti, 2003; Pearlman & Bilodeau, 1999; Riley et al., 1999; Wandersman, 2003). Throughout the study we were mindful of the ramifications of different approaches to technology transfer. For example, how do providers and families view collaboration with academia? What levels of accountability and evaluation would they find acceptable? Are there circumstances that favor some approaches to technology transfer over others? Is the terminology of technology transfer misleading or even offensive to community-based agencies that understand how specific cultural and contextual factors would require substantial alternations to so-called evidence-based programs before the would be accepted and successfully implemented.

We hypothesized that our ability to transfer new technologies to the community hinged on several key issues, including identification of interested partners, access to expertise and resources needed to mount new programs and practices in diverse settings, and effective management – or mitigation – of barriers to collaboration between partnered stakeholders from community and research settings

(Best et al., 2003; Campbell, Dienemann, Kub, Wurmser, & Loy, 1999; Foster-Fishman, Berkowitz, Lounsbury, Jacobson, & Allen, 2001; Foster-Fishman, Berkowitz, Lounsbury, Jacobson, & Allen, 2001; La Piana, 2001; Mattessich & Monsey, 1992; Means, Harrison, Jeffers, & Smith, 1991; Minkler & Wallerstein, 2003; Mintzberg, Dougherty, Jorgensen, & Westley, 1996; Torres & Margolin, 2003; Weiss, Anderson, & Lasker, 2002). It is important to emphasize that our study did not involve the transfer of any one specific type of intervention or program. Rather, the goal was to explore the feasibility of implementing a range of family-oriented programs to a wide variety of settings.

Building Better Partnerships: The Facs Memoranda of Understanding

One tool that we developed to study and test these propositions was the FACS Memorandum of Understanding (MOU). The MOU was a document composed of more than 30 sections that described the terms of the collaborative relationship we sought to initiate with a given agency. Key elements of the academic-community relationship that the MOU helped to clarify were obtaining approval to work together (including approval of a secondary IRB, where required), delineating procedures for patient and family member accrual, addressing staffs' interests and concerns about their roles in the study, and coming to terms about who owned the data and how study results would be used.

Thinking about what happens "before the beginning". Clearly, there is cause to be uncertain about potential audiences for direct transfer of family intervention technology. The usual practice is to enlist convenient and available settings interested in undertaking a given intervention program. However, in Seymour Sarason's (1974) terms, this approach to research glosses over important questions about what happens "before the beginning." How did selected settings come to participate in the research? What convinced them to try? What other settings were approached but chose to decline? What others were considered but not approached? What is the infrastructure necessary to transfer a particular intervention technology into a setting? How can the processes of recruiting settings and transferring technology be replicated? At this stage in the development of interventions for families affected by HIV/AIDS, we cannot afford to ignore these questions. Rather, it is critical that we try to systematically understand factors that might impede or enhance our ability to undertake community-based dissemination.

Several traditions in health promotion and community organizing are relevant to this discussion. The first involves the concept of partnership formation (Currie et al., 2005; Davis, Olson, Jason, Alvarez, & Ferrari, 2006; Nelson, Raskind-Hood, Galvin, Essien, & Levine, 1998; Sebastian, Davis, & Chappell, 1998). Israel, Schulz, Parker and Becker (1998) provide a key review of this concept. They emphasize partnership as a way of conceptualizing relationships between researchers and communities. The partnership model emphasizes collaboration to build upon the knowledge and strengths of both the researchers and the communities, to help partners achieve desired and often common goals. Barriers to establishing partnership include mistrust of academic research, researchers' difficulty communicating with community stakeholders, competing demands, significant power differentials, and limited funding (Nelson et al., 1998; Schensul, 1999).

A second tradition involves capacity building. This concept has been widely applied in areas ranging from international development (Kotellos, Amon, & Benazerga, 1998) to urban planning (Chavis, 1995). Capacity building introduces the possibility of expanding the autonomy and effectiveness of potential community collaborators by increasing their skills, management systems, infrastructure and access to resources (Foster-Fishman et al., 2001; Goodman et al., 1998). Sustainability is a third key concept. Altman (1995) suggests that we view technology transfer or program diffusion as a process. As an intervention is transferred into the field, it may be modified, perhaps significantly. Although changes may violate fidelity with the original design, such modification may be critical for the program to take hold. Ultimately, observing interventions developed in this way leads to new, researchable questions about factors that influence behavioral change and adaptation (Davis et al., 2006).

To be sure, these concepts of partnership are fundamental to the process of technology transfer, whether accomplished through formal Phase IV trials, grass roots participatory research, or any other model. They direct us to consider "before the beginning" questions prior to undertaking program transfer. We need to better understand the range of potential collaborators, their capacities for and interest in adopting emerging family intervention technology, and their needs and challenges. Invariably, researchers in communities encounter and address these phenomena. A deeper understanding of these factors will help us to understand what is required to systematically disseminate and sustain programs for families affected by HIV/AIDS.

Embracing naturalistic inquiry and observational research. It is not possible to study providers' readiness and capacity for academic partnerships without influencing the very phenomena under investigation (Boser, 2006). Asking providers to obtain data on family needs and organizational resources involves overcoming barriers to research and establishing a sufficient working partnership. The act of engaging providers in this sort of study raises the possibility of partnerships and makes them aware of possible programs and resources.

Numerous authors have characterized early stages in partnerships between research and practice. For example, in discussing sustainability of interventions, Altman (1995) refers to a "transfer phase" in which providers and researchers deepen their relationships. In the area of HIV prevention, Sanstad, Stall, Goldstein, Everett and Brousseau (1999) refer to this as "building the collaboration." Lindley (1984) suggests that forming partnerships involves a process of listening, participatory dialogue, envisioning change and acting. Orlandi (1996) and Nutbeam (1998) discuss processes of mutual influences in "problem definition" within organizations such as work places and schools attempting to engage in technology transfer for health promotion such as tobacco use prevention. Freemantle and Stocken (2004) call attention to changes in clinical community oncology programs' organizational ecology, as new, often commercial, network relationships are formed to facilitate the conduct of clinical trials.

In their recent review of the literature on partnerships, Roussos and Fawcett (2000) suggest key steps to partnership formation including action planning, developing leadership, feedback and technical assistance. In contrast with a passive diffusion model, Sibbald and Kossuth (1998) describe research to study adoption of innovations in community-based health delivery through "coordinated implementation." They suggest use of a series of observable indicators of strategies to implement innovations. Brown (2000) calls attention to the need for attitudinal and behavioral measures to study phases in the adoption of innovations.

Overview of Our Multi-Site Research Design

Each site we recruited into the FACS was engaged in a four-step process (see Figure 11.1). Agencies were contacted sequentially, according to our randomly ordered list of providers. Initial contact with an agency representative was made via letters and/or phone.

Information packets providing details about the FACS study were sent to senior program administrators or directors at each selected agency. In the follow-up to the initial recruitment steps, a member of the FACS team met with the agency

Family Access to Care Study Design

1 Introduction and Setup	· Initial meeting · Develop and implement the Memorandum of Understanding
2 Data Collection	· Adult patients/clients (approx. 30 interviews) · Adult family members (approx. 20 interviews) · Providers (up to 4 interviews) · Program profile (description of organization and program of focus for the study)
3 Feedback	· Feedback focus group with staff · Feedback focus group wtih patients/clients · Report on patient/client and family member data customized for your agency
4 Consultation	· Agency-specific networking/program development assistance/ grantwriting opportunities or other possibilities · Periodic summary reports on study findings · Invitation to participate in NIMH Family and HIV conference

FIGURE 11.1 Family Access to Care Study Design

leader to answer additional questions about the study, determine the agency's willingness to participate, and assess whether any special agency approval of the study is required. In general, it was at this meeting that the concept of jointly developing an MOU was introduced to the agency.

Within each agency, providers, PLWHAs (that is, people living with HIV/AIDS who were clients/patients of the agency), and clients' designated family members of participating PLWHA were recruited. In general, PLWHAs who received services from the agency were eligible to participate. PLWHAs who agreed to have one of their family members participate in the study provided contact information for their relatives to facilitate family member recruitment. Generally, flyers were posted at the agencies so that PLWHAs would be able to contact FACS staff directly for information about the study. At some agencies, FACS staff were provided with a random contact list of PLWHAs, from which participants were recruited. Other conditions required that a FACS representative recruit participants on-site at the partnered agency, where potential clients were screened for eligibility, and an interview appointment was scheduled.

On average, the FACS intervention with a given participating agency took approximately 6 months. Due to time limitations we were only able to engage 24 of our original sample of 64 agencies in New York City, 22 of which agreed to participate in all four steps of the study. Incentives were offered to participating agencies, PLWHAs and designated family members. Each participating agency received $500 as compensation for provider time and expenses associated with study activities. PLWHAs and their participating family members were each compensated with $25 and a roundtrip subway metro card after participating in the study. In addition, benefits to agencies included interactive feedback sessions (i.e., 'Focus Feedback Groups') to agency clients/patients and to providers that featured agency-specific analyses. These sessions were used to assess interest in mounting new programs or initiating new research partnerships that address needs of families affected by HIV/AIDS.

Cross-sectional interview data were collected from 22 community-based organizations (CBO), clinics, and hospitals throughout New York City. To date, we have accrued an overall sample of 869 participants, including 626 PLWHAs, 195 family members, and 68 providers. At the time of writing this chapter, feedback and consultation services (i.e., Phases 3 and 4) have been delivered to most, but not all, participating agencies, and we are working with these and other interested agencies to develop an HIV Action Research Network intended to provide an infrastructure to encourage and sustain bi-directional transfer of both strategies and ideas between research and practice settings.

Ethical Principals and Values Embodied in the MOU

In order to specify the ethical issues that arise in community collaboration, it will be useful to outline the range of issues addressed in the MOU. Table 11.1 indicates each of the sections of the MOU, in the order that they are discussed with

TABLE 11.1 Ethical principles and values in working with community agencies embodied in the memorandum of understanding

Sections of the memorandum	Ethics/values
1. Overview of Process to Develop and Complete Memorandum of Understanding	Transparency
2. Sequence of Major Study Activities	Shared Authority
3. Designation of the Agency's Main Contact and Secondary Contact Persons	Shared Authority
4. Agency Study Approval	Shared Authority
5. Agency's Protocol Review Process	Shared Authority, Specificity
6. The Agency Profile	Transparency, Specificity
7. Readiness and Capacity Interview	Specificity
8. Organizing Meetings with Providers to Complete the Readiness and Capacity Interview	Shared Authority
9. Confidentiality of Agency Staff Data	Employees' Rights
10. Client/Patient Interviews	Shared Authority
11. Client/Patient Informed	Human Subjects Consent and Confidentiality
12. Location of Client Interviews	Shared Authority, Human Subjects
13. Client/Patient Eligibility	Human Subjects
14. Client/Patient Recruitment	Shared Authority
15. Family Member Selection by the Study Participant	Human Subjects, Self-Determination
16. Family Member Eligibility	Self-Determination
17. Family Member Consent and Interview Procedures	Human Subjects
18. Arrangements for Client and Family Referrals	Business Practice, Human Subjects
19. Handling of Partner Notification	Human Subjects
20. Handling of Emergency Situations	Business Practice, Shared Authority
21. Waiver of Third-Party Consent	Third-Party Rights
22. Client/Patient–Family Members Feedback Focus Group	Shared Ownership, Specificity
23. Feedback Focus Group Session with Agency Staff	Shared Ownership, Specificity
24. Tape Recording of Interviews and Feedback Focus Group Sessions	Privacy
25. Client/Patient/Family Member Feedback Focus Group Logistics	Shared Authority
26. Agency Staff Feedback Focus Group Logistics	Shared Ownership
27. Follow-up Readiness and Capacity Interview	Shared Authority, Specificity
28. Consultation for Participation in Research Partnership Activities	Shared Ownership, Business Practice
29. Use of Agency-Specific Study Data	Shared Ownership, Business Practice
30. Final Report Summarizing Study Results	Shared Ownership
31. Confidentiality of Agency-Level Data	Business Practice, Shared Ownership
32. Compensation to Agency	Business Practice
33. MSKCC Contact Information	Transparency

agency representatives. In that table, we indicate the ethics and values that dictated our response to common issues that arise in community practice. We outline the thinking behind these principles below:

Transparency is probably the over-arching concern reflected throughout the MOU. Transparency means that the agency has the right to know what the study is about and what they will be required to do. Moreover, transparency also means that the agency has the right to know about our role as researchers, our organizations, and how we will benefit from the study.

Shared authority indicates areas where the study required the collaborating agency to provide some resource, whether it was staff time, space, or access to clients. We wanted to ensure that these expectations were clear and the rationale for them was explicit. At the same time, we needed the agency representative to tell us how to best carry out these activities within the organization.

The MOU clearly supports familiar standards for individual rights of **human subjects**, and acknowledges an agency's responsibilities in that area as well as our own. In the FACS study, this pertained to family participants as well as clients. We also established the client's right to **self-determination**, in terms of who constituted their family and whether and how family members would be approached for the study. This concern for the family also extended to the area of **third-party rights**. FACS study required clients and participating family members to provide information about the range of problems and concerns within the family, which effectively meant that we were obtaining data on other people who had not consented to be in the study. A number of additional steps were included in the review process to ensure protection of these third-party research subjects (see (Lounsbury, Reynolds, Rapkin, Robson, & Ostroff, in press)). We wanted agency representatives to be aware of these issues, and to consider third-party rights in their internal review of the FACS study.

Another issue that arose in the FACS study involved concern with **employees' rights**. Our study protocol involved conversations with agency staff at several levels. This presented a complex situation. On the one hand, if the agency determines that it wants to be part of a collaborative research project, than participating in FACS activities could and should be viewed as part of an employee's job. On the other hand, we did not want employees to feel coerced to participate. In addition, employee participants needed to know that the information that they provided would be kept private and confidential. However, this conflicted with the value of shared ownership of study data, and the desire to ensure that FACS benefited the agency. In the MOU, we required agency leadership to agree that employees' participation in the study would be voluntary. This meant that the leadership would refer several staff for us to interview, but that we would not report whether or not interviews were refused. Further, we stipulated that agencies would not receive reports on agency-specific employee data, as they did about clients and family. Agency staffs were so small and close knit that it would have been impossible to ensure confidentiality at that level. Rather, the MOU stated that we would provide data summarizing staff concerns across all agencies after the conclusion of the study.

A value that we incorporated into the MOU was the notion of **specificity**. This was embodied in our attempt to tailor study activities to the needs and ecology of the agency. Further, the rationale for some of the detailed information that we were collecting at the agency level (which, of course, required more time from agency representatives) was to ensure that we really understood the specifics about the organization and could link study conclusions to information about history and context. In community practice, we believe that it verges on being unethical to disregard unique agency differences in the dissemination of programs, essentially akin to a doctor prescribing the same treatment to every patient that walks through the door without doing a physical or determining the diagnosis. The MOU holds us to the standard, and makes provisions to get data necessary to understand local relationships, resources and concerns.

Shared ownership refers to the view that data generated by the FACS study could be claimed by the agency as well as by the researchers. Of course, this shared ownership had to be bounded by human subject and employee rights concerns. Nonetheless, we felt that it was the responsibility of the researchers to help agencies make use of data for complementary goals. Indeed, information on clients' needs, service preferences, and barriers to care could be used by agencies for their own reports, planning, and development. Thus, we provided agencies with a standard set of summary charts presenting client and family data, and interpreted these charts with them as part of our feedback focus group process. We also offered to conduct further analyses for agencies to examine additional questions not included in the project. Ownership also includes conditions pertaining to co-authorship, in this instance, on case-study papers but not automatically on summative papers across all agencies. Ownership of the data also pertains to confidentiality. We were concerned that agencies may not want their funding sources to know that certain data existed, and that in some instances, it could be used to compare them to their competitors. Thus, we made it clear that we would not identify agencies, nor use it in a way that anyone familiar with agencies could identify them.

Finally, the MOU also included sound **business practice** that we wanted to guarantee. For example, if we identified problems with clients that required follow-up, we felt it necessary for each agency to guide us in how to respond. In most instances, that meant a referral back to the agency, or to the agency's routine service partners. However, agencies also needed to be aware that we would also take individual concerns and preferences into account, and offer other resources if the client or family member did not want to receive services at that agency, Business practice also includes discussion of how we could represent the collaboration. For example, our institution did not want FACS agencies to advertise that they were formally affiliated with our hospital or imply any special access to Memorial Sloan-Kettering simply by taking part in the FACS study. Of course, we would similarly not represent our relationship with any of these agencies without prior additional discussion, beyond the MOU.

How the MOU Affected Quality of Collaboration Among Participating Agencies

To assess how our MOU affected the quality of the FACS teams' partnership with community-based sites, we held a series of 'debriefing sessions' with key members of the MSKCC research team. In addition, we reviewed field notes from agency provider interviews to identify common concerns and interests that the MOU addressed, or could be fashioned to address. We begin with some general reflections about the use of the MOU and our collaborative relationships with agencies and providers, and then we elaborate on our reflections in a series of specific observations about the utility of the MOU as it was employed in the FACS.

Pre-existing conditions. We observed that agencies in which there was a pre-existing interest in the proposed topic of research and in which there was a positive impression of the researchers or the research institution, recruitment of the site into the study was generally faster and easier. We also noted that the level of authority of those we negotiated with to initiate the partnership as well as the professional status of the MSKCC researcher was important. In general, if the agency representatives had the direct authority to decide whether or not to partner with us, recruitment of the site was streamlined. Similarly, if the researcher who first contacted a target site had a doctoral degree or a robust professional title, then obtaining the interest of the agency was often facilitated.

Our experiences recruiting agencies to the FACS reaffirmed our initial assumption that communities and community-based agencies are suspicious of researchers. In general, agency staff assume the researcher will behave in an exploitive manner (i.e., they come in and grab data and run, and they don't come back). This is almost accepted as a truism among agency staff that we encountered. We assert that fact that we incorporated an MOU into our research process put the FACS into a different category from the outset. It served as a way to state that partnering with us would be different, or at least that we wanted to be different, than other community research projects. Research arrangements would be formally articulated and transparent. Results about the agency would be made available in "real time:" and directed to agency stakeholders. We would help agencies make use of FACS data in ways that supported agency needs assessment and/or program planning. Study activities were linked to future steps that could benefit the agency. We researchers were not planning to disappear.

Implementation of the MOU. Our MOU helped us tailor the study to each site, but to ensure that overarching goals and objectives were attained. MOU was not introduced until there was a oral agreement to participate, at which point it would be brought up as tool for reviewing the terms of the partnership. The MOU dialogue was at once an orientation to the study for agency leaders, a fact-finding process for our research team, and a structured negotiation about how we could expect our organizations to relate to one another around the FACS study.

The MOU was finalized more quickly when higher-level staff were involved early on in our process. Also, when higher-level staff were involved, they were more likely to appreciate how they could use the MOU to get information, support, and other resources from us. MOUs were scrutinized more carefully and changed in specific ways when the agency was particularly concerned about a benefit or risk identified in our MOU. In particular, those agencies with more research experience, more interest in data, were part of a hospital (and had an IRB), had a pre-existing interest in family services or mental health services were all more likely to give our MOU more scrutiny.

Although based on the same common template, some MOUs were more detailed than others. The final shape of the MOU was reflective of what the agency wanted and what concerned them. Other factors could add to the complexity of the MOU included the number and varieties of programs and clients an agency served, the number of sites, space and time constraints, and the levels of accountability within the agency to senior leadership and/or to community advisors. Once the MOU was developed, it was generally not changed, and would not be changed unilaterally. Key agency contacts, research activities and milestones were described in the MOU. If there was a need to make changes in procedures, the MOU was helpful because it explained who would be involved in making the changes.

Researcher benefits. MOUs established the authority or permission for an agency to work with the FACS study. The MOU was also an effective way for us to communicate what we wanted to do and how we would do it. MOU helped us dialogue with the agency about how we could carry out our research successfully. Every agency was different. We needed to work with the agency to articulate effective strategies to get approval from the agency, interact with staff, screen participants, conduct interviews, and provider feedback. The MOU helped us elicit information from agency staff about what would work and served as a communication tool to help us listen to what the agency said would work (or not work) in terms of carrying out the research partnership. The MOU process helped the agency staff become more involved and invested in the process. Turnover in staff often challenged our partnership with the agency, but the MOU was useful for helping a new staff person pick up where the previous staff person had left off.

The MOU was helpful not only for laying out the logistics of the process, but also what the agency could expect to get from working with us. It helped us both to anticipate problems that might arise and to communicate the terms of the collaboration clearly. We used the MOU to set boundaries about what we could and could not do with a particular agency. We also relied on what our IRB had approved in our protocol in order to set limits or to explain what we could or could not do with a given agency.

Agency partner benefits. The MOU made it easier for agency staff to approach the agency's executive director or Board in order to get their approval to work with us. It was an assurance that we would not negatively impact on their relationships with their clients or try to take their clients away from them. In general, we believe that MSKCC's strong reputation as a top cancer care and research institution helped. Agencies had confidence that we would make good on what we said we would do for them, and the MOU contributed to their confidence that this was true. Agencies that saw a use for the data we were collecting were

TABLE 11.2 Providers' interests and concerns about participating
in a research partnership

Providers' interests and concerns addressed in the MOU	% of FACS providers (N=68)
To arrange feedback and consultation about the results of the study	20%
To confirm that the agency had control over how data would be used or reported	19%
To ensure that clients would not be referred out or recruited away	16%
To sooth staffs' reluctance to collaborate with researchers	15%
To maintain or improve the agency's reputation in the community	12%
To officially inform staff about the research partnership	9%
To secure assistance from researchers should procedural problems arise	4%
To minimize the potential for disruption as the agency	3%
To protect the confidentiality of all participants	2%

most likely to contact us regarding our progress. They wanted our feedback and/or consultation services, and the MOU helped them keep us accountable.

In a qualitative review of field notes from agency providers, we identified nine common interests or concerns that help validate the purpose and content of our FACS MOU (see Table 11.2). We have confirmed that each theme was, in fact, included in the template we used to flesh out agency-tailored MOUs. The most frequently mentioned themes by providers during their interview with us concerned arranging appropriate and meaningful feedback and consultation about the results of the study (20%) and that the agency would have control over how data would be used or reported (19%). Less frequently mentioned were concerns over the management of unanticipated problems with study procedures (4%), minimization of the potential to disrupt agency activities (3%), and somewhat surprisingly, protection of study participants' confidentiality (2%). This finding is attributed to the fact that informed consents were reviewed and signed as part of our research procedure, for providers as well as participating agency clients and their family members.

Dealing with other Institutional Review Boards. Community-based agencies not linked with a hospital typically agreed to partner with us if their Board approved. Hospital-based programs often involved the added step of communicating and getting approval from their IRB. Seven of 22 programs had a formal relationship an IRB. We found that agency staff were often confused or unsure about whether or not their agency's IRB needed to be involved. Moreover, agency staff were often unsure of how to contact their IRB and were unfamiliar with IRB procedures for research proposal review.

In general, IRBs at other sites slowed partnership formation. For two sites that required IRB approval, the process took so long that we were not able to include them in the study. Typically, the biggest sticking point was whether or not interviews with participants would occur on the agency's site or not. If it was agreed that interviews could be conducted at MSKCC, an option that we were able to offer all sites, then the agency's IRB review process was not required. However, when IRB review was required by the agency, having the MOU helped facilitate approval. We attributed this to the language we employed in the MOU, which did a good job of covering privacy and confidentiality issues for clients and providers, which is a major concern of any IRB.

Specific Observations about the Utility of the MOU. A thematic analysis of information elicited during researcher debriefing sessions showed that the use of the FACS MOU impacted agency participation in a variety of ways.

1) *Facilitates authorized involvement.* The MOU authorized involvement in the study. It allows us to establish a partnership with the agency and it served as proof that the project was a legitimate activity, one that staff could officially spend time working on. Although the MOU was never treated as a traditional contract (i.e., it was not a signed document), it was nonetheless treated as a formal missive. In general, we worked with the program director or other key staff members until the MOU was finalized, then it was passed to the agency director, the agency's authority figure, who gave his/her formal approval to participate. Once this was accomplished, we had little to do with the top authority figures at the agency: we worked almost exclusively with lower level staff, those who would actually help implement the study at the agency.

Ultimately, the effect of the MOU was to give permission to staff to work with the FACS research team. Once signed by the authority figure, we move into the next phase with the agency, that of working with line staff to determine which providers we would interview, how to recruit clients and family members, etc. Lesson learned: Talk about it and write it down. The MOU truly provided a valuable touchstone, by defining the parameters of the relationship in clear and mutually acceptable ways.

2) *The MOU process alone did not lead to effective communication about the purpose (and potential significance) of the study to line staff.* Often lower level staff, those who were actually working with us day-to-day to conduct the study with the agency, those doing the hands-on work, were not told about the very nature of the project, only what they were required to do (i.e., recruit 30 patients for an interview with a member of the FACS research team). It was not used to inform agency staff about the purpose of the study and the nature of the research partnership with MSKCC. It was not used to explain goals of the study to agency staff.

This posed certain challenges regarding our collaborative relationship with the agency, was we implemented the various phases of the project (startup, data collection, feedback, and consultation). For example, at one feedback session with agency staff, it became clear that staff understood the purpose of FACS in a very different way. Staff had not been informed about the explicit family-focus and emphasis on behavioral and mental health interventions, as opposed to medical or clinical HIV-related interventions. In this situation, we learned that this occurred in part due to time constraints on the part of managing staff. Serious time constraints nearly prohibited more in-depth and rounded communication about our involvement at the agency and the purpose of the study. There was little time for directors to spend additional time fully explaining the research project (and, likewise) we had little time to do this, too. We spent more time building collaborative relationship with agency leadership, not much time working with line staff. Lesson learned: Saying it once means that you've said it once. Agencies are complex places and there is no guarantee that communication will reach others, especially about a non-routine matter. In practice, we were able to resolve communication gaps on an as-needed basis, because the FACS procedures were straightforward and time-delimited. In more complicated and demanding studies, mechanisms will be needed to conduct multiple levels of MOU-like discussions, to help different people in the agency express concerns and have input as needed in the process. This can be anticipated in the MOU.

3) *MOU helps set boundaries for working with agencies.* FACS team members used the MOU as a guideline for working inside an agency. The MOU helped us set limits on the work we would do with a given agency. For example, it helped us to ensure confidentiality and privacy for agency clients. Sometimes agency staff would suggest procedures that threatened participants' privacy. MOU spelled out that confidentiality and privacy was an important issue for us. Often agency staff would suggest a strategy for reaching a participant (client or family member) that would violate confidentiality. FACS team members could invoke the MOU (was well as the informed consent documents) as the reason why we could – or could not – use the suggested strategy. Lesson learned: It's a two-way street. As much as we intended the MOU process to benefit and protect agencies, it also served to benefit and protect us. Agency expertise was acknowledged and used to facilitate our work.

4) *Agencies have not used the MOU to demand timely feedback, consultation services or other benefits of participation to the same extent that we have used the MOU to ensure that we got what we wanted (i.e., data from the agency).* The MOU lays out a general timeline for the partnership. Agencies anticipated that our feedback and consultation services would occur soon after data collection was completed. The best we have been able to do was deliver agency reports as early as three months after data collection was completed. Providers and their clients were attracted to the idea that there would be specific activities and meetings set up expressly to report back on findings from the study. This feature was what allowed us to engage the agency and clients so that we could collect data for the study.

In one sense we may have became the 'traditional [ugly] researcher.' We made promises to do more than simply 'come and get the data and leave,' but it may seem to some that this is all we accomplished. To date, more than a year after the study has been officially closed to accrual, not all agencies have received feedback their promised sessions. Also, our promise to report study findings at last year's NIMH Family Conference in New York City did not occur. Although we still have every intention of making good on our promises to organize one-on-one feedback events with each agency and to provide consultation about initiating new research activities and/or mounting new programs, what we will ultimately deliver to many of the agencies and their clients will come some time later, perhaps after the FACS has faded from our partners' memories. Here, our own capacity to carry out the full program of research with each agency we partnered with proved to be more than we would accomplish in a timely manner.

One explanation for why agencies did not hold us to a shorter timeline for the research partnership is that they never expected as much. The narrative may have been different if we had not made good on the promise of $500 per agency to compensate for the agency's expenditure of time and effort to accommodate our data collection activities. Lesson learned: Less is more.

5) *Working with agencies on next steps has been a valuable extension of some (but so far not all) of the relationships established in the FACS study.* Despite our difficulty in delivering tailored feedback and consultation in the FACS study's abbreviated three-year timeframe, we have not ignored our commitments established in the MOU. Several agencies have requested our assistance in identifying research collaborators from the NIMH Family Consortium. We have also presented rounds at one agency. All FACS agencies were invited as guests to the NIMH Family Conference Community Day, to participate in a day-long interactive workshop on agency-research collaborations, and about six attended. Based on the results of this workshop, FACS participants and others were invited to a planning meeting for the HIV Action Research Network. This meeting again included about 25% of FACS sites.

As the HIV Action Research Network takes shape over the next year, we will attempt to reengage other FACS agencies. It is possible that some may choose to wait and see how the Network develops before getting involved, while others choose not to get involved. As the story continues to unfold, we will attempt to see whether agencies concerns, as reflected in the FACS MOU, determined whether, when and how they decide to get involved with further collaborative research. Lesson Learned: Agencies, like people, differ in their willingness to be early adopters of technology. We need to be prepared to foster long term ties that may take time to solidify. We expect that the best way to re-engage some agencies will be to put some of the plans in place with the early adopters.

6) *The values and principals embodied in the MOU process are widely shared and readily scaled up to larger consortia.* Additional insights about the MOU arose in our planning conference for the HIV Action Research Network. Over twenty community agencies participated in this meeting, including several direct "competitors." Nonetheless, agencies unanimously and strongly expressed the need for establishing procedures that ensured mutual benefit, transparency, and constancy, principals that were also important in the MOU. Agencies participants also wanted to find ways to get researchers to study their innovative programs.

After formally inaugurating the Network, our first act was to propose four working groups, to identify opportunities for practice-based research, to create infrastructure for data sharing and for surveillance of needs, and to determine priorities for training and capacity building. Lesson learned: A deliberative process to discuss, negotiate and reach consensus about responsibilities, roles, resources, and remedies is a good early step to start any research collaboration. In a new project that our research team is undertaking with four community partners, we decided to develop a single MOU that simultaneously laid out project roles and activities for all involved. This multilateral negotiation took several long meetings, in part because time was needed simply to clarify how things worked at one another's agencies. This process served to ensure that every party was fully informed and knew what was expected of all of the collaborators.

7) *The MOU formally recognized the research partnership between MSKCC and the agency, which was meaningful to the agency independent of any other perceived or realized benefit of participating in the study.* The notion that agencies could, if they so chose, invoke reference to a formal partnership with a prestigious academic medical center such as MSKCC was of sufficient benefit to agree to participate in our study. The credibility of MSKCC was a selling point to participate. While agencies were willing to partner with us, some may not have valued the specific feedback or consultation services we were offering. In many ways, the FACS study was intentionally an "easy" project for agencies to join. Agencies provided clients referrals, limited use of space, interviews with several staff, and time to receive feedback. There was no direct involvement in carrying out the study, conducting an intervention, or even designing the research in a truly participatory manner. It is very likely that a more demanding study would present greater barriers, but ultimately instill greater interest in those agencies that do take part. Lesson learned: What you put in equals what you get out. Having few expectations of sites may have led some sites to have low investment in the study. More complicated relationships involving collaborative interventions will require more in-depth MOU processes.

8) *The MOU was a planning exercise but not a working document.* The document was not designed to be reviewed or altered as the project unfolded. Indeed, MSKCC's IRB was initially reticent to authorize the use of a flexible document, and wanted us to standardize the MOU. Our institution had difficulty distinguishing the MOU from a "scope of work" contract. Indeed, we have come to see the MOU as an initial step in an on-going process of negotiation that might require additional financial contracts and other documents. Further, concepts included in the MOU can help investigators and agencies monitor progress and problems. After experience with the present study, we would anticipate using the initial MOU to set up on-going mechanisms for revisiting the document. Lessons Learned: Things change and "understanding" should be an evolving standard based on mutual experience and need.

Conclusion

The FACS MOU documented what we intended to do and what principles and procedures we felt were important to the integrity of the study. It was a way for us to inform the agency as well as for us to call specific attention to matters of research ethics and quality assurance. However, it did not give us the power to compel the agency to adhere to or sustain a particular grade of behavior during the partnership. Our success with an agency came by working through the research process with them, dealing with issues as they came up.

In terms of limitations, we found that the MOU was not a useful tool for ensuring full communication about what the study was about or not about to all agency staff and it was not an effective tool for managing the quality of research procedures in a particular community setting. In addition, we noted that the MOU did not always ensure that we were delivering on our promises to agencies in the most timely way. In fact, issues of timing were not made explicit in the FACS MOU. However, the MOU still held us accountable. Rather, the cliché, 'the squeaking wheel gets the oil' often explained our responsiveness to a given site. Those agencies that wanted more attention, faster, got it, in part because those agencies that were less demanding of our attention allowed us to focus on them.

Our field experience related to the multi-site FACS study as well as our follow-up community organizing efforts, suggest that MOU is an effective tool for promoting community research partnerships. When used as part of a process to build collaborative relationships, it can help investigators to identify issues that will effect implementation of research projects and meet these challenges head-on. The data we have analyzed on the feasibility of integrating such interventions has provided insight into how to target dissemination efforts to appropriate providers, and to tailor technology transfer approaches to suit providers' circumstances.

The ethical principles and values embodied in the MOU seem to be widely-shared and widely appreciated by agency representatives and staff. Outlining potentially sensitive issues like employee rights and shared-ownership right from the start help to engender trust and support the collaborative enterprise. These values and practices also form the basis for establishing sound multilateral partnerships. Procedures to revisit and amend an MOU could be built into such a document, especially in more complex, longitudinal studies. Future work must find ways to ensure that the MOU dialogue permeates organizations so that everyone affected by its provisions and its protections is appropriately informed.

References

Altman, D. G. (1995). Sustaining interventions in community systems: On the relationship between researchers and communities. *Health Psychology, 14*(6), 526–536.

Bauman, L., Camacho, S., & Silver, E. J. (2002). Behavioral problems in school-aged children of mothers with HIV/AIDS. *Clinical Child Psychology and Psychiatry, 7*(1), 39–54.

Best, A., Stokols, D., Green, L., Leischow, S., Holmes, B., & Buchholz, K. (2003). An integrative framework for community partnering to translate theory into effective health promotion strategy. *American Journal of Health Promotion, 18*(2), 168–176.

Biggar, H., Forehand, R., Chance, M. W., Morse, E., Morse, P., & Stock, M. (2000). The relationship of maternal HIV status and home variables to academic performance of African American children. *AIDS and Behavior, 4*(3), 241–252.

Blackwell, P., Gruber, D., & vonAlmen, K. (1997). The pediatric AIDS program: A systems approach to services to children and families in New Orleans. *Infant Toddler Intervention: The Transdisciplinary Journal, 7*(4), 251–269.

Bor, R., & Du Plessis, P. (1997). The impact of HIV/AIDS on families: An overview of recent research. *Families, Systems & Health, 15*(4), 413–427.

Boser, S. (2006). Ethics and power in community-campus partnerships for research. *Action Research, 4*(1), 9–21.

Brown, B. (2000). From research to practice: The bridge is out and the water's rising. *Advances in Medical Sociology, 7*, 345–365.

Campbell, J. C., Dienemann, J., Kub, J., Wurmser, T., & Loy, E. (1999). Collaboration as partnership. *Violence Against Women, 5*(10), 1140–1157.

Chavis, D. M. (1995). Building community capacity to prevent violence through coalitions and partnerships. *Journal of Health Care for the Poor and Underserved, 6*(2), 235–245.

Crosby, R. A., DiClemente, R. J., Wingood, G. M., & Harrington, K. (2002). HIV/STD prevention benefits of living in supportive families: A prospective analysis of high risk African-American female teens. *American Journal of Health Promotion, 16*(3), 142–145.

Currie, M., Gillian, K., Rosenbaum, P., Law, M., Kertoy, M., & Specht, J. (2005). A model of impacts of research partnerships in health and social services. *Evaluation and Program Planning, 28*, 400–412.

Davis, M. I., Olson, B., Jason, L. A., Alvarez, J., & Ferrari, J. R. (2006). Cultivating and maintaining effective action research partnerships: The DePaul and Oxford House Collaborative. *Journal of Prevention and Intervention in the Community, 31*(1/2), 3–12.

Diamant, A., Wold, C., Spritzer, K., & Gelberg, L. (2000). Health behaviors, health status, and access to and use of health care: A population-based study of lesbian, bisexual, and heterosexual women. *Archives of Family Medicine, 9*, 1043–1051.

Foster-Fishman, P. G., Berkowitz, S. L., Lounsbury, D. W., Jacobson, S., & Allen, N. A. (2001). Building collaborative capacity in community coalitions: A review of integrative framework. *American Journal of Community Psychology, 29*(2), 241–261.

Freemantle, N., & Stocken, D. (2004). The commercialization of clinical research: Who pays the piper, calls the tune? *Family Practice, 21*(4), 335–336.

Goodman, R. M., Speers, M. A., McLeroy, K., Fawcett, S., Kegler, M., Parker, E., et al. (1998). Identifying and defining the dimensions of community capacity to provide a basis for measurement. *Health Education & Behavior, 25*(3), 258–278.

Green, L. (2001). From research to "best practices" in other settings and populations. *American Journal of Health Behavior, 25*(3), 165–178.

Israel, B., Schulz, A., Parker, E., & Becker, A. (1998). Review of community based research: Assessing partnership approaches to improve public health. *Annual Review of Public Health, 19*, 173–202.

Kelly, J. A., Sogolow, E. D., & Neumann, M. S. (2000). Future directions and emerging issues in technology transfer between HIV prevention researchers and community-based service providers. *AIDS Education & Prevention, 12*(5 Suppl), 126–141.

Kerner, J., Rimer, B., & Emmons, K. (2005). Dissemination research and research dissemination: How can we close the gap? *Health Psychology, 24*(5), 443–446.

Kmita, G., Baranska, M., & Niemiec, T. (2002). Psychosocial intervention in the process of empowering families with children living with HIV/AIDS-a descriptive study. *AIDS Care, 14*(2), 279–284.

Kotellos, K. A., Amon, J. J., & Benazerga, W. M. G. (1998). Field experiences: Measuring capacity building efforts in HIV/AIDS prevention programmes. *AIDS, 12*(S2), S109-S117.

La Piana, D. (2001). *Real collaboration: A guide for grantmakers.* San Francisco: La Piana Associates, Inc.

Lamb, S., Greenlick, M. R., & McCarty, D. (Eds.). (1998). *Bridging the gap between practice and research: Forging partnerships with community-based drug and alcohol treatment.*Washington, DC: National Academy Press.

Lasker, R. D., & Weiss, E. S. (2003). Broadening participation in community problem-solving: A multidisciplinary model to support collaborative practice and research. *Journal of Urban Health: Bulletin of the New York Academy of Medicine, 80*(1), 14–60.

Lindley, C. (1984). Putting "human" into human resource management. *Public Personnel Management, 13*(4), 501–510.

Lounsbury, D. W., Reynolds, T., Rapkin, B. D., Robson, M., & Ostroff, J. (in press). Protecting third party rights in social and behavioral research. *Social Science & Medicine.*

Macaulay, A. C., & Nutting, P. A. (2006). Moving the frontiers foward: Incorporating community-based participatory research into practice-based research networks. *Annals of Family Medicine, 4*(1), 4–6.

Mattessich, P. W., & Monsey, B. R. (1992). *Collaboration: What makes it work.* Saint Paul, MN: Amherst H. Wilder Foundation.

Means, R., Harrison, L., Jeffers, S., & Smith, R. (1991). Co-ordination, collaboration and health promotion: Lessons and issues from an alcohol education programme. *Health Promotion International, 6*(1), 31–40.

Mellins, C. A., Ehrhardt, A. A., Rapkin, B., & Havens, J. F. (2000). Psychosocial factors associated with adaptation in HIV-infected mothers. *AIDS and Behavior, 4*(4), 317–328.

Merzel, C., & D'Afflitti, J. (2003). Reconsidering community-based health promotion: Promise, performance, and potential. *American Journal of Public Health, 93*(4), 557–574.

Minkler, M., & Wallerstein, N. (2003). *Community-based participatory research for health.* San Francisco: Jossey-Bass.

Mintzberg, H., Dougherty, D., Jorgensen, J., & Westley, F. (1996). Some surprising things about collaboration: Knowing how people connect makes it work better. *Organizational dynamics, 25*(1), 60–71.

Nelson, J. C., Raskind-Hood, C., Galvin, V. G., Essien, J. D. K., & Levine, L. M. (1998). Positioning for partnerships: Assessing public health agency readiness. *American Journal of Preventive Medicine, 16*(3S), 103–113.

Nutbeam, D. (1998). Evaluating health promotion: Progress, problems and solutions. *Health Promotion Journal, 13*(1), 27–44.

O'Hara, M. J., & D'Orlando, D. (1996). Ambulatory care of the HIV-infected child. *Nursing Clinics of North America, 31*(1), 179–205.

Orlandi, M. (1996). Prevention technologies for drug-involved youth. In C. B. McCoy, L. R. Metsch & J. A. Inciardi (Eds.), *Intervening with drug-involved youth* (pp. 81–100). Thousand Oaks, CA: Sage Publications, Inc.

Pearlman, S. F., & Bilodeau, R. (1999). Academic-community collaboration in teen pregnancy prevention: New roles for professional psychologists. *Professional Psychology: Research and Practice, 30*(1), 92–98.

Pequegnat, W., Bauman, L., Bray, J., DiClemente, R., DiIrio, C., Icard, L., et al. (2002). Research issues with children infected and affected with HIV and their families. *Clinical Child Psychology and Psychiatry, 7*(1), 7–15.

Riley, D., Sawka, E., Conley, P., Hewitt, D., Mitic, W., Poulin, C., et al. (1999). Harm reduction: Concepts and practice. A policy discussion paper. *Substance Abuse and Misuse, 34*, 9–24.

Rotheram-Borus, M. J., Lee, M. B., Gwadz, M., & Draimin, B. (2001). An intervention for parents with AIDS and their adolescent children. *American Journal of Public Health, 91*(8), 1294–1302.

Rotheram-Borus, M. J., Leonard, N. R., Lightfoot, M., Franzke, L. H., Tottenham, N., & Lee, S. J. (2002). Picking up the pieces: Caregivers of adolescents bereaved by parental AIDS. *Clinical Child Psychology and Psychiatry, 7*(1), 115–124.

Roussos, S. T., & Fawcett, S. B. (2000). A review of collaborative partnerships as a strategy for improving community health [review]. *Annual Review of Public Health, 21,* 369–402.

Sanstad, K. H., Stall, R., Goldstein, E., Everett, W., & Brousseau, R. (1999). Collaborative community research consortium: A model for HIV prevention. *Health Education & Behavior, 26*(2), 171–184.

Schaaf, R. C., Sherwen, L. N., & Youngblood, N. (1997). An interdisciplinary, environmentally-based model of care for children with HIV infection and their caregivers. *Physical and Occupational Therapy in Pediatrics, 17*(3), 63–85.

Schensul, J. J. (1999). Organizing community research partnerships in the struggle against AIDS. *Health Education & Behavior, 26*(2), 266–283.

Sebastian, J. G., Davis, R. R., & Chappell, H. (1998). Academia as partner in organizational change. *Nursing Administration Quarterly, 23*(1), 62–71.

Sherer, R., Stieglitz, K., Narra, J., Jasek, J., Green, L., Moore, B., et al. (2002). HIV multidisciplinary teams work: Support services improve access to and retention in HIV primary care. *AIDS Care: Psychological & Socio-Medical Aspects of AIDS/HIV, 14,* S31–44.

Sibbald, W. J., & Kossuth, J. D. (1998). The Ontario health care evaluation network and the critical care research network as vehicles for research transfer. *Medical Decision Making, 18*(1), 9–16.

Torres, G. W., & Margolin, F. S. (2003). The collaboration primer: Proven strategies, considerations, and tools to get you started. *Health Research and Education.*

Trickett, E., J., & Pequegnat, W. (2005). *Community interventions and AIDS.* New York: Oxford.

Vandehey, M. A., & Shuff, I. M. (2001). HIV infection and stage of illness: A comparison of family, friend, and professional social-support providers over a 2-year period. *Journal of Applied Social Psychology, 31*(11), 2217–2229.

Walkup, J., Satriano, J., Barry, D., Sadler, D., & Cournos, F. (2002). Public health matters. HIV testing policy and serious mental illness. *American Journal of Public Health, 92*(12), 1931–1939.

Wandersman, A. (2003). Community science: Bridging the gap between science and practice with community-centered models. *American Journal of Community Psychology, 31*(3/4), 227–242.

Weiss, E. S., Anderson, R. M., & Lasker, R. D. (2002). Making the most of collaboration: Exploring the relationship between partnership synergy and partnership functioning. *Health Education & Behavior, 29*(6), 683–698.

Westfall, J. M., VanVorst, R. F., Main, D. S., & Herbert, C. (2006). Community-based participatory research in practice-based research networks. *Annals of Family Medicine, 4*(1), 8–14.

Winiarski, M. G., Beckett, E., & Salcedo, J. (2005). Outcomes of an inner-city HIV mental health programme integrated with primary care and emphasizing cultural responsiveness. *AIDS Care, 17*(6), 747–756.

Wood, S. A., & Tobias, C. (2004). Barriers to care and unmet needs for HIV-positive women caring for children: Perceptions of women and providers. *Journal of HIV/AIDS and Social Services, 3*(2), 47–65.

Chapter 12
HIV Research with Children

There are in excess of 3.2 million children who have acquired HIV infection through vertical transmission, and 90% of those cases reside in resource-constrained settings—mostly Africa, but as the epidemic migrates, India, Southeast Asia, and China as well (Mofenson, 2003). According to UNAIDS, children now account for 14% of all new infections and 16% of all AIDS-related deaths, despite the fact that perinatal acquisition of HIV is almost entirely preventable, and that the costs of such prevention, in relative terms, are negligible (UNAIDS, 2006).

Despite overwhelming evidence of excellent clinical outcomes for children receiving medical monitoring and standard-of-care medical management of HIV illness, there remain considerable challenges in pediatric HIV/AIDS treatment. Infant diagnosis is still problematic, and there are developmentally-related biological and behavioral limitations imposed by the size and age of the child. There are differences in drug synthesis, and long-term management is complicated by dosing changes required as the child grows from infancy, to childhood, to adolescence. Pediatric formulations are still limited; reduced-dose adult formulations are still commonly used, which may also affect palatability. For many in the health professions, pediatric HIV/AIDS remains a specialty within a specialty: there are simply not enough professionals with strong clinical expertise in treatment of HIV infection in children. And some would contend—though there has been meaningful progress in the last several years—that global scale-up for treatment of pediatric AIDS remains a low priority, in part because children can seldom advocate for themselves in a manner affecting social policy in the same way that other HIV-impacted populations or communities can (such as men who have sex with men in the United States) (Institute of Medicine, 2005).

To cite one example, it has been recently reported that pediatric HIV care has lagged behind adult HIV services in Kenya due to inadequately trained personnel, the lack of appropriate pediatric antiretroviral preparations, and lack of appropriate capacity for children under 18 months of age. As a result, by early 2004 there were less than 400 children receiving antiretrovirals, only increasing, as a result of intensive scale-up, to 1,600 children in June of 2005, and 4,300 children a year later (Ojoo et al., 2006).

Though difficult to verify, a universal commitment to high-quality medical management of HIV infection in children may also be attenuated because of an underlying (and callous) perception that providing treatment to children would only delay childhood mortality. Rarely (it is assumed), will treated children have the opportunity to grow into productive adulthood, and medical resources, therefore should be apportioned to those for whom stable health will result in the greatest overall social and collective gain. Such a perception would only be rarely uttered publicly, of course. Nevertheless, it is important to continue to ask questions about allocation of health resources and perceived "social value" in any disease context.

Each of these challenges presents ethical dilemmas, and the research itself related to each of these specific issues generates still more ethical challenges. These, in turn, can be overlaid onto a host of fundamental and fluidly-articulated and addressed dilemmas related to the bioethics and children in general, dilemmas such as vulnerability, consent/assent and autonomy, surrogate and proxy decision-making, confidentiality and disclosure, paternalism and individual volition or agency, beneficence, and distributive justice. Finally, the moral complexity of each of the issues is, in turn, compounded by one of the most heartbreaking aspects of the HIV/AIDS epidemic: that so many of the world's children now living with HIV infection, whether treated or not, are themselves orphaned by the death of one or more parent due to AIDS, leaving millions of children in foster care or orphanage systems (or nowhere at all) that are not only poorly resourced and staffed, but also inadequately supported with experience and expertise in the ethical conduct of proxy decisionmaking and in foundational or operational ethics in general.

Background and Issues

The reality of unethical research involving children and other "convenient" subject populations—convenient because they are centralized in orphanages, care homes, and the like, and often, due to tragic circumstances, outside of parental control—is longstanding. In 1966 distinguished Harvard faculty member Henry Beecher outlined 22 examples (some involving children) of unethical research, carefully documenting the extent to which human subjects were unethically involved in trials, usually without their knowledge or without knowledge of the risks they were exposed to (Beecher, 1966). The commonalities between these abuses included, most importantly, failure of informed consent, use of vulnerable ("convenient") populations, and an unfair distribution of risks and benefits. Although the Nuremberg Code (1946) established the ethical necessity of informed consent, it wasn't until nearly 20 years later that the Declaration of Helsinki established the necessity of proxy consent, thus addressing the question of autonomy in decisionmaking for vulnerable adults, and children.

In the years since then, ethicists have tackled the tensions between autonomy and third-party decisionmaking, protection from harm and protection from paternalism, and other questions through an ongoing process of necessary, and often only temporarily resolving, ethical balancing acts. In general, there is near-consensus (there are some dissenting voices) on the right of children to access clinical trials, and the need to conduct them in order to determine effective medical strategies for ill children. Certainly, current NIH policies mandate inclusion: "It is the policy of NIH that children (i.e., individuals under the age of 21) must be included in all human subjects research, conducted or supported by the NIH, unless there are scientific and ethical reasons not to include them" (NIH, 1998). NIH policies also outline levels of information to be provided to both child and parent, required before consent/assent can be said to be met, based on levels of risk and benefit incurred in the proposed research. Levine takes a more explicitly rights-based approach: "If we consider the availability of drugs proved safe and effective through the devices of modern pharmacology a benefit, then it is unjust to deprive classes of persons, e.g., children and pregnant women, of this benefit" (Levine, 1988: 240). Over a decade later, he and other scholars reiterated this same message, this time specifically in relation to HIV/AIDS: "No group should be categorically excluded, on the basis of age, gender, marital status, place of residence or incarceration, or other social or economic characteristic from access to clinical trials or other mechanisms of access to experimental therapies" (Levine, Dubler and Levine, 1991: 14).

Consent

But of all the slippery questions related to research in children (especially since the age span runs from infancy to the legal age of adulthood, and therefore encompasses a wide range of developmental stages), few are as perpetually vexing as the questions related to consent/assent for research, and the interdeterminancy of absolute autonomy upon which such questions rest. According to Levine, children do not have legal capacity to consent and many are unable to comprehend sufficiently to meet the standards of consent to research that are enunciated in relevant documents, such as the Nuremberg Code (Levine, 1988). Parents, or their legal surrogates, are thus conferred the responsibility for consent, and the obligation to advocate on behalf of their children. But there are at least two complicating factors: 1) the capacity of parents or other proxy decision-makers to provide truly informed consent on behalf of the child, and 2) the capacity of children, especially at different stages of cognitive and ethical development, to offer assent.

There are, of course, legal conditions which may create exceptions to the need for parental or proxy consent, including Emancipated Minor and Mature Minor regulations, and those situations where it is legal for a minor to access health services without parental consent, such as certain sexual health services. There are considerable variations across state/provincial and national jurisdictions about whether, under what circumstances, and with what parameters such laws and regulations apply. With regard to adolescent consent in such cases, there is a need for clearer guidelines and practice protocols for their application. As Levine and colleagues assert, "States must establish clear rules to enable mature and emancipated minors to provide legally adequate consent for counseling, test-ing, treatment, and research" (Levine, Dubler, and Levine, 1991: 15). This may carry additional weight given the experience of those HIV+ teenagers who were infected at birth, and who may have "acclimated" to a lifetime of gradually more complex medical decisionmaking on their own behalf.

There are other exceptions to the inviolability of parental or proxy consent. As embodied in U.S. law, parental or guardian consent must precede a child's partic-ipation in research, even in those cases where all parties may agree that the child may be old enough to be of sufficient capacity to understand the nature of research and thereby able to consent to his or her own participation. But, as Nairn points out, "by demanding parental permission, the law does not thereby give parents absolute discretion regarding proxy consent. Traditionally, there have been two limits on proxy consent for a minor, (1) the 'best interest' judgment . . . and (2) the understanding of a 'reasonable person' or a 'reasonable parent.' Thus, for example, there are certain medical procedures, such as a blood transfusion, that must be performed upon a child even if the procedure is against a parent's religious beliefs (Nairn, 1993).

Parental or proxy decision-maker capacity to provide informed consent is legally assumed in most cases, but may not always be effectively verified (an argument that can be made about all forms of "informed consent.") A 2004 paper by Kodish and colleagues studied parental understanding of the concept and process of randomization—one of the core scientific principles upon which research is founded, and therefore a common conceptual component of consent. The study, focusing on childhood leukemia clinical trials, raises concerns about comprehension, and therefore, fully-informed consent. Cases were recruited from six U.S. hospitals associated with academic medical centers from July 1, 1999 to December 31, 2001. Participants included parents and members of health care teams who participated in 137 informed consent conferences for children with newly diagnosed acute leukemia. Randomization was explained by physicians in 83% of cases, and a consent document was presented in 95% of cases. Interviews conducted after each conference indicated that over half of parents did not under-stand the concept of randomization (Kodish et al., 2004). In fact, our verified understanding of whether parents or proxy decision-makers are making judg-ments based on a sense of be truly *informed* is inadequate, and in the absence of more research, we can only assume that comprehension is not universally sufficient.

Consent and/or Assent

Informed consent in pediatrics is an inaccurate phrase: informed consent equals parent permission and, in a number of cases, assent of the child. Kodish and col-leagues, in their 2004 study, make a number of recommendations for improving parental understanding of randomization, including having parents participate in the informed consent conference, including question-asking; describing different types of randomized clinical trials, including the standard group; explaining differences between treatment in a randomized clinical trial and off-study therapy; discussing the right to withdraw from the trial; stressing the importance of reading the consent document and giving parents time to do so; providing formal communication training, including cultural sensitivity training, to enrollment staff; promoting nurse attendance at the consent conference; giving parents and patients as much time as they need to make decisions; providing parents with additional emotional support; assessing parental understanding of randomization; and providing further explanation of randomization until understanding is achieved. These recommendations, it would seem, would apply to understanding of any of the concepts embedded in consent (Kodish et al., 2004).

Assent means that the child is aware of the medical condition, comprehends what she or he can expect, understands the nature of the procedure or treatment, and expresses voluntary support for involvement in the study. Elements of assent include helping the child achieve a developmentally appropriate awareness of their condition; disclosing the nature of the proposed intervention and the child's likely experience with that intervention; assessing the child's understanding; and soliciting the child's willingness to participate.

By and large, it seems that children and youth want *involvement* in decision-making about research participation, even when they recognize that they may not be the final decision-makers.

As an example, small survey of HIV+ children in the Democratic Republic of the Congo indicated that minors believed that they should receive information about a research study in which they might participate, and that they could be asked to sign an assent form. Both minors and their adult caregivers, however, believed that caregiver decisions about whether or not the child should participate are final (Corneli et al., 2006). And a study by Burke, Abramovitch, and Zlotkin (2005), which sought to maximize the amount of information children and adolescents understanding about the risk and benefits of participating in a biomedical research study, found that by creating age-appropriate modules of information, children as young as six could understand potentially difficult and complex topics such as research risks and benefits. Indeed, the conceptualizations and motivations children and youth bring to a discussion of possible involvement in research may be much more complicated and nuanced than commonly acknowledged. In a study on children's' assent to non-therapeutic research, Wolthers (2006) surveyed 1,281 healthy Danish children and adolescents aged 6–16 to evaluate their motives for assent or dissent and their understanding of the information provided. He found that sociodemographic factors had little influence on healthy children's decision to volunteer for non-therapeutic research. Children who assented reported higher altruistic and educational motives, whereas children who dissented were more likely to express worries about medical procedures, such as blood draws.

Consent and assent are also expressed and achieved through the filters of cultural experience, a point often raised in the current literature. Nussbaum takes a common approach, supporting a universalist, trans-cultural approach to ethics in general, but arguing also for a "delicate balancing act between general rules and a keen awareness of particulars, in which . . . the particular takes priority, in the sense that a good rule is a good summary of wise particular choices, and not a court of last resort" (Nussbaum, 1993: 257). If the question of individual consent may be somewhat clouded in collectivist societies where multiple levels of "consent" (such as community leaders) may be required, the question of proxy consent—parents who consent to research participation for their children—can be equally clouded: "One may ask whether the mothers of the children [in an HIV vaccine study in Zaire] were even the proper people from whom to obtain proxy consent. This question is itself multifaceted, eventually raising a final question regarding the very nature and possibility of proxy consent in an African consent" (Nairn, 1993: 239). While there is something needlessly essentialist about this contention on the face of it—it seems to reduce all of the African continent to a single set of cultural values, without acknowledging the extraordinary diversity of values from north to south and east to west—it does highlight an important consideration: proxy, after all, is merely instead-of, and in whatever way cultural variations may influence the meaning of consent in an individual context, it will also influence the meaning of the proxy.

Special Issues

In addition to the fundamental issue of consent/assent, a number of other issues related to ethical inclusion of children in research, and particularly HIV/AIDS research, continue to surface.

As with adults, numerous analysts have asked whether involvement of minority children in research is representational. A study by Walsh and Ross (2003) reviewed 192 pediatric research trials with findings published between July 1999 and June 2000 to assess representation of minority children in trials. Overall, their findings noted an overrepresentation of African American children and an underrepresentation of Hispanic/Latino and white children, with significant variation depending on the type of research. Both African American and Hispanic/Latino children were overrepresented in "politically stigmatizing research"—research focused on topics such as child abuse and neglect, HIV infection, psychiatric illness, and high risk behaviors. While this contrasts with under-representation of adult African Americans in clinical trials, and can therefore could be viewed as a positive finding in that in it helps correct problems of equality of access to clinical trials, it can just as easily be viewed as highly problematic, since communities are highly sensitive to use of populations as research "guinea pigs," and minority communities may be concerned that their children are disproportionately burdened with risk in the implementation of a research agenda. These differences in ways of viewing such disproportionality may speak as much to effective community consultation and management of community expectations and concerns as they do to actual proportionality.

A second recurring issue is the involvement of foster children in clinical research trials. The example of the outcry raised over the involvement of HIV+ foster children, particularly foster children of color, in HIV/AIDS clinical trials in New York City (described in a separate chapter) and reported both in the *New York Times* and the BBC highlights the sensitivity of the issue (Scott and Kaufman, 2005). C. Levine (1991: 234) raises particular concerns about consent for children in foster care, noting that "A determination that a protocol is acceptable for foster children is not the same as a determination that a protocol is the best option for a particular foster child." Given the history of research abuses with "convenient" study populations such as foster children, ethical scrutiny is critical. Again, Levine articulates the salient conclusion about the inevitable weakness of a paternalistic kind of neoliberalism, emphatically to the reality of "how easily ethical considerations can be swept aside under the rubric of 'humanitarianism' or 'compassion'" (Levine, 1991: 236).

Another issue is IRB preparedness to address the complexities and subtleties of decisions involving ethical participation of children in research. Wolf, Zandecki, and Lo (2005) reviewed guidance on pediatric research from 39 IRB websites, finding that few IRBs discuss important ethical issues on which regulations are relatively silent, and that some IRBs actually provide inaccurate information on the regulations. They specifically concluded that more detailed IRB guidance, in the form of checklists and itemizations of "things to consider," could help pediatric investigators think through ethical issues and better protect children in clinical research. (For a discussion of how complex informed consent issues were addressed in the context of one HIV-related study involving children, see the case study by Boyd and colleagues that follows this chapter.)

Finally, the human rights of children participating in research—to the extent that children and youth are recognized as eligible for the spectrum of human rights commonly accorded to those of legal age—remains a complicated question. One example, out of many possibilities, will help illustrate. A Human Rights Watch report written by Jonathan Cohen and Helen Epstein, and based on interviews with children affected by HIV/AIDS in Kenya, South Africa, and Uganda indicates that governments and schools may be acting in subtle ways to deprive children affected by HIV/AIDS, especially those already orphaned by HIV/AIDS, from exercising an equal right to education (Cohen and Epstein, 2005). Anecdotal evidences from around the world over the last 25 years of the AIDS epidemic points to countless examples of HIV+ and HIV-affected children being denied access to public and private education. Researchers working with HIV+ children have a special obligation to work with parents and caregivers to help ensure that educational access and equality of provision of educational resources is in no way compromised for HIV+/HIV-affected children with whom they are working as research participants. Separately, Fanelli and Mushunje (2006) have argued for a rights-based approach in the development of child-friendly communications strategies for national AIDS policies, contending that HIV+ and HIV-affected children can and should have meaningful participation in the policy process.

Approaches

Emanuel, Wendler, and Grady (2000: 2701) have proposed 7 requirements that "systematically elucidate a coherent framework for evaluating the ethics of clinical research studies" in general, and they are worth mentioning here for their depth and their particular applicability to research involving children. The seven requirements include 1) value, or the reasonable supposition that "enhancements of health or knowledge must be derived from the research"; 2) scientific validity, or methodological rigor; 3) fair subject selection, or the requirement that "scientific objectives, not vulnerability or privilege, and the potential for and distribution of risks and benefits, should determine communities selected as study sites and the inclusion criteria for individual subjects"; 4) favorable risk-benefit ratio, or the requirement that the potential benefits to individuals and knowledge gained for society must outweigh the risks; (5) independent review, or the requirement that "unaffiliated individuals must review the research and approve, amend, or terminate it"; (6) informed consent; and 7) respect for enrolled subjects, as evidenced through privacy protections, guarantees of opportunities to withdraw, and monitoring of participant well-being. The authors conclude that meeting *all* of the requirements is necessary to make clinical research ethical, and that the requirements are universal, though they must be adapted to the health, economic, cultural, and technological conditions in which clinical research is conducted. The obligation to monitor well-being may carry particular weight in the case of children involved in HIV/AIDS research, given multiple vulnerabilities. Kennedy and associates (2006), for example, found extraordinarily high rates of trauma, post-traumatic stress disorder, and suicidal tendencies in the HIV+ youth, aged 13–21, attending an inner-city clinic, with 44% meeting the diagnostic criteria for PTSD, and 38% reporting at least one suicide attempt (lifetime). Nearly 15% reported having no one to talk to about their problems.

In addition to general recommendations about the ethical management of clinical trials, there are special considerations for children and youth. The experience of researchers and practitioners who have developed and implemented prevention and medical scale-up programs for adolescents may be informative in this regard. Tiffany, et. al. made recommendations about involvement of youth in HIV prevention research that could well be applied to a full range of research studies or initiatives in which HIV+ and HIV- youth are participants. Preliminary recommendations from experience with a community-based participatory research project include

1) Provide tangible incentives to youth participants at the same time as fostering their intrinsic interest in the research process; 2) Develop infrastructure to allow rapid preliminary data analysis; 3) Organize timely community-based interpretation sessions; 4) Develop processes for timely translation of research findings into program activities; 5) Encourage understanding and appropriation of the research process by participants; 6) Design multiple, flexible modes of participation for youth and frontline program staff; and 7) Facilitate dialogue among

stakeholder groups (Tiffany, et al., 2006). While the research partnership the researchers outline worked with teens, little has been done to consider translating the basic principles of such recommendations—which essentially involve skills and knowledge transfer, and establishment of structures and mechanisms to foster research participant partnership in the research process—to younger youth, in a manner that is developmentally appropriate and manageable. This is a potentially difficult, but also equally potentially fruitful, task, and worth undertaking.

More specifically, researchers must also be sensitive to patterns of HIV disclosure and silence within families where a child is living with HIV. A recent report from Romania estimated that about 20% of 600 HIV-positive children reviewed had not had their HIV status disclosed to them (Ionescu, 2006). While the situation in Romania is unique because of specific modes of HIV transmission, non-disclosure of HIV status to HIV-infected children is common in other countries as well (Mellins et al., 2002; Nöstlinger et al., 2004; Thorne, Newell, and Peckam, 2000).

Researchers should address not only the consent-related educational needs of parents and the assent-related educational needs of children and youth, but the training needs of research team members as well, and the ways in which each of the constituent "groups" phrases or voices its concerns about ethics. Twomey (1994) interviewed physician and nurse researchers, IRB members, and parents of HIV+ children, at four separate institutions conducting pediatric HIV research, about their views on the ethical issues involved in the conduct of clinical drug research that uses as subjects HIV+ children. Two issues emerged as paramount ethical concerns: 1) factors that affect research subject access to care, and 2) determination of appropriate risk for HIV+ children in research. A third factor was family status; Twomey noted the relative lack of access of children in foster care to clinical trials. Most significantly, Twomey found that IRB members and physicians, on the one hand, and nurses and parents, on the other, 'voiced' their ethical concerns in different ways: the former adopted a markedly principled approach to risk/benefit assessments; the latter tended to place those appraisals within the context of the child's overall existence. Understanding the differing vocabularies each constituent group adopts will aid in the development of a common ethical framework, and a common understanding of that framework, to regulate research (Twomey, 1994).

Finally, given that a significant number of research studies are now being sponsored directly by pharmaceutical companies, and that there is growing concern about the possibility of bias under such conditions, it is fair to ask whether parents or proxy decision-makers have the right to know about sponsorship of a study that could negatively—or positively—affect their child's well-being. To be sure, this is a question that should be posed about all research protocols, but given the history of "convenience" employment of children as research subjects, it may be of particular relevance to parental or proxy decision-makers, who will have the right to know who, beyond the child himself or herself, is taking risks, and can expect gains, as a result of the research.

Notes

1. Some would argue that true informed consent is a fallacy in current practice. In an opinion published in 2005 in *Applied Clinical Trials*, Hochhauser (2005) asserts that "Truly informed consent is impossible," noting that one study of oncology consent forms found that the typical form included 14 topics and 2,700 words in 11 pages, not including HIPPA language. Hochhauser based his "impossibility" conclusion on the fact that the human brain, even when fully developed, is simply not capable of absorbing that many sets of information at one time (he suggests no more than four topics). Whether or not the claim of true consent impossibility is an intrinsic condition of the legal and ethical demands that consent must satisfy, Hochhauser makes important points about human capability under reasonable circumstances.

References

Beecher, H.K. (1966). Ethics and clinical research. *New England Journal of Medicine, 274*, 1354–1360.

Burke, T.M., Abramovitch, R., & Zlotkin, S. (2005). Children's understanding of the risks and benefits associated with research. *Journal of Medical Ethics, 31*, 715–720.

Cohen, J. & Epstein, H. (2005). *Letting Them Fail: Government Neglect and the Right to Education for Children Affected by AIDS*. New York: Human Rights Watch.

Corneli, A., Rennie, S., Vaz, L., Dulyx, J., Omba, S., Kayumba Baye, J. et al. (2006). Perspectives of minors and caregivers on the rights of minors to be involved in decision making about research participation: Findings from formative research on assent and disclosure for HIV-positive children in Kinshasa, Democratic Republic of the Congo." Presented at the XVI International AIDS Conference, Toronto, Canada, August 13–18 [abstract no. MOAD0204].

Emanuel, E. J., Wendler, D., & Grady, C. (2000). What makes research ethical? *Journal of the American Medical Association, 283*, 2701–2711.

Fanelli, C.W. & Mushunje, M. (2006). Child-friendly communications: A critical step in facilitating child participation in national HIV/AIDS policies. Presented at the XVI International AIDS Conference, Toronto, Canada, August 13–18 [abstract no. CDE0411].

Hochhauser, M. (2005). Memory overload: The impossibility of informed consent. *Applied Clinical Trials*, November 1. Last accessed January 10, 2007; Available at http://actmagazine.com/appliedclinical trials

Institute of Medicine. (2005). *Scaling Up Treatment for the Global AIDS Pandemic: Challenges and Opportunities*. Washington, D.C.: National Academies Press.

Ionescu, C. (2006). Romanian parents keep HIV secret from infected children. *Lancet, 367*: 1566.

Kennedy, C.A., Botwinick, G., Johnson, D., Ruranga, R., & Johnson, R.L. (2006). High rates of suicidal ideation and attempts to characterize HIV+ minority youth. Presented at the XVI International AIDS Conference, Toronto, Canada, August 13–16 [abstract no. CDB1030].

Kodish, E., Eder, M., Noll, R.B., Ruccione, K., Lange, B., Angiolillo, A. et al. (2004). Communication of randomization in childhood leukemia trials. *Journal of the American Medical Association, 291*, 470–475.

Levine, C. (1991). Children in HIV/AIDS clinical trials: Still vulnerable after all these years. *Law, Medicine & Health Care, 19(3–4)*, 231–237.

Levine, C., Dubler, N.N., & Levine, R. (1991). Building a new consensus: Ethical principles and policies for clinical research on HIV/AIDS". *IRB: A Review of Human Subjects Research, 13(1–2)*, 1–17.

Levine, R.E. (1988). *Ethics and Regulation of Clinical Research*, 2nd Edition. New Haven: Yale University Press.

Mellins, C.A., Brackis-Cott, E., Dolezal, C., Richards, A., Nicholas, S.W., & Abrams, E.S. (2002). Patterns of HIV-status disclosure to perinatally HIV-infected children and subsequent mental health outcomes. *Clinical Child Psychology and Psychiatry, 7*, 101–114.

Mofenson, L. (2003). Tale of two epidemics: The continuing challenge of preventing mother-to-child transmission of human immunodeficiency virus. *The Journal of Infectious Diseases, 187*, 721–724.

Nairn, T.A. (1993). The use of Zairian children in HIV vaccine experimentation: A Cross-cultural study in medical ethics. *The Annals of the Society of Christian Ethics, 1993*, 223–243.

National Institutes of Health. (1998). *NIH Policy and Guidelines on the Inclusion of Children as Participants in Research Involving Human Subjects*. Last accessed January 19, 2007; Available at http://grants.nih.gov/grants/guide/notice-files/not98–024.html

Nöstlinger, C., Jonckheer, T., de Belder, E., van Wijngaerden, E., Wylock, C., Pelgrom, J. et al. (2004). Families affected by HIV: Parents' and children's characteristics and disclosure to the children." *AIDS Care*, 16, 641–648.

Nussbaum, N. (1993). Non-relative virtues: an Aristotelian approach. In Nussbaum, N. & Sen, A. (Eds.). *The Quality of Life* (pp. 242–269). New York: Oxford University Press.

Ojoo, S., Tanui, I., Koech, E., Gathandu, B., Odera, D., & Muiva, C. (2006). Scaling up pediatric HIV care and treatment services in Kenya. Presented at the XVI International AIDS Conference, Toronto, Canada, August 13–16 [abstract MOPE0637].

Scott, J. & Leslie Kaufman, L. (2005). Belated charge ignites furor over AIDS Drug trial. *New York Times*, July 17, Sec. 1, Col. 1, Metropolitan Desk, at 1.

Thorne, C., Newell, L.M., & Peckam, C.S. (2000). Disclosure of diagnosis and planning for the future in HIV-affected families in Europe. *Child Care Healthy Development, 26*, 29.

Tiffany, J.S., Brea, M., Eckenrode, J.J., Friedrichs, E.K., Peters, R.M., Richards-Clarke, C. et al. (2006). Developing a community-based participatory research project with New York City Youth to explore participatory HIV prevention strategies, social connectedness, and HIV risk reduction. Presented at the XVI International AIDS Conference, Toronto, Canada, August 13–18 [abstract no. CDD1417].

Walsh, B.S. & Ross, L.F. (2003). Are minority children under- or over-represented in pediatric research? *Pediatrics, 112(4)*, 89–95.

Twomey, J.G., Jr. (1994). Investigating pediatric HIV research ethics in the field." *Western Journal of Nursing Research, 16(4)*, 404–413.

UNAIDS. (2006). *AIDS Epidemic Update, December, 2006*. Last accessed January 19, 2007; Available at http://data.unaids.org/pub/EpiReport/2006/2006_EpiUpdate_en.pdf.

Wolf, L.E., Zandecki, J, & Lo, B. (2005). Institutional review board guidance on pediatric research: missed opportunities. *Journal of Pediatrics*, 147(1), 84–89.

Wolthers, D.O. (2006). A questionnaire on factors influencing children's assent and dissent to non-therapeutic research. *Journal of Medical Ethics, 32*, 292–297.

Case Study Six
Informed Consent and Assent with Juvenile Detainees

Andrea Boyd, Ph.D., Geri Donenberg, Ph.D., and Kristi Jordan, Ph.D.

Background

The process of informed child assent and parental permission to engage in research can be more complex than simply explaining the procedures and risks of a study and obtaining agreement to participate. Depending on the population, the researcher may face several ethical, and in some cases, legal challenges. Juvenile detainees in particular have unique needs, because they are vulnerable on multiple levels; many detained teens have significant mental health disorders that impair their decision making process, histories of abuse and neglect, minimal or no family involvement, and little education, and they are especially susceptible to coercion if they believe that participating in research will favorably affect their legal disposition. In most cases, youths under age 18 years may not consent to participate in research without permission from a legal guardian. However, there are circumstances in which it may be difficult to obtain parental consent (e.g. run away teens, estranged teens, state wards, etc.), or where obtaining parental consent may put youths at undue harm (e.g., HIV prevention intervention for gay youth who have not disclosed their sexual preference to parents). In each case, it is the researcher's responsibility to identify and adapt their procedures to the specific needs and circumstances of the population under investigation.

This case study describes a collaborative process to design an informed consent and assent procedure for an HIV and substance prevention program for adolescents detained at the Cook County Juvenile Detention Center in Illinois. Three primary considerations influenced our understanding, development, and implementation of the assent process with detainees, each of which is described below. Of note, at each step along the way, we consulted stakeholders and legal advisors with diverse expertise with detained youth. In brief, we examined the feasibility, strengths, and risks related to seeking parental permission for youth to participate in the intervention, strategies to ensure that teens' best interests were represented, and approaches to communicate and emphasize to youths' their rights. We believe our final approach reflects a meaningful collaboration among invested parties and a thoughtful consideration of how best to involve adolescent detainees in intervention-oriented research while protecting their rights, avoiding coercion, and potentially affecting change in a public health epidemic, HIV/AIDS.

The Research Study

The goals of our exploratory federally-funded study are to develop and pilot an HIV and substance abuse prevention program tailored to the specific needs of juvenile offenders. Participants range in age from 12 to 17 years old and are detained at a large temporary juvenile detention center in Chicago. Days prior to the intervention, youths are asked to complete a computerized assessment asking about their personality characteristics, family relationships, peer relationships, sexual behavior, substance use, and psychiatric symptoms. Next, four groups, each consisting of 8 to10 youths, attend a 6-session intervention conducted over one week. The timeframe for the intervention is intended to maximize retention and minimize attrition due to short stays at the detention center. The program is designed to increase knowledge about HIV/AIDS transmission and prevention, reduce and eliminate substance use and risky sexual behavior, enhance perceptions of vulnerability, develop safer sex behaviors and assertive communication skills, and teach effective emotion regulation. Additionally, youths are asked to release confidential health information regarding their sexually transmitted infections (STIs) and psychiatric history. These data will inform intervention development, illuminate mediators and moderators of risk, and evaluate behavioral outcomes.

Ethical Issues

Obtaining Parental Consent for Research with Juvenile Offenders

A important caveat of conducting research with detained youths is that all research procedures and progress are inextricably tied to the cooperation and collaboration of multiple agencies and stakeholders, most importantly the detention center's administration and staff. As a result, one of our first steps involved approaching the detention center administration and engaging key members in a dialogue about the study.

Initially, we proposed enrolling detainees after obtaining parental permission and child assent, but we were advised that this would exclude a substantial number of needy and highly deserving youths. Namely, two segments of the detention population do not have a parent, and together they represent half of all detained youths: (1) wards of the state (i.e., those in the custody of the Department of Child and Family Services), and (2) estranged youth (i.e., those no longer in the care of a parent or guardian due to strained or non-existing relationships).

We evaluated the implications of this reality and decided that juveniles lacking familial involvement were among the most in need of HIV and substance use prevention services. We determined that excluding them would present unfair treatment to half of the detained population, thereby limiting generalizability and posing an important ethical dilemma. In addition, our collaborators at the detention center felt that all detained youth regardless of parental involvement should have the same opportunities to participate in the study as any other youth. We consulted the guidelines for research involving prisoners (CR 46.305 a), and they state that selection for any services should be fair to all prisoners. In sum, we faced a critical ethical dilemma: Requiring parental permission to enroll youth in the study would exclude a significant number of teens, unfairly penalize detainees who lacked family involvement, and fail to evenly distribute the opportunity among youths. In the end, we decided that every detainee should be afforded the same opportunity to participate in the research study, and we began to explore alternatives to parental consent.

Seeking Alternatives to Parental Consent for Special Populations

As a guide for alternatives to parental consent, we consulted the Department of Health and Human Services (DHHS) research guidelines (see 45 CRF 46.116 (c) or 46.177 (c). The guidelines indicate that parental or guardian permission can be waived if the study focuses on a condition for which parental permission is not a reasonable requirement to protect the children (e.g., treatment of STIs, mental health care), or if the study includes youths for which parental permission is not a reasonable requirement to protect the children (e.g., abused or neglected children). In both situations, an alternative to parental permission must be available to ensure that the rights of juveniles are protected and does not interfere with federal, state, or local law. (See 45 CRF 46.408 (c)).

We pursued an alternative to parental consent at two levels, (1) minors' ability to receive specialized services without parental approval, and (2) employing a "participant advocate" to ensure youths' rights and welfare. To the first point, minors as young as 12 years old in Illinois are able to consent for treatment or services related to reproductive health, HIV/sexually transmitted infections, substance use, and mental health. These laws do not specifically include or exclude prevention interventions, but the focus of our program on HIV and substance use led us to inquire as to whether our intervention might qualify within the boundaries of adolescents' ability to consent for services without parental permission.

Second, we investigated the terms for securing a waiver of parental permission by putting in place an alternative mechanism to ensure participant decisional capacity and the protection of youths' rights. Similar to previous research involving detained youths, we employed a "participant advocate" as a substitute for parental permission. The participant advocate is a neutral third party whose role is to protect the rights of the participant, verify the participant's comprehension of the study's purpose and procedures, and help ensure that participants are willing volunteers (and not coerced into study enrollment). Child advocacy groups commonly serve in this capacity.

To determine the best course of action, we sought the advice of stakeholders and agencies involved in decision making around detained youth (e.g., attorney, therapists, social workers, guardian ad litem). All detainees have legal representation whose sole purpose is to ensure their client's legal rights are protected. As such, we contacted the Public Defenders Office (PDO) and met with the Chief Juvenile Public Defender to discuss the purpose and goals of the research and to request feedback about the best way to obtain informed consent and assent for detainee participation. We also invited the PDO to ask questions and express concerns regarding study procedures. The Chief Juvenile Public Defender was highly favorable toward the goals and purpose of the study and reiterated the significant need for HIV and substance use prevention programs for detained youth. However, she also expressed concern about the confidentiality of information provided by youths, and whether anyone outside the study would have access to detainee's data. She noted that youths' responses to questions and potential disclosures might place them at risk for additional criminal charges leading to negative consequences related to their legal disposition.

In all of our work, we use several strategies to protect participant confidentiality. These procedures are required prior to any research activity, and they are overseen by the University Institutional Review Board and the National Institutes of Health. In addition to these efforts, we obtained a Certificate of Confidentiality from the federal government that explicitly states that no information gathered as a result of research participation can be released. What our current procedures did not address was the likelihood that detained youth might disclose information during the study that required mandatory reporting by health practitioners, such as in the case of child abuse. As an example, if a teen reported a sexual offense typically reportable under the law, would the researcher report it, thereby forcing the youth to be charged anew and causing negative outcomes (including a lifetime registration on the national list of sex offenders) as a consequence of his/her participation in the study? The additional conviction of a sexual offense would permanently alter the youth's life trajectory.

This issue raised an important ethical dilemma for us. On the one hand, if we were to concede and agree to employ non-mandated reporters in all research activities (thereby circumventing the technicalities of the reporting law), we would lose an opportunity to stop injurious behaviors, provide services for the offending youth, and protect potential victims. We may also inadvertently condone sexual offending by not addressing it. On the other hand, if we were to stand firm, the study would not proceed because the PDO would withdraw its support. Moreover, it became unclear to us whose rights we were most responsible to protect – our research participants or their potential victims?

Several considerations influenced our decision to adapt study procedures in order to go forward with the study. First, although we might have been able to proceed without the cooperation of the PDO, we opted not to do so. We did not want to alienate the PDO or create an adversarial relationship with potential collaborators. In addition, it was clear that the PDO would advise their youth clients against participation in the study, thereby significantly affecting the utility and generalizability of the findings. Second, we decided our first obligation was to our research participants in accordance with the Hippocratic Oath to "First, do no harm." The real potential for enrolled youths to receive additional charges as a consequence of their participation in the study was contrary to our intended goals and against human subjects guidelines for the protection of research subjects. We decided to adapt our procedures to minimize as much as possible any likelihood that participation in our study could lead to negative consequences. Third, we wanted to be sensitive to the context in which we were conducting the study. Our intervention includes a discussion of illegal behavior (i.e. drugs, sex, antisocial acts), and a participant could inadvertently implicate him or herself in a crime or disclose information that we would have to report to Child Protective Services.

After extensive consideration and discussion with the PDO, both parties agreed on the following strategies to protect youth and yet permit the ethical conduct of our research. (1) Prior to data collection, participants would be reminded of topics that required mandated reporting (child abuse/neglect, elder abuse, homicidal thoughts, and sex with a minor). The researcher would emphasize the potential implications of such disclosures on teens' legal dispositions. (2) While no specific questions inquire about sexual partners' identifying information, the researcher would instruct youths explicitly NOT to share demographic information about their sexual partners (name, age, address, etc.). Reports require such information, and without it, there is no way to proceed. (3) If youths initiate a discussion that may provide incriminating information, research staff would stop the discussion and remind the youth of the limitations of confidentiality. (4) If a teen opted to disclose the information despite our precautions, the researcher would then make a report as mandated by law.

While these procedures met the legal requirements and were approved by the PDO, we felt they came up short on our ethical obligations. Thus, we decided to implement two additional strategies in order to provide an opportunity for the youth to obtain needed services and clinical intervention related to the offense. First, research staff would encourage youths to speak with their attorney about the issue. In this way, a responsible yet non-mandated reporting adult could address the issues more openly and follow up on the offending behavior. The attorney could also address the issue with the family and arrange services for the youth. Second, research staff would ask minors for permission to inform their attorney about some topics that emerged during the study. In this way, we hoped to ensure continuity and follow-up about the offense. The PDO worked closely with us to design these procedures, and we all agreed with the proposed plan.

Conducting applied clinical research within the community has important implications for subject consent and assent, especially for participants with vulnerabilities (e.g., juvenile detainees). Community based research requires investigators to navigate contexts with their own set of rules, regulations, agendas, and politics. Our discussions with detention center officials and the Chief Juvenile Public Defender were extremely helpful in shaping our assent procedures. Their input directly influenced our decisions to incorporate alternative approaches to assent in lieu of parental consent and to implement safeguards that minimize potentially significant negative consequences for detained youths. Collaborating with the PDO proved to be instrumental in creating both a legal and ethical process for informed consent/assent to participate in research.

Chapter 13
HIV-Related Research
with Cognitively Impaired Persons

Introduction

Issues related to cognitive impairment may arise in the context of HIV research in several ways. First, individuals who have a pre-existing diagnosis of mental illness and may episodically suffer from diminished cognitive capacity may be at increased risk of HIV or may be HIV-infected. A study by Carey et al. (1995) found from an examination of aggregated studies that the prevalence of HIV among severely mentally ill persons in the U.S. is 5% among females and 10% among males, compared to a prevalence of 0.24 to 0.35% in the general population (Rosenberg, 1995). Reports indicate that the rate of HIV among severely mentally ill persons ranges from 3% to 23%, or between 8 and 70 times higher than the national rate (Carey et al., 1995; Cournos and McKinnon, 1997).

This increased risk of HIV infection among severely mentally ill persons has been attributed to various factors. Lower levels of disease severity appear to increase the likelihood of being sexually active. Although greater symptom severity, and sexual dysfunction associated with psychotropic medications (Fuller and Sajatovic, 2005) may reduce individuals' desire for sexual activity, they may be more likely to engage in survival sex due to an inability to manage resources; as a result, they may be increasingly vulnerable to coercion (McKinnon, Cournos, Herman, 2002). Individuals' levels of affective instability and behavioral impulsivity are also relevant; a correlation between higher excited symptoms and number of sexual episodes has been noted (McKinnon, Cournos, Herman, 2002).

Previous studies of the risk of HIV among severely mentally ill persons have found that many individuals may be at increased risk due to little fear of HIV (Sacks et al., 1990) and a low perception of personal risk (Carey et al., 1997). Cognitive impairments associated with illnesses such as schizophrenia or bipolar disorder may negatively affect an individual's ability to make decisions in their own best interest or appropriately weigh the risks vs. benefits of some types of sexual behaviors (Young, Zakzanis, Bailey et al., 1998). The prevalence of homelessness among severely mentally ill persons may be as high as 45% (Kalichman, 1994), while the prevalence of HIV infection among severely mentally ill homeless persons has been found to be as high as 19% (Susser et al., 1993). Sexual relationships in these circumstances are often transient and casual, and may be offered in exchange for shelter and/or food (Kalichman et al. 1996).

HIV-related research may be undertaken with individuals with a pre-existing diagnosis of mental illness for a variety of reasons. Investigators may wish to understand why mentally ill persons are at elevated risk of HIV transmission or to develop an efficacious and effective HIV prevention intervention.

Second, individuals who are HIV-infected may experience a diminution in cognitive ability and consequent impairment due to a condition or disease that post-dates the onset of their HIV infection and that is independent of the HIV. For instance, an individual may develop a mental illness, as defined by the *Diagnostic and Statistical Manual* (American Psychiatric Association, 2002), or experience a trauma or illness, such as a traumatic brain injury resulting from an accident. In some cases, an individual may have already been participating in HIV-related research prior to the onset of the cognitive impairment, such as might result from a traumatic injury. In this situation, significant ethical issues may arise as to whether the individual should be or can be continued as a participant in the research and whether the balance of the risks and benefits of study participation has changed so as to warrant or mandate the discontinuation of his or her participation.

In yet other situations, an HIV-infected individual may lack capacity to make decisions due to other personal circumstances. This might include being under the influence of alcohol or another substance, or as the result of fear or shock upon learning of their HIV status (Wolf and Lo, 2004).

Finally, as a result of the progression and impact of HIV disease, HIV-infected individuals may develop HIV-associated disease, such as HIV dementia, that affects their capacity to make decisions (Wolf and Lo, 2004). In some cohorts the prevalence of HIV dementia continues to be high, and there is some indication in the literature that women may be at an increased risk of developing HIV-associated cognitive impairment (McArthur, 2004). It may be important to involve such individuals in HIV-related research in order to investigate interventions to reduce the impact of the cognitive impairment on the individual's functional ability or interventions to reduce the progression of the impairment.

Regardless of the origin of the individual's cognitive impairment, significant ethical issues exist in the context of research with cognitively impaired persons. Many of these issues relate to the informed consent process, such as the capacity of the prospective research participant to provide or withhold his or her consent to participation, the ability of the individual to understand the information that is provided to him or her by the research team, the designation of a surrogate for consent to participate and the standard to be used by the surrogate in providing or withholding that consent, and the standard by which to assess and balance the risks and benefits that may result from participation. The resolution of these issues in a specific context may be rendered even more difficult as individuals age and develop additional conditions that may impact their ability to understand and process information, such as stroke, dementia, hearing loss, and vision loss, and/or encounter circumstances, such as placement in a nursing home or an assisted living situation, that reduce their actual or perceived ability to participate in research free of coercion or duress.

Conducting research with individuals suffering from forms of cognitive impairment is critical if we are to improve our understanding of how best to reduce HIV risk and HIV transmission among individuals with cognitive impairment; how to reduce the risk of developing HIV-associated cognitive impairment; and how to ameliorate the impact of HIV-associated cognitive impairment when it does occur. Yet past history in the United States (*Kaimowitz v. Michigan Department of Health*, 1976; *Valenti v. Prudden*, 1977; Lubasch, 1982; *Scott v. Casey*, 1983; Bein, 1991; Rothman, 1991; Advisory Committee on Human Radiation Experiments, 1996; Garnett, 1996) and in other countries (Bloch, Chodoff, and Green, 1999) demonstrates the vulnerability of cognitively impaired, mentally ill, and institutionalized persons to abuse in research. An outright prohibition against the participation of mentally ill individuals in research would shield them from the potential for such abuse, but would also result in a loss of their individual autonomy and possibly exacerbate their societal isolation and stigmatization. Such a prohibition would also deprive future generations of important scientific knowledge critical to the amelioration and/or prevention of the disease and the improvement of care. Consequently, individuals suffering from cognitive impairment may face the dangers of both exploitation and overprotection; our challenge is to foster such research while simultaneously protecting mentally ill research participants from potential exploitation and abuse.

The Requirement of Informed Consent

Ethically, researchers are required to obtain the informed consent of an individual in order to enroll an individual into a study. This ethical requirement derives from several international documents, including the Nuremberg Code and the Helsinki Declarations, and has been incorporated into research guidelines promulgated by the Council for International Organizations of Medical Sciences (2002, 2005). In addition, the provisions of various international conventions and protocols arguably prohibit the conduct of research involving human participants without their informed consent. For instance, the Universal Declaration of Human Rights (1948) provides in Article 5 that "No one shall be subjected to torture or to cruel, inhuman or degrading treatment or punishment."

These documents, however, provide relatively little specific guidance to researchers who wish to conduct studies with cognitively impaired individuals, including those whose decisionmaking abilities are diminished due to mental illness. Scholars have delineated four elements that must be present for an consent process to be informed and valid: (1) the individual from whom consent is to be obtained must be given the information necessary to make a decision; (2) the individual must understand the information; (3) the prospective participant must have the capacity to consent; and (4) the consent of the individual to participate must be voluntary (Faden & Beauchamp, 1986; Meisel, Roth, & Lidz, 1977). It cannot be emphasized enough that informed consent is a process that continues from the time of recruitment and enrollment throughout the study; it is not and should not be construed as the mere presentation to and signing of a document by the prospective research participant.

Enhanced procedures during this informed consent process may be ethically required to ensure that research participants who have a cognitive impairment are able to provide valid informed consent. Still other enhancements may be required to protect those who are suffering from cognitive impairment due to their mental illness, resulting in their heightened vulnerability. Vulnerable participants are those individuals with "insufficient power, prowess, intelligence, resources, strength or other needed attributes to protect their own interests through negotiations for informed consent" (Levine, 1988: 72). However, the mere fact of having been diagnosed with a mental illness should not serve as the basis for automatically assuming that the individual lacks capacity (National Bioethics Advisory Commission, 1998). These additional enhancements and protections are discussed below in the context of assessing capacity, providing information, ensuring understanding, and voluntariness. Additional considerations of confidentiality and the balancing of risks and benefits that are relevant informed consent are also discussed.

Assessing Capacity to Consent

The terms *capacity* and *competence* are often used synonymously, but they actually represent distinct concepts. The term *capacity* is used here to refer to an individual's decisionmaking ability. In contrast, the term *competence* reflects a legal judgment that an individual has a minimal level of mental, cognitive, or behavioral functioning to perform or assume a specified legal role (Bisbing, McMenamin, & Granville, 1995; Loue, 2001). It is important to recognize that being diagnosed with a particular condition is "relevant to, but not determinative of, incapacity for informed consent" (High et al., 1994). For instance, the course of schizophrenia may fluctuate, so that there may be periods of time during which an individual is able to understand and to give legally valid consent.

In general, it is presumed at the commencement of research studies that a prospective participant has capacity to consent, unless there is some reason to believe that he or she does not or that the capacity to give consent may be limited in some way. However, if a study focuses on a disorder involving either permanent cognitive impairment, such as mental retardation, progressive impairment, such as Alzheimer's disease, or fluctuating impairment, such as bipolar disorder, an assessment of capacity should be conducted when an individual begins to participate in a study. Longitudinal studies with mentally ill participants may find two or more forms of impairment even within the same individual. For instance, individuals may experience fluctuating impairment due to the progression of their schizophrenia but, as they age, they may develop Alzheimer's disease, resulting in additional levels of progressive impairment. Because capacity and decisionmaking ability may vary during the course of the study, depending upon the length of the study and the progression of the disorder or disease, it is also recommended that assessments of capacity and decisionmaking ability be conducted periodically during the course of an individual's participation in research, unless that participation is of very short duration.

It is critical that the conditions under which capacity is to be assessed maximize the likelihood that an accurate finding will be achieved. First, it is important that the individual who is to assess capacity be matched appropriately with the prospective research participant (Kennedy, 2000). For instance, an HIV-infected woman with a history of sexual abuse as a child and partner violence as an adult may continue to be intimidated by males in a position of authority and power and may be less forthcoming when interviewed by a male research team member than she might be with a female. It has been suggested by some commentators that the assessment and monitoring of an individual's capacity to consent and to participate in a study is best done by the research team of a study in collaboration with family members (Keyserlingk et al., 1995). Four exceptions to this basic premise have been noted: (1) when project staff does not have the requisite skill to assess or monitor the participants' capacity; (2) when there is a strong danger of conflict of interest; (3) when the individual had previously executed an advance directive for research while he or she still had capacity, but the document requires interpretation; and (4) when the protocol does not have the potential to confer a direct benefit on the participant and it involves more than minimal risk.

An individual's ability to respond to questions posed or to perform well on a test of cognitive ability may also be impacted by iatrogenic and institutional factors (Kennedy, 2000). The individual's ability to concentrate or his or her level of awareness may be affected by his or her medications. Individuals accustomed to the regimentation associated with institutionalization may become confused or frightened with a change in routine; absent a careful assessment, signs of that confusion may be mistaken for signs of diminished capacity. Environmentally induced stress, such as sleep deprivation and recent bereavement, resulting in depression and a decline in functional ability, may also adversely impact the individual's decisionmaking ability (American Psychiatric Association, 2000). Physiological causes, such as fluctuations in the blood sugar of individuals with diabetes, sodium deficiency, and electrolyte imbalances, such as might occur with HIV-associated chronic diarrhea, can also affect cognition. Because a determination of (in)capacity is so complex, it has been suggested that a determination of (in)capacity be verified through reliance on second opinions or the services of individuals who are consent specialists (Bonnie, 1997).

Providing Information

The following information should be provided to all research participants during the informed consent process: (1) a statement that the study involves research, an explanation of the purposes of the research, the expected duration of the subject's participation, a description of the procedures required for participation, and the identification of any procedures which are experimental; (2) a description of any reasonably foreseeable risks or discomforts to the research participant; (3) a description of any benefits from the research that may be reasonable expected for the research participant or others; (4) a disclosure of appropriate alternative procedures or courses of treatment, if any, that might be advantageous to the research participant; (5) a statement describing the extent to which confidentiality of records identifying the research participant will be maintained; and (6) for research involving more than minimal risk, an explanation as to whether any compensation or any medical treatments are available if injury occurs and, if so, what they consist of, or where further information may be obtained; (7) an explanation of whom to contact for answers to pertinent questions about the research and the rights of research participants, and whom to contact in the event of a research-related injury to the research participant; and (8) a statement that participation is voluntary, that a refusal to participate will not involve any penalty or loss of benefits to which the research participant is otherwise entitled, and the participant may discontinue participation at any time without penalty or loss of benefits to which the subject is otherwise entitled.

In addition to these disclosures, it may be advisable to provide the following to research participants: (1) a statement that the particular treatment or procedure may involve risks to the research participant (or to the embryo or fetus, if the subject is or may become pregnant) which are currently unforeseeable; (2) anticipated circumstances under which the participation of a research participant may be terminated by the investigator without regard to the subject's consent; (3) any additional costs to the subject that may result from participation in the research; (4) the consequences of a participant's decision to withdraw from the research and procedures for orderly termination of participation by the subject; (5) a statement that significant new findings developed during the course of the research which may relate to the subject's willingness to continue participation will be provided to the subject; and (6) the approximate number of subjects involved in the study.

Ensuring Understanding

It is critical that individuals understand that they are participating in research and that the procedures that they will undergo may not yield any direct benefit to them. A number of studies have found that many research participants may not understand either that they are participating in research rather than receiving clinical care, or the nature of the procedures that they will undergo in conjunction with their participation (Park, Slaughter, Covi, & Kniffin, 1966; McCollum & Schwartz, 1969; Fletcher, 1973; Gray, 1975; Hassar & Weintraub, 1976; Howard, DeMets, & The BHAT Research Group, 1981; Riecken & Ravich, 1982). Research suggests that among severely mentally ill individuals, the ability to understand is related both to the level of psychopathology and the quality of the information that is presented (Benson et al., 1988).

The National Bioethics Advisory Commission of the United States has recommended that the informed consent procedure be tailored to the specific abilities of each individual participant to receive and process information (National Bioethics

Advisory Commission, 2001). For instance, in addition to cognitive impairment, some HIV-infected participants may have hearing or vision impairments that either pre-dated the onset of their HIV infection or that occurred subsequently as a result of any of a variety of circumstances, including traumatic injury or an opportunistic infection such as cytomegalovirus retinitis (CMV) (Henderly and Jampol, 1991). These sensory limitations may further impede the individuals' ability to understand the information in the form in which it might be presented; accommodations must be made for these limitations in order to ensure that potential research participants understand the substance of the information being presented.

A number of suggestions have been made to maximize understanding, including the use of a clear and simple presentation format for the information (Bergler, Pennington, Metcalfe, & Freis, 1980), the provision of sufficient time to enable the individual to process the information given to him or her (Morrow, 1978), and discussion of the information with the investigator or member of the research team (Williams et al., 1977). The individual may be asked to restate or summarize in his or her own words the information provided in order to confirm that he or she understood. Tailored questions, whether in multiple choice, true-false, or essay format, may be asked of the participant following the presentation of the information, to ascertain whether and how much the prospective participant understood of the information presented (McCollum & Schwartz, 1969; Hassar & Weintraub, 1976; Williams et al., 1977; Flanery et al., 1978; Roth et al., 1982; Bonnie, 1997). One commentator has suggested that a family member participate with the cognitively impaired member in the informed consent process to ensure understanding and provide concurrent consent (Bonnie, 1997).

Voluntariness

The life situation of many individuals with cognitive impairment may affect their ability to consent or to refuse consent to participate in research. One research study in the United States found that 21% of adults with serious mental illness live below the poverty threshold, compared with 9% of the general adult population (Barker et al., 1992). Many homeless individuals suffer from mental illness (Isaac and Armat, 1990). A lack of adequate medical care may be associated with the poverty and lack of stable housing that they experience (Douaihy, Stowell, Bui, Daley, and Salloum, 2005). Consequently, the possibility of participation in research with its attendant psychiatric and medical care may represent an otherwise unavailable and unattainable resource, leading individuals to disregard the risks that may be inherent in participation and to overemphasize the likelihood that they will obtain a direct benefit from their participation (National Bioethics Advisory Commission, 1998).

Some individuals may be dependent on others for their physical care, for attention to their personal needs, or for their medical care. This may occur, for instance, if an individual at risk for HIV has a cognitive impairment or in situations in which an HIV-infected individual is physically dependent on others due to the severity of the HIV disease and/or the HIV-associated cognitive impairment. In such circumstances, individuals may fear that a refusal to participate in a particular research study will result in the withdrawal of such assistance, a diminution in the quality of this assistance, or complete abandonment. This may be of particular concern to individuals living in institutions, such as nursing homes, hospice facilities, or mental hospitals (Annas and Glantz, 1997). Individuals may also be concerned that they will disappoint their caregiver or care provider if they refuse to participate (Sachs and Cassel, 1989). Some individuals may also believe that they would not have been offered the possibility of participation in a study unless the researcher believed that their participation would yield some clinical benefit to them personally. They may believe this despite all assertions by the research team that they may not receive any personal benefit from their participation and only future patients will derive any benefit from the new-found knowledge gained through the study. This misconception is known as the "therapeutic misconception" (Grisso and Appelbaum, 1998).

The provision of some kind of payment or other reward in exchange for an individual's participation in research may also raise concerns if the payment or reward interferes with the individual's ability to act autonomously in deciding whether or not to participate in a study. For instance, if the individual has experienced homelessness, an excessively large monetary payment offered in exchange for participation may be so great as to induce the individual to participate in a study even if he or she might prefer not to do so; the individual might believe that he or she could not refuse and had to do it (Beauchamp, Jennings, Kinney, & Levine, 2002). It has been suggested that, if a monetary incentive or reward is to be offered, the value should be sufficiently high so as not to exploit the participants by underpayment, and sufficiently low so as not to create an "irresistible inducement" (Beauchamp, Jennings, Kinney, & Levine, 2002).

Other Considerations

Confidentiality of the Data

The level of confidentiality protection of the information that is disclosed to the researcher may be of concern for a number of reasons. First, confidentiality may be difficult to maintain if interviews or other procedures are conducted in the context of an institutional residence, such as a hospice facility, group home or nursing home, due to the physical layout of the institution, a scarcity of private space, and the possibility that the participant may have impaired hearing ability, thereby requiring that the researcher to speak at a level that is audible to others (Cassel, 1985, 1988).

Assessing and Balancing Risks and Benefits

A decision relating to participation in a research protocol requires that the decision-maker, usually the prospective participant him- or herself, balance the risks and benefits of participation. A balancing of risks and benefits must also be done by the researcher proposing the study prior to its initiation and must also be conducted by the institutional review board (IRB) of the researcher's institution in its initial and continuing reviews of the research protocol.

Commentators have identified four categories into which research protocols may be classified: (1) research in which there is the potential for a direct therapeutic benefit to the participant and minimal risk is involved; (2) research in which the participant may obtain some direct therapeutic benefit, but more than minimal risk is involved; (3) research in which there is no expected benefit for the individual participating, but there is no more than minimal risk; and (4) research in which there is no expected therapeutic benefit to the participant and there is more than minimal risk (Kapp, 1998; LeBlang & Kirchner, 1996). "Minimal risk" is often interpreted to mean that the risks of participation are no greater than those that would be experienced in the everyday course of living (Levine, 1988). Examples of risks include the physiological effects of an experimental drug or procedure and increased levels of anxiety associated with study questions or procedures (Dresser, 2001).

Participation in a study may involve emotional risks for all individuals, but particularly for individuals suffering from a severe mental illness. Depending upon the nature of the study and duration of engagement between the research team and the participants, the participants may form strong emotional bonds to study staff. This could result in a blurring of boundaries, whereby the participant comes to believe that one or more staff members are friends (Ensign, 2003). A failure by the team member to fulfill the participant's expectations of a "friend" and/or the termination of the relationship due to termination of the study, could potentially lead to severe distress for the participant. It is incumbent upon the research team to foresee this possibility and to provide for adequate reminders of the research nature of the engagement and sufficient time for the participant to disengage. The case study by Méndez and Loue that follows this chapter described the involuntary termination of the participation of a woman in an HIV-related observational study as a result of her inability to understand and acknowledge boundaries with the study team and her resulting increased risk of traumatization.

Emotional pain may also result from interviews conducted as part of a study, in which participants may be asked to recount particular painful experiences, such as childhood abuse or violence. The question has been raised as to whether "the objectification of highly charged emotional events [is] itself a form of violence" (Harvey & Gow, 1994: 2). It is critical that appropriate safeguards for the participants be in place to help participants deal with resulting emotional pain and/or trauma if they are to be asked to speak about painful or traumatic events (cf. Romero-Daza, Weeks, & Singer, 2003). Such safeguards could include referral to a community-based agency that can provide counseling services, with follow-up to assure that the agency has provided requested services, or referral to a colleague unrelated to the study for appropriate follow-up, at no cost to the participant.

Direct benefits may include short- or long-term improvement in the individual's condition, an improvement in the individual's symptoms, and the slowing of the degenerative process (Keyserlingk et al., 1995). Indirect benefits may include enhanced opportunities for social interaction, increased attention from health and ancillary health professionals, and a feeling of contributing in a way that may help others.

Research suggests that even when risks of study participation are divulged to prospective participants, individuals may have difficulty comprehending the risks. In one clinical trial of a drug, respondents were found to be well-informed about the study design and general risks of participation, but 39% were unable to enumerate specific minor side effects of the drug and 64% were unable to identify the serious risks of the medication that had been divulged to them (Howard, DeMets, & The BHAT Research Group, 1981). In yet another study, few of the respondents recognized the possibility of unknown risk, meaning that there could be risks that had not been anticipated prior to the initiation of the study (Gray, 1975). These data underscore the importance of assessing understanding during the initial informed consent process. Because a participant's individual situation may change during the course of a study, the risks and benefits of participation to a particular participant may also change during the course of a study. Consequently, it becomes important to reassess the risks and benefits of participation on an ongoing basis. However, there is no formula that will dictate how the benefits and risks of a particular individual's participation are to be weighed against each other. There is also no consensus among researchers or ethicists as to the level of risk or benefit that must be present for a surrogate decision maker to be able to consent to research participation by a cognitively impaired individual (Dresser, 2001).

Mechanisms for Expressing Choice During Incapacity

Surrogate Consent

Some have suggested that adults who lack capacity to consent should be able to participate in research through the consent of a surrogate. A question arises as to which individual(s) are best suited to be appointed as the surrogate decision makers. The National Alliance for the Mentally Ill (NAMI) in the United States has proposed family members as the most appropriate surrogates in a research context (Flynn, 1997). Many IRBs in the U.S. allow family members or friends to give consent (LeBlang & Kirchner, 1996).

The appointment of a surrogate decision-maker for HIV-infected individuals may be even more difficult, depending upon their personal circumstances and the legal and political context in which the potential surrogate must function. A number of states have adopted legislation which deprives all unmarried opposite-sex and/or same-sex couples of legal recognition as couples (Cox, 2005; Krotoszynski & Spitko, 2005). As a result, in some such jurisdictions, unmarried couples may not be able to rely on legislation that arguably confers benefits only on married individuals, such as the ability to designate an individual as an agent to make health care decisions, if that agent is not related by blood or marriage. This may be a particular problem in the context of HIV because of the high proportion of HIV-infected individuals whose primary relationships are with individuals of the same sex (Centers for Disease Control and Prevention, 2006).

Even if the appointment of a surrogate decision-maker is possible, there are dangers associated with having decisions made by a surrogate. First, family members may be inappropriate due to their own lack of capacity, unavailability, or inattention to the needs of the cognitively impaired individual (High, White-house, Post, & Berg, 1994). Second, the surrogate may act in his or her own interest, rather than that of the individual (Sachs, 1994). This may be an issue of special concern in the context of HIV-related illness because some HIV-infected individuals may have been estranged from their families for extended periods of time due to disagreements related to sexual orientation, sexual behavior, and/or substance use. In such situations, the family member may have little understanding of the individual's personal circumstances or preferences and may decide upon a particular course of action out of anger, guilt, or fear. Accordingly, it has been suggested that an appropriate surrogate be an individual who (1) is chosen, known, and trusted by the individual; (2) participates with the cognitively impaired individual in the informed consent process; (3) is familiar with the individual's medical and psychiatric history; (4) is familiar with the prodromal signs and symptoms indicative of a relapse, in the case of a mental illness diagnosis; (5) is informed about and is willing to assume the responsibilities of a surrogate decision maker; (6) is willing to overrule the individual's previously expressed desire to participate in research if the participation could adversely affect the individual; and (7) is willing and able to ensure appropriate medical and/or psychiatric follow-up care if needed (Backlar, 1998).

Assuming that a surrogate is able to decide for the individual who lacks capacity to decide for him- or herself, there remains the question of how the surrogate should make that determination. Two processes have been suggested in the United States: the best interest test and the substituted judgment test. The best interest test requires an assessment of what is in the individual's best interest at the time that the decision by the surrogate is to be made. This perspective allows a surrogate to more easily disregard any previously expressed desire or intent of the cognitively impaired individual because what was once expressed may no longer be in his or her best interest, as determined by the surrogate. The substituted judgment test requires that the surrogate decide the issue of research participation in a manner consistent with what the individual would have chosen for him- or herself if he or she had remained able to do so. This perspective allows the surrogate to preserve to a greater degree the psychological continuity between the once-capable then-self and the now-self. In situations in which an IRB permits reliance on the substituted judgment test, the IRB may require in addition to the surrogate's consent, the assent of the cognitively impaired individual to participate, meaning that, to the best of their ability, they must indicate some preference, although that indication does not rise to the level of legal consent (cf. Sachs et al., 1994).

Acknowledgements. This work was supported in part by the National Institute of Mental Health R01 MH-63016.

References

Advisory Committee on Human Radiation Experiments. (1996). Final Report. Washington, D.C: The Committee.

American Psychiatric Association. (2000). *Diagnostic and Statistical Manual of Mental Disorders, Fourth Edition, Text Revision.* Washington, D.C.: Author.

Annas, G.J., & Glantz, L.H. (1997). Informed consent to research on institutionalized mentally disabled persons: The dual problems of incapacity and voluntariness. In A.E. Shamoo (Ed.), *Ethics in Neurobiological Research with Human Subjects: The Baltimore Conference on Ethics* (pp. 55–79). Amsterdam: Gordon and Breach Publishers.

Backlar, P. (1998). Anticipatory planning for research participants with psychotic disorders like schizophrenia. *Psychology, Public Policy, and Law, 4,* 829–848.

Barker, P.R. et al. (1992). Serious mental illness and disability in the adult household population: United States, 1989. In R.W. Manderscheid, M.A. Sonnenschein (Eds.), *Advance Data from Vital and Health Statistics of the National Center for Health Statistics,* No. 218. Washington, D.C.: Department of Health and Human Services.

Beauchamp, T.L., Jennings, B., Kinney, E.D., & Levine, R.J. (2002). Pharmaceutical research involving the homeless. *Journal of Medicine and Philosophy, 27(5),* 547–564.

Bein, P.M. (1991). Surrogate consent and the incompetent experimental subject. *Food, Drug, & Cosmetic Law Journal, 46(5),* 739–771.

Benson, P.R., Roth, L.H., Appelbaum, P.S., Lidz, C.W., & Winslade, W.J. (1988). Information disclosure, subject understanding, and informed consent in psychiatric research. *Law and Human Behavior, 12(4),* 455–475.

Bergler, J.H.., Pennington, A.C., Metcalfe, M., & Freis, E.D. (1980). Informed consent: How much does the patient understand? *Clinical Pharmacology & Therapeutics, 27,* 435–440.

Bisbing, S., McMenamin, J., & Granville, R. (1995). Competency, capacity, and immunity. In ACLM Textbook Committee (Ed.), *Legal Medicine,* 3rd edition (pp. 27–45). St. Louis, Missouri: Mosby-Year Book.

Bloch, S., Chodoff, P., & Green, S. (Eds.). (1999). *Psychiatric Ethics.* New York: Oxford University Press.

Bonnie, R.J. (1997). Research with cognitively impaired subjects: Unfinished business in the regulation of human research. *Archives of General Psychiatry, 54(2),* 105–111.

Carey, M.P., Carey K.B., & Kalichman, S.C. (1997). Risk for human immunodeficiency virus (HIV) infection among persons with severe mental illnesses. *Clinical Psychology Review, 17,* 271–291.

Carey, M.P., Weinhardt, L.S., & Carey, K.B. (1995). Prevalence of infection with HIV among the seriously mentally ill: Review of the research and implications for practice. *Professional Psychology: Research and Practice, 26,* 262–268.

Cassel, C. (1985). Research in nursing homes: Ethical issues. *Journal of the American Geriatrics Society, 33,* 795–799.

Cassel, C. (1988). Ethical issues in the conduct of research in long term care. *Gerontologist, 28,* 90–96.

Centers for Disease Control and Prevention. (2006). Basic statistics. Available at http://www.cdc.gov/hiv/topics/surveillance/basic.htm#exposure. Last accessed April 3, 2006.

Cournos, F. & McKinnon, K. (1997). HIV seroprevalence among people with severe mental illness in the United States: A critical review. *Clinical Psychology Review, 17*, 259–269.

Cox, S.E. (2005). Interjurisdictional recognition of civil unions, domestic partnerships, and benefits: article: red states, blue states, marriage debates. *Ave Maria Law Review, 3*, 637–656.

Douaihy, A.B., Stowell, K.R., Bui, T., Daley, D., & Salloum, I. (2005). HIV/AIDS and homelessness: Part 1: Background and barriers to care. *The AIDS Reader, 15(10)*, 516–520.

Dresser, R. (2001). Dementia research: Ethics and policy for the twenty-first century. *Georgia Law Review, 35*, 661–690.

Ensign, J. (2003). Ethical issues in qualitative health research with homeless youths. *Journal of Advanced Nursing, 43(1)*, 43–50.

Faden, R. & Beauchamp, T. (1986). *A History and Theory of Informed Consent*. New York: Oxford University Press.

Flanery, M., Gravdal, J., Hendrix., P. et al. (1978). Just sign here . . .*South Dakota Journal of Medicine, 31(5)*, 33–37.

Flynn, L.M. (1997). Statement. *Issues Concerning Informed Consent and Protections of Human Subjects in Research: Hearings before the Subcommittee on Human resources of the House Committee on Government Reform and Oversight*, 105th Cong.

Fuller, M. & Sajatovic, M. (2005). *Drug Information Handbook for Psychiatry*. 5th ed. Cleveland, Ohio: Lexi-Comp., Inc.

Garnett, R.W. (1996). Why informed consent? Human experimentation and the ethics of autonomy. *Catholic Lawyer, 36*, 455–511.

Gray, B. (1975). *Human Subjects in Medical Experimentation: A Sociological Study of the Conduct and Regulation of Clinical Research*. New York: Wiley.

Grisso, T. & Appelbaum, P. (1998). *Assessing Competence to Consent to Treatment: A Guide for Physicians and Other Health Professionals*. New York: Oxford University Press.

Harvey, P., Gow, P. (1994). Introduction. In P. Harvey, P. Gow (Eds.). *Sex and Violence: Issues in Representation and Experience* (pp. 1–17). New York: Routledge.

Hassar, M. & Weintraub, M. (1976). "Uninformed" consent and the wealthy volunteer: An analysis of patient volunteers in a clinical trial of a new anti-inflammatory drug. *Clinical Pharmacology & Therapeutics, 20*, 379–386.

Henderly, D.E. & Jampol, L.M. (1991). Diagnosis and treatment of cytomegalovirus retinitis. *Journal of Acquired Immune Deficiency Syndromes, 4(1)*, S6-S10.

High, D.M., Whitehouse, P.J., Post, S.G., & Berg, L. (1994). Guidelines for addressing ethical and legal issues in Alzheimer disease research: A position statement. *Alzheimer's Disease & Associated Disorders, 8 (Supp. 4)*, 66–74.

Howard, J.M., DeMets, D., & The BHAT Research Group. (1981). How informed is informed consent? *Controlled Clinical Trials, 2*, 287–303.

Isaac, R.J., Armat, V.C. (1990). *Madness in the Streets: How Psychiatry and the Law Abandoned the Mentally Ill*. New York: Free Press.

Kaimowitz v. Michigan Department of Mental Health. (1973). *U.S. Law Week, 42*, 2063 (Circuit Court, Wayne County, Michigan).

Kalichman, S.C., Carey, M.P., & Carey, K.B. (1996). Human immunodeficiency virus (HIV) risk among the seriously mentally ill. *Clinical Psychology: Science and Practice, 3*, 130–143.

Kalichman, S.C., Kelly, J.A., Johnson, J.R., & Bulto, M. (1994). Factors associated with risk for HIV infection among chronic mentally ill adults. *American Journal of Psychiatry, 151(2)*, 221–227.

Kapp, M. (1998). Decisional capacity, older human research subjects, and IRBs: Beyond forms and guidelines. *Stanford Law & Policy Review, 9*, 359–365.

Kennedy, G.J. (2000). *Geriatric Mental Health Care: A Treatment Guide for Health Professionals*. New York: Guilford Press.

Keyserlingk, E.W., Glass, K., Kogan, S., & Gauthier, S. (1995). Proposed guidelines for the participation of persons with dementia as research subjects. *Perspectives in Biology & Medicine, 38*, 319–361.

Krotoszynski, R.J. Jr. & E. Gary Spitko, E.G. (2005). Navigating dangerous constitutional straits: A prolegomenon on the federal marriage amendment and the disenfranchisement of sexual minorities. *University of Colorado Law Review, 76*, 599–652.

LeBlang, T.R. & Kirchner, J.L. (1996). Informed consent and Alzheimer disease research: Institutional review board policies and practices. In R. Becker & E. Giacobini (Eds.), *Alzheimer's Disease from Molecular Biology to Therapy* (pp. 529–534). Boston: Birkhauser.

Levine, R. J. (1988). *Ethics and Regulation of Clinical Research*. New Haven, Connecticut: Yale University Press.

Loue, S. (2001). Elder abuse and neglect in medicine and law: The need for reform. *Journal of Legal Medicine, 22*, 159–209.

Lubasch, A.H. (1982).Trial ruled in 1953 death case. *N.Y. Times, Sept. 14*, A-14.

McArthur, J.C. (2004). HIV dementia: An evolving disease. *Journal of Neuroimmunology, 157*, 3–10.

McCollum, A.T. & Schwartz, A.H. (1969). Pediatric research hospitalization: Its meaning to parents. *Pediatric Research, 3*, 199–204.

McKinnon K, Cournos F, & Herman R. (2002). HIV among people with chronic mental illness. *The Psychiatric Quarterly, 73*, 17–31.

Meisel, A., Roth, L.H., & Lidz, C.W. (1977). Toward a model of the legal doctrine of informed consent. *American Journal of Psychiatry, 134*, 285–289.

Morrow, G., Gootnick, J., & Schmale, A. (1978). A simple technique for increasing cancer patients' knowledge of informed consent to treatment. *Cancer, 42*, 793–799.

National Bioethics Advisory Commission. (1998). *Research Involving Persons with Mental Disorders That May Affect Decisionmaking Capacity*. Rockville, Maryland: National Bioethics Advisory Commission.

Park, L.C., Slaughter, R.S., Covi, L., & Kniffin, H.G., Jr. (1966). The subjective experience of the research patient: An investigation of psychiatric outpatients' reactions to the research treatment situation. *Journal of Nervous & Mental Disease, 143*, 199–206.

Riecken, H.W., & Ravich, R. (1982). Informed consent to biomedical research in Veterans Administration hospitals. *Journal of the American Medical Association, 248(3)*, 344–348.

Romero-Daza, N., Weeks, M., & Singer, M. (2003). "Nobody gives a damn if I live or die": Violence, drugs, and street-level prostitution in inner-city Hartford, Connecticut. *Medical Anthropology, 22*, 233–259.

Rosenberg, P.S. (1995). Scope of the AIDS epidemic in the United States. *Science, 270*, 1372–1375.

Roth, L.H., Lidz, C.W., Meisel, A., Soloff, P.H., Kaufman, K., Spiker, D.G., et al. (1982). Competency to decide about treatment or research: An overview of some empirical data. *International Journal of Law and Psychiatry, 5*, 29–50.

Rothman, D.J. (1991). *Strangers at the Bedside: A History of How Law and Bioethics Transformed Medical Decision Making*. New York: Basic Books.

Sachs, G. & Cassel, C. (1989). Ethical aspects of dementia. *Neurologic Clinics, 7*, 845–858.

Sacks, M.H., Perry, S., Graver, R., Shindledecker, R., & Hall, S. (1990). Self-reported HIV-related risk behaviors in acute psychiatric inpatients. *Hospital & Community Psychiatry, 41,* 1253–1255.

Scott v. Casey. (1983). 562 F. Supp. 475 (N.D. Ga.).

Susser, E., Valencia, E., & Conover, S. (1993). Prevalence of HIV infection among psychiatric patients in a New York City men's shelter. *American Journal of Public Health, 83,* 568–570.

Universal Declaration of Human Rights. (1948). G.A. Res. 217A (III), U.N. Doc. A/810 at 71.

Valenti v. Prudden. (1977). 58 A.D.2d 956, 397 N.Y.S.2d 181.

Williams, R.L., Rieckmann, K.H., Trenholme, G.M., Frischer, H., & Carson, P.E. (1977). The use of a test to determine that consent is informed. *Military Medicine, 142,* 542–545.

Wolf, L.E. & Lo, B. (2004). Informed consent in HIV research. *FOCUS, 19(4),* 5–6.

Young, D.A., Zakzanis, K.K., Bailey, C., Davila, R., Griese, J., & Sartory, T.A. (1998). Further parameters of insight and neuropsychological deficit in schizophrenia and other chronic mental disease. *Journal of Nervous and Mental Disease, 186(1),* 44–50.

Case Study Seven
Balancing Risks and Benefits in Research with Severely Mentally ill Participants

Sana Loue, J.D., Ph.D., M.P.H., Martha Sajatovic, M.D.,
Nancy Méndez, and Ingrid Vargas

Introduction

This observational study was designed to improve our understanding of the context of HIV risk among severely mentally ill Mexican and Puerto Rican women. Hispanics have been disproportionately impacted by HIV/AIDS. Although comprising 13.7% of the U.S. population in 2003, Hispanics accounted for 20.3% of new AIDS cases and 21.7% of all AIDS-attributable deaths that year (Centers for Disease Control and Prevention, 2003a, b, 2004a). In 1996, HIV became the second leading cause of death for Latinas aged 25 to 44 (United States Department of Health and Human Services, 1998). Additionally, a trend has been observed of increasing AIDS cases among foreign-born Hispanic men and women and heterosexual U.S.-born Hispanics (Klevens, Diaz, Fleming, Mays, and Frey, 1999). HIV infection among Latinos has been attributed primarily to heterosexual transmission and injection drug use (Centers for Disease Control and Prevention, 2002).

Numerous barriers to HIV risk reduction and prevention have been identified. Hispanics have consistently demonstrated lower levels of HIV knowledge compared to other groups (Aruffo, Cloverdale, and Vallbona, 1991; Flaskerud and Nyamathi, 1989; Hingson et al., 1989; Kroliczak, 1989). Barriers to HIV education among Hispanics include language and cultural differences (Marin, 1989), low literacy levels (Rivera, Petty, Krepcho, and Haley, 1989), strong anti-homosexual (Burgos and Perez, 1986), and machista attitudes (Burgos and Perez, 1986; Marin, Tschann, Gomez, and Kegeles, 1993), a denial or minimization of risk (Mata and Jorquez, 1988), and an emphasis on traditional gender roles and differences (Sabogal, Perez-Stable, Otero-Sabogal, and Hiatt, 1995). Condom use by Hispanic women is typically low (Marin, Gomez, and Hearst, 1993; Marin and Marin, 1992; Marin, Tschann, Gomez, and Kegeles, 1993), particularly among foreign-born (Deren, Shedlin, and Beardsley, 1996) and less acculturated persons (Hines and Caetano, 1998; Marin and Marin, 1992). Compared to non-Hispanic white and African American women, Latinas have been found less likely to use condoms in the context of casual sexual relationships (Aversa, McCoy, Randall, and McBridge, 1998), although Puerto Rican women specifically have been

236

found less likely to use condoms in relationships of longer duration and greater emotional investment (Saul et al., 2000). The relative lack of condom utilization has been attributed to cultural norms (Marin and Marin 1990; May and Cochran, 1988) and the reduced likelihood that Hispanic men will use condoms with their primary sexual partners (Caetano and Hines, 1995; Marin, Gomez, and Hearst, 1993; Sabogal, Faigeles, and Catania, 1993). Among Puerto Rican women, low resource power as measured by level of education and (un)employment appear to be related to lower levels of condom use (Saul et al., 2000). Hispanic men have also been found to be more likely to have sexual contact with prostitutes, further increasing their risk for HIV (Kim, Marmor, Dubn, and Wolfe, 1993).

HIV prevalence among individuals with serious mental illness (SMI) has been found to be 5% among women and 10% among men, compared to 0.24% to 0.35% in the general population (Carey, Weinhardt, and Carey, 1995; Rosenberg, 1995). Individuals with SMI have chronic or relapsing mental disorders such as schizophrenia, bipolar disorder and major depression. The rate of HIV among individuals with SMI ranges from 3% to 23%, or between 8 and 70 times higher than the national rate (Carey, Weinhardt, and Carey, 1995; Cournos and McKinnon, 1997). Among individuals with SMI, the prevalence of HIV among Hispanics has been found to be three times that among whites (Cournos and McKinnon, 1997), reflecting the increased risk of HIV among Hispanics (Centers for Disease Control and Prevention, 2004b).

Lower levels of disease severity appear to increase the likelihood of being sexually active. Even though greater symptom severity and sexual dysfunction associated with psychotropic medications (Fuller and Sajatovic, 2005) may reduce individuals' desire for sexual activity, they may be more likely to engage in survival sex due to an inability to manage resources; as a result, they may be increasingly vulnerable to coercion (McKinnon, Cournos, and Herman, 2002). Individuals' levels of affective instability and behavioral impulsivity are also relevant; a correlation between higher excited symptoms and number of sexual episodes has been noted (McKinnon, Cournos, and Herman, 2002).

Increased risk may also be attributable to little fear of HIV and a low perception of personal risk (Carey, Carey, and Kalichman, 1997; Sacks, Perry, Graver, Shindledecker, and Hall, 1990). Cognitive impairments associated with SMI may negatively affect individuals' ability to make decisions in their own best interest or appropriately weigh the risks versus benefits of specific sexual behaviors (Young, Zakzanis, Bailey, Davila, Griese, and Sartory, 1998). The prevalence of homelessness among SMI persons may be as high as 45% (Kalichman, Kelly, Johnson, and Bulto, 1994), while the prevalence of HIV infection among SMI homeless persons has been found to be as high as 19% (Susser, Valencia, and Conover, 1993). Sexual relationships in these circumstances are often transient, casual, and may be offered in exchange for shelter and/or food (Kalichman, Carey, and Carey, 1996). Additionally, homeless women with SMI may be at increased risk for sexual abuse, assault, and rape, further increasing their risk of HIV transmission (Fisher, Hovell, Hofstetter, and Hough, 1995).

The Research Study

This group of investigators is conducting an on-going research study that focuses on HIV risk among Hispanic women with SMI, and have begun to identify preliminary critical issues in the context of HIV status/vulnerability among this population. Eligibility for participation in the study required a diagnosis of major depression, bipolar disorder, or schizophrenia; Puerto Rican ethnicity and residence in any of six enumerated counties of northeastern Ohio; Mexican ethnicity and residence in San Diego County or any of six enumerated counties of northeastern Ohio; and age between 18 and 50 years at enrollment. Information about the study was disseminated within these communities through presentations conducted in diverse settings (e.g., language and vocational classes) and the distribution of flyers to a broad spectrum of venues (e.g., churches, nightclubs, government assistance offices). We also contacted clinicians and counselors serving mentally ill Mexican and Puerto Rican women to apprise them of the study. The presentations and flyers advised individuals that we were conducting a study to understand how to reduce HIV risk among Puerto Rican and Mexican women who were *deprimida* (depressed), who had suffered from *ataques de nervios* (nervous or panic attacks), or who had emotional troubles, and provided contact information for additional information about the study. We used these terms to avoid stigmatizing or embarrassing interested individuals who might signify their interest in a public forum.

Interested individuals were provided with details about the study and were asked for their consent to a baseline interview to assess eligibility for study participation (see Methods, below). Eligible individuals were invited to join the study and were asked for their informed consent for an additional two-part baseline interview, a follow-up interview each year for two years, and 100 hours of shadowing. Shadowing required that the study ethnographers observe the participants in a variety of situations including interactions with mental health care providers, family members, and friends; everyday activities, such as grocery shopping; and, in some cases, hospitalizations. No situations involving questionable capacity to provide informed consent to study participation arose. Our study procedures were approved by the IRB.

Interviews were conducted by one of several trained bilingual female interviewers in English and/or Spanish, as indicated by the participant. Interviewers received training on questionnaire administration and the conduct of interviews, which were designed to maximize privacy and confidentiality. Individuals' qualifying diagnosis was made using the Structured Clinical Interview for Axis I DSM-IV Diagnoses (SCID). The two-part baseline interview following this assessment focused on acculturation level, migration history, demographic information, sexual and drug use histories, and HIV knowledge. Observations during shadowing supplemented these data and allowed us to understand participants' viewpoints regarding key concepts, such as risk, religiosity, spirituality, love, and commitment.

Shadowing required that the interviewer accompany each participant in the course of her life's activities. Interviewers observed similar situations among all study participants, including: interactions with romantic and/or sexual partners, children, and other family members; social situations; appointments with medical and mental health care providers; inpatient hospitalization episodes; everyday activities such as church attendance, grocery shopping, and interactions with children's school personnel; and interactions in the course of obtaining publicly funded services, e.g., Social Security. Some participants were shadowed while using drugs and/or approaching men to exchange sex for money or drugs.

All participants were aware when shadowing occurred, as were most third parties. In order to minimize safety risks to the participants and the interviewers, interviewers' true purpose was not revealed to persons selling drugs to the study participants or to those with whom they were engaging in other illegal activities. Interviewers were instructed not to participate in any way in observed illegal activities and to leave any situation in which their safety was at risk.

The Ethical Issue and Analysis

During the course of the study, two major issues arose that relate to participant risk: (1) the ethical obligations of the research team to intervene with participants who they see engaging in high HIV risk behaviors and (2) the circumstances under which the risks to the participant will be deemed to outweigh the benefits of participation so as to ethically warrant the involuntarily termination of that individual's participation in the study.

Intervening with Participants

The majority of our research participants engaged in high risk behaviors prior to their recruitment and enrollment into the study, and most continued these behaviors during the course of the study. These included unprotected sexual intercourse with multiple partners, unprotected sexual intercourse with one or more individuals known to be HIV seropositive, and/or shared use of injection equipment.

Attempts to intervene with their behaviors would transform the study from an observational study to an intervention study. However, such an "intervention" would be rendered on an ad hoc basis, absent a conceptual framework, a protocol, and an evaluation. Indeed, such an approach could unwittingly result in increased harm to the participants due to a lack of understanding of their life situations and the demands placed on them. For instance, advising some of our participants to demand that their partners use a condom could have resulted in an increased risk of partner violence.

In deciding not to intervene each time that we saw or learned of a participant's high risk behavior, we relied on the provisions of the *International Guidelines for*

Ethical Review of Epidemiological Studies (Council for International Organizations of Medical Sciences, 1991), which provide:

23. Disruption of social mores is usually regarded as harmful. Although cultural values and social mores must be respected, it may be a specific aim of an epidemiological study to stimulate change in certain customs or conventional behaviour to lead through change to healthful behaviour – for instance, with regard to diet or a hazardous occupation.

24. Although members of communities have a right not to have others impose an uninvited "good" on them, studies expected to result in health benefits are usually considered ethically acceptable and not harmful. Ethical review committees should consider a study's potential for beneficial change. However, investigators should not overstate such benefits, in case a community's agreement to participate is unduly influenced by its expectation of better health services.

As Guideline 24 observes, if we were to attempt intervention for each participant, individuals in the community might be influenced to participate because of an expectation that they would receive some form of counseling or health benefit. This unwanted result would not be unlikely, in view of the low socioeconomic level of many prospective participants and their difficulty obtaining gainful employment as a result of their mental illnesses.

However, we also felt that it would be unethical to stand by and watch our participants, many of whom did not possess adequate knowledge about HIV and its transmission, engage in behavior that could expose them to risk. Our solution to this dilemma was to provide each participant at enrollment with basic HIV prevention information and a list of resources within the community for additional information and HIV testing. During the course of the study, we provided to each participant on request these same referrals, plus any others that might be needed for medical care, counseling, and substance use treatment. In this way, we provided information to our participants that they could use to reduce their risk and obtain counseling, testing, and medical treatment, without intervening in a manner that could constitute coercion or increase risk in another aspect of their lives.

Involuntary Termination of Participation

As noted in the preceding chapter, it is critical that the balance of risks and benefits to the participants be evaluated on an ongoing basis throughout the course of the study. The balance between the risks and benefits may change over time as the participant's circumstances evolve.

Esperanza (a fictitious name), one of our participants, had been diagnosed with schizophrenia. Although we did not conduct intelligence tests as part of our screening or assessment, Esperanza's mannerisms and affect led our team to believe that her intelligence level was in the low normal range.

Over the course of our shadowing experiences with Esperanza, we learned that she had extreme difficulty in recognizing and acknowledging interpersonal boundaries. She developed extremely intense attachments to several members of the research team. Initially, Esperanza would phone the office on a daily basis to speak to one of these team members if she did not have a shadowing appointment with them. Over time, the frequency of her calls increased to several times a day, and she would panic if her phone call was not returned immediately. These attempts at frequent phone contact evolved into sudden appearances at our offices, with a demand to see a specified member of the research team.

It was explained to Esperanza on numerous occasions that the research team was responsible for shadowing and maintaining contact with over 100 women and could not spend all of their time with Esperanza. During meetings of the research team, we discussed in great depth the possibility of terminating Esperanza's participation in the study, due to the frequent disruptions of study routine. Each time we concluded that Esperanza was one of our participants at greatest risk of HIV transmission as a result of her desperate need to connect to others, regardless of the nature of the relationship, and resolved to closely monitor her while continuing her participation in the study.

Later, one of our study team members required an extended leave of absence from work for personal reasons. In accordance with our study procedures, we advised each of the participants being followed by this staff member of her absence, and indicated which team member would be available instead for shadowing. Shortly after receiving this information, Esperanza indicated to us that she was suffering increasingly frequent episodes of suicidal ideation.

At this point, we re-evaluated the balance of the risks and benefits to Esperanza and determined that the risks greatly outweighed any of the potential benefits due to the degree of emotional trauma that she seemed to be experiencing. However, Esperanza objected vociferously to the termination of her participation and pleaded with us to give her "another chance." Together with her community-based therapist, we negotiated with her a behavioral contract that limited her contact with the study team. Esperanza understood and agreed that she could remain in the study if and only if she was able to adhere to the agreement whose terms she had helped to develop. We reported the increased suicidal ideation to our IRB and our decision to continue Esperanza's participation if she could adhere to the behavioral contract.

Unfortunately, Esperanza was unable to abide by the terms of the behavioral contract, which would have limited her to one phone call per day with study team members. Ultimately, we made a decision to terminate her participation in order to avoid further emotional trauma to her. We reported to our IRB the series of events and our decision to terminate Esperanza's participation in the study.

Conclusions

The events surrounding Esperanza's participation underscore both the need to involve severely mentally ill persons in research to reduce their risk of HIV and the need to continuously monitor their mental and emotional state to assess the balance of risks and benefits to them of participation. A base-line assessment of cognitive capacity that identifies an individual with borderline intellectual functioning might indicate a need for either more intensive follow-up, or might support a decision for non-inclusion should study resources not permit such adaptations. And, if we had realized at the commencement of Esperanza's participation the extent of her dependence issues, we may have decided to assign staff differently, for example, having multiple "primary" staff assigned to shadow her.

In this situation, the balance of risks and benefits changed over time, necessitating termination in order to avoid the foreseeable potential of increased harm to the participant.

References

Aruffo, J., Cloverdale, J., & Vallbona, C. (1991) AIDS knowledge in low income and minority populations. *Public Health Reports, 106,* 115–119.

Aversa, S., McCoy, H.V., Randall, L., & McBridge, D.C. (1998). Socio-demographic influences on condom use: Rural, chronic drug users and their main and other sexual partners. *Abstracts of the Twelfth International Conference on AIDS,* 215 [abstract no. 14164].

Burgos, N. & Perez, V. (1986). An exploration of human sexuality in the Puerto Rican culture. *Journal of Social Work and Human Sexuality, 4,* 135–150.

Caetano, R. & Hines, A. (1995). Alcohol, sexual practices, and risk of AIDS among Whites, Blacks, and Hispanics. *Journal of Acquired Immune Deficiency Syndromes and Human Retrovirology, 10,* 554–561.

Carey, M.P., Carey, K.B., & Kalichman, S.C. (1997). Risk for human immunodeficiency virus (HIV) infection among persons with severe mental illness. *Clinical Psychology Review, 17,* 271–291.

Carey, M.P., Weinhardt, L.S., & Carey, K.B. (1995). Prevalence of infection with HIV among the seriously mentally ill: Review of the research and implications for practice. *Professional Psychology: Research and Practice, 26,* 262–268.

Centers for Disease Control and Prevention. (2004a). *HIV/AIDS among Hispanics, 2004.* Atlanta, Georgia: Author.

Centers for Disease Control. (2004b). *HIV/AIDS among Hispanics in the United States.* Atlanta, Georgia: Author.

Centers for Disease Control and Prevention. (2003a). *HIV/AIDS Surveillance Report, 15,* 12, Table 3.

Centers for Disease Control and Prevention. (2003b). *HIV/AIDS Surveillance Report, 15,* 12, Table 7.

Centers for Disease Control and Prevention. (2002). *The State of Latinos in HIV Prevention in Community Planning.* March 18.

Council for International Organizations of Medical Sciences. (1991). *International Guidelines for Ethical Review of Epidemiological Studies.* Geneva, Switzerland: Author.

Cournos, F. & McKinnon, K. (1997). HIV seroprevalence among people with severe mental illness in the United States: A critical review. *Clinical Psychology Review, 17,* 259–269.

Deren, S., Shedlin, M., & Beardsley, M. (1996). HIV-related concerns and behaviors among Hispanic women. *AIDS Education and Prevention, 8,* 335–342.

Fisher, B., Hovell, M., Hofstetter, C.R., & Hough, R. (1995). Risks associated with long-term homelessness among women: Battery, rape, and HIV infection. *International Journal of Health Services, 25(2),* 351–369.

Flaskerud, J.H. & Nyamathi, A. (1989). An AIDS education program for black and Latina women. *Abstracts of the Fifth International Conference on AIDS*, June 4–9, 701 [abstract no. T.D.O.17].

Fuller, M.& Sajatovic, M. (2005). *Drug Information Handbook for Psychiatry*. 5th ed. Cleveland, Ohio: Lexi-Comp., Inc.

Hines, A.M. & Caetano, R. (1998). Alcohol and AIDS-related sexual behavior among Hispanics: Acculturation and gender differences. *AIDS Education and Prevention, 10,* 533–547.

Hingson, R., Strunin, L., Craven, D.E., Mofensen, L., Mangione, T., Berlin, B. et al. (1989). Survey of AIDS knowledge and behavior changes among Massachusetts adults. *Preventive Medicine, 18,* 806–816.

Kalichman, S.C., Carey, M.P., & Carey, K.B. (1996). Human immunodeficiency virus (HIV) risk among the seriously mentally ill. *Clinical Psychology: Science and Practice, 3,* 130–143.

Kalichman, S.C., Kelly, J.A., Johnson, J.R., & Bulto, M. (1994). Factors associated with risk for HIV infection among chronic mentally ill adults. *American Journal of Psychiatry, 151(2),* 221–227.

Kim, M.Y., Marmor, M., Dubn, N., & Wolfe, H. (1993). HIV risk-related sexual behaviors among heterosexuals in New York City: Associations with race, sex, and intravenous drug use. *AIDS, 7,* 409–414.

Klevens, R.M., Diaz, T., Fleming, P.L., Mays, M.A., & Frey, R. (1999). Trends in AIDS among Hispanics in the United States, 1991–1996. *American Journal of Public Health, 89,* 1104–1106.

Kroliczak A. AIDS among Hispanics in the United States. *Abstracts of the Fifth International Conference on AIDS*, June 4–9, 1989: 748 [abstract no. WDP33].

Marin, B.V., Gomez, C.A., & Hearst, N. (1993). Multiple heterosexual partners and condom use among Hispanics and non-Hispanic whites. *Family Planning Perspectives, 25,* 170–174.

Marin, B.V. & Marin, G. (1990). Effects of acculturation on knowledge of AIDS and HIV among Hispanics. *Hispanic Journal of Behavioral Sciences, 12,* 110–121.

Marin, B.V. & Marin, G. (1992). Predictors of condom accessibility among Hispanics in San Francisco. *American Journal of Public Health, 82,* 592–595.

Marin, B.V., Tschann, J.M., Gomez, C.A., & Kegeles, S.M. (1993). Acculturation and gender differences in sexual attitudes and behaviors: Hispanic vs. non-Hispanic white unmarried adults. *American Journal of Public Health, 83,* 1759–1761.

Marin, G. (1989). AIDS prevention among Hispanics: Needs, risk behaviors and cultural values. *Public Health Reports, 104,* 411–415.

Mata, A. & Jorquez, J. (1988). Mexican-American intravenous drug users. Needleshot practices: Implications for AIDS prevention. *National Institute on Drug Abuse Monograph Services, 81,* 40–58.

May, V.M. & Cochran, S.D. (1988). Issues in the perception of AIDS risk and risk reduction activities by Black and Hispanic/Latina women. *American Psychologist, 43,* 949–957.

McKinnon, K., Cournos, F., & Herman, R. (2002). HIV among people with chronic mental illness. *The Psychiatric Quarterly, 73,* 17–31.

Rivera, Y., Petty, P.A., Krepcho, M.A., & Haley, C.E. (1989). AIDS knowledge and attitudes of adult Latino students in English-as-a-second-language classes. *Fifth International Conference on AIDS*, June 4–9, 894 [abstract no. E552].

Rosenberg, P.S. (1995). Scope of the AIDS epidemic in the United States. *Science, 270,* 1372–1375.

Sabogal, F., Faigeles, B., & Catania, J.A. (1993). Data from the National AIDS Behavioral Surveys. II. Multiple sexual partners among Hispanics in high-risk cities. *Family Planning Perspectives, 25,* 257–262.

Sabogal, F., Perez-Stable, E.J., Otero-Sabogal, R., & Hiatt, R.A. (1995). Gender, ethnic, and acculturation differences in sexual behaviors: Hispanic and non-Hispanic white adults. *Hispanic Journal of Behavioral Sciences, 17,* 139–159.

Sacks, M.H., Perry, S., Graver, R., Shindledecker, R., & Hall, S. (1990). Self-reported HIV-related risk behaviors in acute psychiatric inpatients. *Hospital & Community Psychiatry, 41,* 1253–1255.

Saul, J., Norris, F.H., Bartholow, K.K., Dixon, D., Peters, M., & Moore, J. (2000). Heterosexual risk for HIV among Puerto Rican women: Does power influence self-protective behavior? *AIDS and Behavior, 4,* 361–371.

Susser, E., Valencia, E., & Conover, S. (1993). Prevalence of HIV infection among psychiatric patients in a New York City men's shelter. *American Journal of Public Health, 83,* 568–570.

United States Department of Health and Human Services, Offices on Women's Health. (1998). *Hispanic Women and HIV/AIDS.* Rockville, Maryland: Author.

Young, D.A., Zakzanis, K.K., Bailey, C., Davila, R., Griese, J., & Sartory, T.A. (1998). Further parameters of insight and neuropsychological deficit in schizophrenia and other chronic mental disease. *Journal of Nervous and Mental Disease, 186(1),* 44–50.

Chapter 14
Studies with Minority Populations

Background

The challenge of HIV/AIDS has, in and of itself, necessitated accelerated clinical, epidemiologic, behavioral, community, and policy research to learn more about how HIV/AIDS impacts, and can be effectively addressed within, a wide range of minority populations. This demand has raised of host of ethical issues, which will be tentatively, rather than exhaustively, surveyed in this chapter.

At the outset, however, we must acknowledge the elusiveness and inexactitude of terminology, especially in an international context. "Minority" is often taken to mean non-white (or specifically European), non-LGBT, non-English-speaking, non-Christian, and non-poor (among others). In many U.S. settings, however, African Americans or Latinos, or Jews, or the poor, or even LGBT persons may constitute the functional or numerical majority; in cities devastated by the HIV/AIDS epidemic, from Washington, D.C. to Cleveland to west coast urban centers, African Americans or Latinos are in the majority. The term "minority" has less and less useful meaning as our understanding of the dynamics of diversity and difference evolves, but shorthand descriptors have not kept pace with our understandings.

We therefore use the term "minority" in this chapter to mean, specifically, African Americans, Hispanics/Latinos, Asian-Pacific Islanders, and Native Americans/Alaskan Natives in terms of race and ethnicity, because, even combined, they constitute a minority of the overall population of the United States; and lesbian/gay/bisexual/transgender persons, because they are a clear minority within any given overall society. It is also these two broad sets of "minorities" that have been fundamentally devastated by the HIV/AIDS epidemic in the United States, and that are likely to remain disproportionately affected in the coming years. Even this particularization, however, loses any meaning when talking about people of African or Asian or Spanish/Indian descent outside of the United States, where in any given country, they may constitute the majority culturally, ethnographically, and politically.

In some ways, however, the experience of African Americans and Black Africans—just to focus momentarily on one of the "minority" populations under consideration—is not fundamentally different with respect to HIV/AIDS. Roberts and colleagues (2004) noted that, despite their many cultural differences, African Americans in the U.S. and Blacks living in Africa share many similar experiences in the context of HIV prevention and care. For instance, 70% of all HIV infections in the world occur among Blacks in sub-Saharan Africa, while Blacks in America represent at least 70% of all AIDS cases in many southern states, including Alabama, Georgia, Louisiana, Maryland, Mississippi, North Carolina, and South Carolina. U.S. African Americans and Blacks in Africa also share historical legacies characterized by oppression, racism, and multiple threats to the well-being of their communities (Roberts, McNair, and Smith, 2004).

There are, too, structural similarities with other racial minority groups in the United States, and with Lesbian, Gay, Bisexual, and Transgender persons. All live with the historic and ongoing reality of diminished political, cultural, and generally economic power in relation to majority culture. All suffer real and often legally-supported discrimination, rejection, restricted employment and educational opportunities, and abuse at the hands of social institutions—including medical and behavioral health institutions. All have endured, with some inter-racial variation and inter-ethnic variation, disproportionate burdens as a result of the HIV/AIDS epidemic—burdens that have, in a number of cases, only served to compound preexisting burdens that are themselves also the result of health disparities. And many share the experience of struggling with other health disparities—disparities that pre-date the current HIV/AIDS, and/or that continue in parallel with the HIV/AIDS epidemic (Institute of Medicine, 2005).

The disparate burden of the HIV/AIDS epidemic on racial minorities in the United States is longstanding and well-documented. In 2004, African Americans made up but 13% of the U.S. population, but accounted for 50% of HIV/AIDS cases. African American males had over 8 times the AIDS rate as non-Hispanic white males; African American females had over 22 times the AIDS rate as non-Hispanic white females; African American men were over nine times more likely to die from HIV/AIDS than non-Hispanic white men; and African American women were over 21 times more likely to die from HIV/AIDS than non-Hispanic white women. Similar, though somewhat less stark, disparities in HIV/AIDS case rates and morbidity exist for Hispanics/Latinos, and American Indians/Alaskan Natives. The only notable exception to prevailing racial/ethnic disparity in the case of HIV/AIDS is Asian/Pacific Islanders, who in general have lower HIV/AIDS cases rates than their non-Hispanic white counterparts, and are less likely to die of HIV/AIDS (Office of Minority Health, 2006). Nevertheless, HIV epidemiology within data certain subsets of the U.S. Asian/Pacific Islander population, such as more recent immigrants from highly-impacted regions such as southeast Asia and India, is poorly documented for a variety of reasons. Language, literacy, and cultural barriers, in addition to tenuous residency status that heightens fears of institutional settings, may translate into low testing rates and unknown infection.

Despite the persistent evidence of disproportionality, analysts have yet to articulate a fully satisfying, durable explanation. A recent report by Robert

Fullilove (2006: 7), published by the National Minority AIDS Council, offers a cogent response:

Each year, we ask the same question: Why is AIDS hitting black Americans hardest? While much of the existing literature focuses on quality of care, health care access or individual risk behaviors, we believe that the HIV/AIDS epidemic in African American communities results from a complex set of social, individual, and environmental factors.

In other words, initiatives designed to prevent new infections and maximize medical outcomes for people with HIV/AIDS have consisted largely of what are often called "downstream" approaches, when the available evidence suggests that "midstream" and "upstream" approaches are much more urgently needed. "Downstream approaches" refer to "interventions aimed at individuals; upstream interventions are aimed at large systems effects such as laws, policies, and poverty; and midstream interventions are aimed at local or organizational levels such as school- or community-based interventions" (Beatty, Wheeler, and Gaiter, 2004; Smedley & Syme, 2001).

Research Disproportionality

One could argue, without much qualification, that 1) for some minority groups in the United States, especially some dual-minority groups, the current profile of the community's HIV/AIDS epidemic is epidemiologically, economically, politically, and culturally analogous to the HIV/AIDS profile that now exists in the world's hardest-impact regions or countries; and 2) that since many of the research academy's "accredited" representatives are still European American, heterosexual, and male, HIV/AIDS research conducted with minority community participants in the United States has many of the same features (and problems and ethical challenges) of trans-nationality research conducted by Western researchers in developing-world settings—a transnationality that is significantly marked by unequal allocation of resources, and a sometimes relentless drive toward a protected hegemony of Western, European, male, heterosexual, and Christian values and power. In other words, if the HIV/AIDS epidemic were to be mapped out functionally rather than geographically, many HIV+ minority community members in the United States would more logically reside outside the U.S.—Africa, Asia, Central America— with research resources, and the individuals and institutions that allocate and deploy them, still residing in the United States and a handful of European countries.

The challenges of Western-sponsored and managed HIV/AIDS research conducted in international settings, and HIV/AIDS research conducted about or within minority populations in the United States, may therefore have a great deal in common. Part of the problem is due to the relatively short-term focus of research and publicly-sponsored health and health promotion interventions, when the disparate realities of relatively poor health—not just in terms of HIV/AIDS, but in relation to a wide and troubling array of health and disease variables— is entrenched, long-standing, and supported not only by individual behavior or inaction, but institutional behavior and inaction as well. A recent observation by Laurie Garrett in *Foreign Affairs*, though made specifically about the long term public health needs of developing countries, applies equally well to the realities of African Americans, Hispanic/Latinos, American Indian/Alaskan Natives, Asian/Pacific Islanders, and LGBT persons in the United States as well:

[I]n all to many cases, aid is tied to short-term numerical targets such as increasing the number of people receiving specific drugs, decreasing the number of pregnant women diagnosed with HIV, or increasing the quantity of bed nets handed out to children to block disease-carrying mosquitoes. Few donors seem to understand that it will take at least a full generation (if not two or three) to substantially improve public health—and that efforts should focus less on particular diseases than on broad measures that affect populations' general well-being (Garrett, 2007: 15).[1]

Change a few qualifiers, and it becomes an effective description of what needs to take place for minority populations in the United States: long-term, integrated resource allocations that focus not merely on disease, but the intricately tangled web of economic, political, social, and health management forces that engender inequity in the first place.

Beatty, Wheeler and Gaiter (2004) arrive at similar conclusions, but focus more specifically on African Americans in the U.S. The note that research findings relating to HIV prevention among African Americans are limited due to researchers' exclusion of African Americans in these studies, underlying conceptual and methodological weaknesses of the studies that do include African Americans, and a continuing focus of the studies on modification of individual behaviors.

Problems related to HIV/AIDS research and minorities in the U.S. are not limited to overly narrow foci, inadequate understanding of minority communities, and the inapplicability of HIV prevention research to, at minimum, African Americans. Representation in clinical trials is problematic as well. African Americans and Hispanics/Latinos are under-represented in U.S. clinical trials. A 2002 study indicated that while HIV+ African Americans constituted 33% of those receiving care, they represented only 23% of those participating in clinical trials. Fractionally, Hispanics/Latinos fared little better; HIV+ Hispanics/Latinos made of 15% of those receiving care, but only 11% of those in clinical trials (Burchard, Ziv, Ciyle, & Gomez, 2003; Gifford, et al., 2002). Commonly-identified barriers to minority participation in clinical trials include mistrust of medical institutions, economic limitations (since trial participation requires time and commitment, thus making it difficult for low-income working people or parents who need assistance with child care), cross-cultural communication issues, language, immigration status, and others (Mariner, 2005).

Among these factors, distrust of medical "establishments" is particularly relevant, for it will help explain not only reduced involvement in clinical trials, but greater reluctance to find out one's HIV serostatus, and to seek out and utilize medical and behavioral health services if HIV positive. In a widely-cited 1991 analysis, Thomas and Quinn suggested that the enduring power of intra-community memory of the infamous Tuskegee Syphilis Study within the African American community has significant impact on trust of medical institutions and, therefore, the effectiveness of HIV prevention programs designed and delivered by those institutions. More recently, Bogart and Thompson (2005) found significant endorsement of conspiracy theories about the origin of HIV/AIDS among African Americans, and a strong correction between acceptance of conspiracy theories and distrust of and reduced use of condoms among men. And in an abstract presented at the 2004 AmFAR National Update Conference, Jerry Nessel and Beny Primm of the Addiction Research and Treatment Corporation found high levels of distrust in medical institutions *among individuals trained as HIV peer educators*, with 83.9% of individuals reporting at least one distrustful answer as to the origins of HIV among over 2,000 individuals surveyed. Racial and ethnic minorities reported substantially higher rates of distrust.

These and other findings point to a number of problems in relation to HIV/AIDS research and minority populations, including 1) significant and persistent disparities in health access and health outcomes; 2) the narrow focus of research and care/preventions interventions; 3) under-representation by minorities in medical research; 4) inadequate research and medical institution understanding of minority communities and intra-community diversities; 5) disconnectedness between research and interventions focused on HIV/AIDS, and research and interventions focused on an array of other individual and collective challenges facing minority communities in the U.S.; and 6) a tendency to limit the field of analysis to individual knowledge, attitudes, and behavior, which may, in itself, reinforce prevailing assumptions that individuals are entirely responsible for personal embodiments of history, culture, and inequality. To these we would add, especially in relation to lesbian/gay/bisexual/transgender identity and practice, inadequate and sometimes harmfully-conceived notions of the diversity of human sexuality, a diversity that multiplies at the intersections of race, sexual orientation and practice, and gender in ways that few in the research establishment fully understand, and fully appreciate and respect as viable expressions and manifestations of desire, practice, affection, and love. As Cindy Patton (1986: 142) put it early on in the epidemic,

The paradoxical message about lesbians and gay men voiced by straight society as well as gay liberation ideology is that gay men and lesbians are different *because of sexuality*: sexual objects are different even if people are bisexual across their life experience, styles and range of sexual practice are different, and the identity and community constructed by lesbians and gay men have different value and support difference. (Emphasis added.)

Ethics

Given the profound parallels between Western-sponsored research conducted in international settings, and much of the HIV/AIDS research conducted in the United States about and within minority communities, the ethics of transcultural equality and reciprocity must be fully appreciated and respected. Christakis outlines four possible ethical models to guide transcultural biomedical research. The first is to simply prohibit use of subjects from one culture when researchers are from another. The second is to require transcultural research to independently adhere to the ethical cultures of the two cultures involved—the investigator's, and the subject's. The third involves abstracting a system of research ethics through cross-cultural examination of systems of medical ethics. And the fourth requires that Western—or in this case, dominant-culture—research ethics be adopted as the universal standard (Christakis, 1992).

The problem with the first two models is conflation: they conflate multiple, overlapping cultural behaviors and values from a specific location into a unitary, seamless "culture." The distillation of the formula to (only) two cultures, with two individuals as actors, effaces pre-existing cultural transposition, permeability, and multiplicity, and ignores the degree to which polyculturalism exists in many cases—such as is the case for many African American gay men in large U.S. cities. And the problem with the second two is reductionism: ethics, in the models, are fixed-point thought models. A cross-cultural examination of medical ethics will reveal content, but not the evolutionary trajectory of that content, before or after the examination. As we have seen over the past 50 years, "Western research ethics," as a body of ideas, prescriptions, and proscriptions, is neither fixed nor univocal.

It might be better to propose a set of negotiable values, informed by Western research ethics as they are defined at any particular point in time, which can help guide HIV/AIDS research in relation to minority communities. The Association of Black Psychologists (1997) has proposed a framework for the development of prevention interventions for African Americans that is founded on cultural values associated with the African American community, such as consubstantiation, interdependence, egalitarianism, collectivism, transformation, cooperation, humanness, and synergism. While the degree to which each of these values undergirds and directs the life of any one particular local African American community varies, they do have strong and enduring historical and cultural power. That said, it is worth asking whether a recognition of those values can help inform research design, implementation, and ethics when such research is carried out in African American communities.

Beatty, Wheeler, and Gaiter (2004: 52–53) address similar themes in articulating new directions. However, they also add additional recommendations, which

include (a) using and developing, if necessary, well-articulated theories appropriate to the culture and experiences of the African American population to guide the research; (b) acknowledging and controlling for the diversity of the African American population, especially in regard to factors that are associated with HIV/AIDS and health issues (e.g., gender, age, country of origin, education, sexual orientation); (c) conducting more studies on structural interventions that investigate how sociopolitical and environmental factors shape and can effectively change health behaviors in African American communities; (d) conducting more studies across the upstream-downstream continuum that focus on the most vulnerable African American populations (i.e., MSM, women, residents of the Southeast); (e) clarifying the use of culture whenever it is used in research; and (f) investigating the use of stigmatization.

Recommendations

While useful as broad recommendations, there are a number of more specific recommendations that can be made. Recommendations made in a separate chapter
on Researcher-Community Relations are valid here as well:

1. Research committees can be strengthened by ensuring representative community
   participation; conducting research committee deliberations as openly as possible;
   developing and consistently following protocols for reporting back to the
   community; and ensuring that the research committee has developed lines of
   communication with all key community constituents and stakeholders.
2. Community leadership can be engaged, formally or informally, to gain support
   for research objectives, and to serve as a conduit through which community
   concerns can be relayed to researchers.
3. Researchers, in collaboration with community leadership, can develop
   community-focused educational materials and campaigns, which will attempt
   to communicate the nature of the proposed research, its short- and long-term
   goals and objectives, its specific protocols, its eligibility criteria and exclusions, its ethical architecture, and its sponsorship, financing, the credentials of
   the research team, and other key facts back to the community.
4. Researchers can establish formal or recognizable structures that can offer
   opportunities for community-decision about the design and implementation of
   research, and ownership of the research process and its findings.
5. Researchers should clearly plan for, and communicate to the participant
   community about, the inevitable conclusion of a particular study, the effects
   on the community and research participants, and strategies for addressing those
   effects.

In addition, we encourage adoption of the following approaches and strategies
to help ensure the ethical conduct of HIV/AIDS research in minority communities.

1) Researchers and sponsors, with active involvement from affected minority
community members, should work collaboratively to develop a research agenda
based on a multiplicity of causes and effects, and on interlocking individual and
structural minority community issues, as they are lived and experienced, in
minority communities. Most evidently that agenda should focus on downstream,
mid-stream, and upstream points of intervention and analysis, and should more
carefully research the interconnections between HIV/AIDS, other minority
community health issues (such as diabetes and cardiovascular disease), and the
degree to which both social/economic institutions support or impede progress
toward healthier communities.

2) Researchers should make proactive efforts to engage and interact with
minority cultures—and cultural institutions, processes, practices, and activities—
in ways that go beyond the specific spheres related to the research study under
consideration or being implemented. Research on HIV risk behavior of African
American gay men, for example, requires researcher familiarity with venues
and settings in which Black gay men are often found. Even further, researcher

familiarity with larger cultural institutions affecting the lives of Black gay men, such as faith institutions, minority media, popular culture in local African American communities, local African American politics and non-elected community leaders, will most certainly aid the research team in implementing the most community-sensitive, and community-knowledgeable, research possible.

3) Researchers should advocate, and support advocacy, around the specific needs of HIV-infected and HIV-affected people of color, and elimination of disparities in care access and health outcomes. This may require active promotion of policies, initiatives, funding allocations, services requirements, and system monitoring to reduce such disparities.

4) Researchers should advocate, and support advocacy around, those specific conditions—high rates of incarceration and racial disparities in arrests and sentencing; disproportionate injection drug use; higher rates of homelessness and substandard housing in urban centers; lower rates of public and private economic investment in city cores; inadequately funded and failing public school systems; inadequate access to health insurance and a living wage; legally-sanctioned discrimination and exclusion from entitlements for LGBT persons; and others— that directly or indirectly contribute to disproportionate rates of HIV infection in minority communities. A recent report from the National Minority AIDS Council has carefully outlined the links between poverty, high rates of incarceration, homelessness, and HIV infection, and called for a more global analysis of contributing factors, and a more coordinated responses to the current array of challenges. That call should be heeded, and embraced not merely by policy advocates and elected officials, but by the HIV/AIDS research community as well (Fullilove, 2006).

5) Research initiatives should endeavor to ensure adequate minority community representation on the research team, and commitment to the development of cultural competence at all levels for all professionals and support staff working on the study. Representation should not be viewed as merely an obligation to match research staff with the target community for the purposes of a particular research study (though it remains an important consideration), but an ongoing effort to nurture the polycultural identity and skills of research teams, research institutions, and the field as a whole.

6) Researchers must address the dynamics of HIV stigma and intra-community communications and visibility. Numerous researchers, especially Wilson (1986, 1996), have explored the degree to which poverty itself creates and reinforces marginalized social status, or stigma, in communities of color. In relation to HIV and minority communities, the National Minority AIDS Council Report (Fullilove, 2006: 12) notes that "Stigma is also a part of the pattern of marginalization that affects and influences patterns of morbidity and mortality among African Americans." And while many African Americans know that HIV can be spread through unprotected intercourse (99%) and that increasing condom usage can help reduce risk of acquisition and transmission (89%), a significant number of African Americans believe that HIV can be spread through kissing (38% of African Americans, compared to 33% of Hispanics/Latinos and 26% of whites), sharing a drinking glass with someone who has HIV (25% of African Americans,

17% of Hispanics/Latinos, and 15% of whites), or touching a toilet seat (13% of African Americans, 15% of Hispanics/Latinos, and 8% of whites) (Fullilove, 2006; Kaiser Family Foundation, 2006).

These data suggest that there will be higher levels of fear of people living with HIV/AIDS in African American communities and, therefore, greater vulnerabilities for African Americans living with HIV/AIDS. Given that in many minority communities—especially in mid-sized cities, smaller towns, and rural areas—community knowledge of individual situations is high and rapidly communicated through informal networks, disclosure of an individual's HIV+ status effectively terminates the individual's option to remain private, and may lead to harm. Researchers therefore have a special duty to develop research protocols that will protect research participant confidentiality in every respect.

7) Researchers must directly address the profound and enduring distrust of the medical establishment that underlies and contaminates so many interactions between minority community members and researchers as authorized representatives of medical and academic institutions. The prevalence of such attitudes should be assessed within the target community, and researchers should devise ways to address specific attitudes without negating the historical legacy of medical abuse, and without resorting to a paternalistic posture that concludes that such attitudes are mere ignorance that can be rectified with accurate, scientific information. In particular, researchers should work in collaboration with trusted minority community leadership to explore the depth and breadth of such distrust, articulate appropriate messages that can help restore trust, and designate effective messengers who can assist in the process. In the field of HIV/AIDS overall—research, prevention, and care and treatment—too little has been done to address the painful origins of distrust, and develop social marketing initiatives that can speak to both the reality of that legacy, and the facts about HIV/AIDS transmission and the psychosocial effects of living with HIV/AIDS.

8) Researchers should craft, with input from community leaders, effective plans for representational *involvement* of minority community participants in the research study. Research-related exclusions—such as the need to study dermatological manifestations of HIV among African Americans, which inherently excludes Hispanics/Latinos and members of other minority communities—must be explicitly stated, along with the scientific basis for exclusion. Outreach for participation in community trials should recruit minority community peers, and should be based on true *outreach* protocols—that is, reaching out, often in person, to prospective candidates where they live, work, congregate, pray, learn, and recreate—rather than simply relying on advertising in minority media, or posted announcements at minority-serving organizations.

9) Researchers should advocate for enhanced resources to support training and education to broaden the available pool of minority ethicists and researchers, and should create opportunities within the study for skill development and career progression for minority researchers. Researchers should particularly advocate for support mechanisms that encourage HIV+ minority students to enter the research field.

10) It goes nearly without saying, and echoes discussions in other chapters of this volume, that researchers should integrate the highest standard of care and best proven preventative methods into the research design. The absence of such implies a concession to an inequality of resources and power:

The implicit ethical imperative to provide the 'best proven preventive' methods to trial participants, however, should include social and behavioral interventions to reduce HIV risk behavior such as needle exchange programs, for example. The absence of the best proven preventive methods in a community is often an indication that persons at high risk for HIV infection do not have the political power they need to obtain those services (des Jarlais, 1998: 1431).

Finally, we need to bear always in mind the devastation that HIV/AIDS has caused in minority communities in the United States, a devastation that often goes unrecognized in the larger culture and its media outlets. Many reacted with alarm at the release of the June 1, 2001 *Morbidity and Mortality Weekly Report*, which provided preliminary data on a seroprevalence study conducted among young men who have sex (MSM) with men in seven U.S. cities. While there were limits to the universality of the findings—men were screened in gay community venues frequented by young MSM, and could not therefore be said to accurately represent all young MSM in the United States—the results were nevertheless astonishing: of 2,942 individuals tested, HIV prevalence was 7% among whites, 14% among Hispanics/Latinos, and 32% among African Americans (Centers for Disease Control and Prevention, 2001). By comparison, current national HIV infection rates in Africa are, as of October 2006, are 20.1% in Zimbabwe, 18.8% in South Africa, 19.6% in Namibia, 23.2% in Lesotho, 24.1% in Botswana, and 33.4% in Swaziland—just to mention a handful of the most-impacted countries (AVERT, 2006). This has led some observers to note that if young African American men who have sex with men in the United States constituted a separate nation, they would rival Swaziland for the dubious distinction of having the highest HIV infection rate in the world. Among young African American MSM in the U.S., that fact is well known, and knowledge of it circulates rapidly. The emotional effects, both for individuals and communities, are well-known as well, and are characterized by enormous grief, frustration, anger, hopelessness, and panic—as well as resurgent activism and community-building in response. To fail to keep the scale of the disaster in front of us at all times is to ignore that the fates of entire communities hang in the balance, communities that are desperately looking to researchers, the medical community, behavioral scientists, elected officials, and the American public as a whole for answers.

Notes

1. Related points are made by Parker, Easton, and Klein (2000: S23) in relation to the available research on structural barriers and facilitators in HIV prevention. "During the past decades, researchers have documented a number of structural factors that facilitate HIV transmission and its concentration within particular geographic areas and populations. Most of these factors can be grouped into three analytically distinct but interconnected

categories: (i) economic (under)development and poverty; (ii) mobility, including migration, seasonal work, and political instability; and (iii) gender inequalities. This research reveals that despite the uniqueness of each local HIV/AIDS epidemic, the same general structures and processes are at work in Africa, Asia, Latin America and the Caribbean, and certain groups and communities in North America." One could make a compelling case for the addition of other categories—to degree to which, for example, suppression of gay, lesbian, bisexual and transgender identity and sexual practice is an almost universal phenomenon—but that does not negate the authors' fundamental argument.

References

Association of Black Psychologists. (1997). *An Introduction to the African Centered Behavior Change Model for the Prevention of HIV/STDs.* Cooperative Agreement Number U22/CCU309881. Atlanta, GA: CDC.

AVERT, at www.avert.org/aafrica.htm. Last updated November 2, 2006; accessed December 31, 2006.

Beatty, L.A., Wheeler, D., & Gaiter, J. (2004). HIV prevention research for African Americans: current and future directions. *The Journal of Black Psychology, 30(1),* 41–58.

Bogart, L.M, & Thorburn, S. (2005). Are HIV/AIDS conspiracy beliefs a barrier to prevention among African Americans? *Journal of Acquired Immune Deficiency Syndromes, 38(2),* 213–218.

Burchard, E.G., Ziv, E., Ciyle, N., & Gomez, S.I. (2003). The importance of race and ethnic background in biomedical research and clinical practice. *New England Journal of Medicine, 348,* 1170–1175.

Centers for Disease Control and Prevention. (2005). Health disparities experienced by black or African Americans—United States. *Morbidity and Mortality Weekly Report, 54(1),* 1–3.

Centers for Disease Control and Prevention. (2001). HIV incidence among young men who have sex with men—Seven U.S. cities, 1994–2000. *Morbidity and Mortality Weekly Report, 50(21),* 440–444.

Christakis, N.A. (1992). Ethics are local: Engaging cross-cultural variation in the ethics for clinical research. *Social Science and Medicine, 35(9),* 1079–1091.

des Jarlais, D.C. (1998). Letter. *Science, 279 (March 6),* 1431.

Fullilove, R.E. (2006). *African Americans, Health Disparities and HIV/AIDS: Recommendations for Confronting the Epidemic in Black America. A Report from the National Minority AIDS Council.* Washington, D.C.: National AIDS Minority Council.

Garrett, L. (2007). The challenge of global health. *Foreign Affairs, 86(1).* Last accessed Jan. 4, 2007; Available at http://www.foreignaffairs.org/20070101faessay86103 /laurie-garrett/the-challenge-of-global-health

Gifford, A.L., Cunningham, W.E., Heslin, K.C., Anderson, R.M., Nakazono, T., & Lieu, D.K. (2002). Participation in research and access to experimental treatments by HIV-infected patients. *New England Journal of Medicine, 346(28),* 1373–1382.

Institute of Medicine. (2002). *Unequal Treatment: Confronting Racial and Ethnic Barriers in Health Care.* Washington, D.C.: National Academies Press.

Kaiser Family Foundation. (2006). *2006 Kaiser Family Foundation Survey of Americans on HIV/AIDS.* Last revised May 2006; Last accessed December 31, 2006; Available at http://www.kff.org/kaiserpolls/upload/7513.pdf.

Mariner, D. (2005). HIV/AIDS Clinical Trials & People of Color. Last accessed January 1, 2007; Available at http://www.researchadvocates.org/article003.htm.

Nessel, J.J. & Primm, B.J. (2004). Distrustful thinking about the origins of HIV. 2004 AmFAR National Update Conference, March 27–30, Miami, Florida. [abstract no. S011.]

Office of Minority Health. (2006). HIV?AIDS and Asians and Pacific Islanders. Last modified June 8, 2006; Last accessed January 1, 2007; Available at http://www.omhrc.gov/templates/ content. aspx?ID = 3062

Parker, R.G., Easton, D., & Klein, C.H. (2000). Structural barriers and facilitators in HIV prevention: A review of international research. *AIDS, 14 (supp. 1)*, S22-S32.

Patton, C. (1986) *Sex and Germs: The Politics of AIDS*. Montreal-New York: Black Rose Books.

Roberts, G.W., McNair, L.D., & Smith, D.K. (2004). Introduction. *The Journal of Black Psychology, 30(1)*, 5–10.

Smedley, B.D. & Syme, S.L. (2001). Promoting health: Intervention strategies from social and behavioral research. *American Journal of Health Promotion, 15(3)*, 149–166.

Thomas, S.B. & Quinn, S.C. (1991). The Tuskegee Syphilis Study, 1932–1972: Implications for HIV education and AIDS risk reduction programs in the Black community. *American Journal of Public Health, 81(11)*, 1498–1504.

Watney, S. (1987). *Policing Desire: Pornography, AIDS and the Media*. Minneapolis, Minnesota: University of Minnesota Press.

Wilson, W.J. (1987). *The Truly Disadvantaged: The Inner City, the Underclass, and Public Policy*. Chicago: University of Chicago Press.

Wilson, W.J. (1996). *When Work Disappears*. Chicago: University of Chicago Press.

Case Study Eight
Ethical Issues: A Case Study Involving a Research Study Logo and African American Families

Kristi Y. Jordan, Ph.D., Geri R. Donenberg, Ph.D., and Andrea Boyd, Ph.D.

Introduction

An important aspect of many community-based research studies is creating a logo that appeals to both study participants and their community. Logos can serve many purposes such as: 1) introducing the study to members of the target population, 2) providing insight into the basic principles that buttress the work, 3) promoting (or deterring) the recruitment of families, and 4) facilitating community buy-in into the research, which is especially relevant when involving ethnic minority groups. This paper describes an ethical dilemma involving the creation of a research study logo that was considered offensive by some members of the community. We discuss the ethical considerations behind this controversy, describe factors that may have influenced it, and review our strategies to resolve the issue.

The Case Study

Community Consultation and the Logo

GIRL TALK (Growing Into Responsible Leaders by Talking About Love and Kinship) is a federally funded basic prevention research project that examines mother/daughter relationship factors that contribute to HIV risk behavior among African American teenage girls seeking mental health services. The study targets African American families, and thus, we sought to develop a logo that would appeal to research participants by representing African American culture in a constructive and optimistic light. We took several steps to achieve this goal. First, because the study focused on African American girls in psychiatric care, we invited a teenage girl receiving mental health services to create it. The teen drew a picture of an African American mother and daughter facing one another in profile and talking to each other. Both depictions incorporated facial features characteristic of some African Americans (e.g., larger lips, prominent noses). In addition, the mom was portrayed as wearing her hair in a short afro while the daughter had a pony tail and bangs. The picture included unique characteristics

257

often promoted in the African American culture (e.g., mother and daughter are wearing red and green shirts because these colors are identified as Pan-African colors). The young girl's mother gave permission to use the picture as our research logo.

The next step was to obtain feedback from a broader audience regarding the logo. The picture was presented to the research team (over half are African American women) for feedback and comments. The team overwhelmingly approved the logo, noting its cultural sensitivity and positive depiction. In the third step, the picture was distributed to the research project's Advisory Board, consisting of national and local African American researchers, community leaders, community members, and clinicians working with the study population. Similar to the research team, members of the Advisory Board expressed considerable enthusiasm for the picture's utility as the project logo.

The GIRL TALK logo was applied to all study incentives and materials (water bottles, t-shirts, key chains, project letterhead, and flyers advertising the study) given to participants (mothers and daughters) and community mental health agency collaborators. We informed the agencies and participants that the logo was created specifically for GIRL TALK by an African American teenage girl in mental health treatment. During the first two years of the study, we consistently received positive responses to the logo from families who participated in the research and our community collaborators. On several occasions during follow-up appointments, participants requested extra incentives (e.g., t-shirts) to share with family members and friends who expressed positive feedback about the logo.

Ethical Issues

Two years into the study, GIRL TALK's project director was asked to meet with the executive director of one of the nine mental health agencies collaborating on the study. During the meeting, the executive director (an African American woman) expressed concern about the depiction of the mother and daughter in the logo, stating that she found it offensive and stereotypical. The project director (also an African American woman) reminded her that the logo was drawn by an African American teenage girl specifically for GIRL TALK, and that families expressed considerable enthusiasm for it. Moreover, staff members at the agency, the majority of whom were African American, previously indicated positive responses to the logo. Nevertheless, the executive director required us to remove the picture from all of the incentives provided to participants recruited from her agency, or she threatened to withdraw her agency's collaboration.

This issue may be best understood in the context of two ethical principles. The first principle, the intent "to do no harm," provided the rationale for our extensive efforts to evaluate the acceptability of the logo prior to its usage. Indeed, we followed several steps to ensure that the logo favorably represented African American women, and we invited opinions and feedback from a variety of stakeholders and community members. Despite these efforts, the executive director's complaint prompted us to re-examine our decision to use the logo (see below).

The second principle, the intent to respect the rights and welfare of all people and communities, guided our actions following the complaint. We took three steps to safeguard the well-being of our participants and our collaborating agencies. First, we conducted a brief survey of ten research participants (mothers and daughters) recruited from the other eight collaborating agencies and directly queried their views about the logo. We showed them the logo and asked mothers and daughters to comment on what they thought of it, whether it depicted African American women in a positive way, and whether or not they approved its use for GIRL TALK. All 10 families expressed very positive reactions to the logo and unanimously approved using it.

Second, we shared the executive director's concerns with the research team, and this prompted extensive discussion concerning the role of stereotypes in our work and our obligation to respond thoughtfully and respectfully to our community collaborators. At this point, many of our staff was hired after the logo was integrated into the study, and had not participated in the earlier conversations about its acceptability. Several newer staff members reported that they understood the complaint and noted that the logo could be viewed as stereotyping African Americans (e.g., depictions of big lips and large eyes). Following extensive discussion, and learning of the participants' responses to the survey, the research team voted unanimously to retain the logo.

Finally, we informed the Advisory Board of the community agency's concerns and requested renewed consideration of the logo's acceptability and appropriateness for the study. Board members were responsive to the issue, discussed the concerns, and ultimately voted to retain it. However, a compromise was reached whereby the logo was removed from all recruitment materials and incentives given to participants recruited from the concerned agency.

Reflecting Back: An Update and Analysis

GIRL TALK is now in its fourth year and has enrolled 170 families from 11 mental health agencies. To date, no family has expressed concern about the logo. Moreover, our staff has received numerous compliments over the years by participants who express pride in the logo and report positive feedback from neighbors, relatives and friends who request obtaining their own t-shirts. Similarly, all of our other community collaborators are consistently favorable about the logo.

Our efforts to understand and address the concerns prompted considerable discussion about potential factors that may have influenced the executive director's complaint. Her concerns may reflect the history and broader context of clinical research initiated by members of academic institutions and involving ethnic minority populations.

1) **City-wide gentrification and the University of Illinois at Chicago (UIC):** UIC is located in an urban, low-income community and has recently acquired large amounts of surrounding land to develop and expand its campus. As a consequence,

many low-income, African American families have been uprooted from their homes. Their former communities are being rebuilt to house middle-income individuals and businesses. The displacement of these families has created outrage in the African American community and targeted at UIC, which is criticized for taking advantage of the city's gentrification efforts. GIRL TALK, by expanding into the community and working with local mental health agencies, may be viewed as an extension of a university that exploits rather than preserves the African American community.

2) **Racism and myths of sexuality:** Our focus on HIV risk behavior among African American females may be interpreted by those outside of the academic/ research community as promulgating racist myths of overly sexualized, licentious African American women. African Americans unfamiliar with the goals of research studies and influenced by the egregious betrayal of research like the Tuskegee Syphilis Experiment may believe GIRL TALK reinforces negative images of their culture. In this context, the logo may be viewed as perpetuating similarly degrading stereotypes.

3) **The Principal Investigator of GIRL TALK is not African American.** Since GIRL TALK began, the principal investigator has encouraged ongoing conversations with research staff about the impact of her race (she is Caucasian) on the study's community collaborators, participants, and ability to reach the target population. Despite her efforts, her leadership of this research study may still bring to light several emotional responses among the African American community: 1) How much can a white woman know about African Americans? 2) What is her ulterior motive for "helping us"? 3) Is the logo an indication that she lacks awareness of the population and therefore, is unable to study them effectively? These concerns underscore the need to attend to the role that race of the researcher plays in conducting studies involving minority groups. Of note, the study's project director and most of the staff are African American women.

4) **The executive director's need to "protect" her community.** While meeting with the project director, the executive director indicated that we could not ask families from her agency to evaluate the logo's acceptability or allow them to see the logo during their enrollment in GIRL TALK. By doing so, she implied that the families could not decide for themselves whether or not the logo was offensive. Many of our staff members interpreted her decision to "protect" the community as infantilizing and demeaning. In fact, the team reflected on the irony that the executive director sought to prevent racist stereotypes by potentially engaging in behaviors that stereotype her community.

In the end, the research team and the advisory board reached a consensus about how to address the complaint about the logo; GIRL TALK currently has two sets of incentives: one with the logo and one without, and families are offered a choice. To date, the majority of families have chosen to receive incentives that include the logo.

Acknowledgements. Funded by the National Institute of Mental Health (R01 MH65155)

Chapter 15
Research in International Settings

The ethics of American researchers or funders conducting research in resource-poor environments is challenging beyond words

(Faden & Kass, 1998: 4).

Introduction

In the September 18, 1997 edition of the *New England Journal of Medicine*, executive director Marcia Angell, M.D., sharply criticized government-funded placebo-controlled research studies on the perinatal transmission of HIV, and compared these studies to the notorious Tuskegee Syphilis study carried out in the US between 1932 and 1972—noting that in both cases, medical interventions known to be effective were withheld. Her letter sparked vigorous controversy and debate. That letter was fueled by a letter written by Peter Lurie and Sidney Wolfe (1997) to then-Secretary of Human and Human Services Director Donna Shalala earlier in the year in which they argued that placebo-control trials were unethical because effective treatments for prevention of perinatal transmission of HIV were available. In the study in question, which had enrolled 12,000 HIV+ women in nine countries, preliminary data demonstrated that 9% of pregnant women receiving treatment had transmitted HIV to their babies, while 19% of the pregnant women in a placebo-control group had transmitted HIV to their babies—even though it was widely known and accepted that AZT was highly effective in reducing the likelihood of vertical transmission. It was estimated that more than 1,000 babies acquired HIV vertically because of the placebo treatment given to their mothers. (Roman, 2001).

The controversy was and remains notable for reasons that go beyond obvious narrow ethical issues. First, the critique originated from outside the research community: Lurie and Wolfe were representatives from Public Citizen's Health Research Group. If there were ethical violations, the research community had failed in its responsibility to self-police. And second, the controversy quickly took on an unusually *public* character: two members of the editorial board of the *New England Journal of Medicine*, who are recognized experts in HIV/AIDS, David

Ho and Catherine Wilfert, publicly resigned to protest Angell's letter. It haunted David Satcher, then Director of the Center for Disease Control and Prevention, who had just been nominated to the position of Surgeon General; and the New York Times carried it as a front page story that quickly circulated through the global news media (French, 1997; Hane, 1997; Stolberg, 1997a, 1997b, 1998).

In their critique, Lurie and Wolfe called for a single, international standard of research on human subjects. Angell echoed the call, asserting that investigators should adhere to the "highest ethical standards, no matter where the research is conducted." In their response, Satcher and Harold Varmus, Director of the National Institutes of Health, asserted their position that ethical standards need to take into account the social, economic, and scientific conditions of research (Varmus and Satcher, 1997). Therein the debate between a kind of nuanced relativism, and ethical fundamentalism, was renewed.

To some, the issue is unambiguous: "To allow for research studies in developing countries that are not permitted in the United States appears to make the Third World equivalent to a 'research sweat shop'" (Clark, 2002: ED14). But research studies in the developed world are often carried out in a particular economic and political context, one that, in relative terms, generally ensures greater access to resources necessary for survival, and access to rights and legal codes that at least in principle help ensure freedom from coercion or violations of civil rights. Clark's contention may very well be correct, but deserves elaboration: research trials carried out in developing countries cannot meet a test of permissibility and equality if they simply transplant clinical protocols, even ethically-managed clinical protocols, to another setting, because the settings, themselves, differ greatly. "Take these pills with breakfast, and two large glasses of water," is one prescription in most developed countries, and another prescription altogether in the poorest countries of the world, where clean water, itself, may be unavailable.

Carlos Del Rio poses the foundational question, asking whether it is even possible to conduct ethical research in developing countries—whether it is possible to conduct research within the ethical framework of CIOMS and WHO guidelines— and concludes that at the very least, communities where new drugs are tested must be given "affordable access to the newly developed and approved drugs." On at least one point, he is unequivocal: "any trials conducted in developing countries to test interventions primarily applicable to developed countries are unethical, whether or not a placebo group is included" (Del Rio, 1998: 330).

And Thomas appropriately contextualizes research emanating from developed countries, but carried out in developing countries, within the largely market-driven mechanisms of drug development. He that researchers seek opportunities for the conduct of research that pose a minimum of obstacles and developing countries provide such an opportunity and permit, as a consequence, the maximization of profit (Thomas, 1998). While the U.S. has detailed regulations governing bioethical research, many countries do not have such laws or regulations, and some have the barest minimum. Such non-uniformity, by itself, would—in the absence of any other regulatory factors—tend to drive research toward the least-regulated environments. Again, the question of investments—the physical bodies of research participants—and benefits (drug profit) is again raised, and poorly answered.

And in the resolution of these questions, impatience is not necessarily a virtue. The development of applied bioethics in the case of AIDS is sometimes shaped by the urgency of the epidemic, in which the lives of millions are at stake. (Barry, 1988). The question has to be raised whether such urgency puts pressure on bioethical practice to "cut corners" in a perceived attempt to serve the greater good (Fortin, 1991).

Research in Developing Countries

Gladys Mabunda (2001) has identified six major ethical questions related to HIV research in developing countries: (1) the user of placebos in clinical trials, (2) comparability of standards of care, (3) approval of research protocols by host countries and participant communities, (4) the potentially dual role of the health care provider as research participant recruiter, (5) the issue of informed consent, and finally, (6) the lingering question of whether HIV research should be conducted in developing countries at all. Some of these are addressed elsewhere in this volume; in this chapter, we address the underlying philosophical or moral question about universality of ethical standards, and the role of standards of care in the attainment of universality. The question is not fundamentally about ethical science, but about ethical relationships:

Increasingly, there is an awareness that the success of North-South research collaboration should not be judged solely on the results of scientific research activities. This awareness must be coupled with a learning approach to create a sustainable, mutually beneficial working relationship, that aside from advancing science must address inequity and put local priorities first, develop capacity with a long term perspective and preserve the dignity of local people by ensuring that the benefits of research will truly uplift their status (Edejer, 1999: 440).

Universalism versus Relativism

The core philosophical or moral tension underlying the ethics of research in international settings is between those who subscribe to universalist standards of ethics, as represented by legal codes and professional standards largely established in the US, and those who argue for a relativistic approach. But the distinction is seldom absolute. For one thing, ethicists must acknowledge the homogenizing force of Western medicine, which, combined with the force of globalization, means that a particular kind of thought is more likely to influence communities scattered across the globe than other competing, perhaps local, forms of thought: "Despite the cultural specificity of Western medicine—product that it is of a particular cultural tradition—it has been extraordinarily widely diffused throughout the world. In non-Western settings, Western biomedicine typically comes to form one part of a heterogeneous collection of medical systems and competes for patients with the others" (Christakis, 1992: 1079).

Hellsten (2005) reflects much current thinking in his support of uniformity in general in international guidelines, but calls for more discussion about their applicability in different economic and social conditions, arguing that present debates can too often be employed to defend double standards of practice. Hellsten (2005: 265) recommends that, in the larger African context, we explore bioethical questions in the following areas of debates:

1. the distribution of all resources and the problem of global and local inequity;
2. conflicting rights and duties, in relation to a) individual rights versus (collective) public health, b) individual rights versus the social responsibilities of individuals, and c) individual rights versus collective group rights;
3. the universality of codes of medical ethics and medical research in relation to cultural values and practices, and in relation to available resources.

For some, absolutist or fundamentalist rigidity must be dethroned and the nature of international bioethics as a matter of negotiated conventions must be acknowledged (Baker, 1998). But for others, excessive relativism sets exactly the kinds of conditions that have permitted ethical abridgements: after all, "local custom" can be used to justify all manner of iatrogenic abuse. The issue is, indeed, "challenging beyond words:"

More broadly, if modern public health depends on citizens' self-policing and their adherence to the standards of risk avoidance, how uniform does this require conduct to be throughout the world? Must other cultures now adopt behavioral [and research] norms determined in, and possibly dictated by, the industrialized world? If so, is this a form of cultural imperialism or merely a part of the benign globalization that is proceeding apace anyhow? Do American civil liberties make sense in developing nations? (Baldwin, 2005: 288).

Orentlicher (2002) believes that although universality in biomedical ethics is the fundamental principle, but there are limitations and variations in the application of those principles due to differing local circumstances. Indeed, he notes that studies conducted in the U.S. may be considered unethical in other countries, just as studies conducted elsewhere could be considered unethical in the U.S. But this gives us little more than an overly-broad hope that universalism and relativism, as two unique perspectives, can be comfortably integrated into a meaningful whole; it does not tell us what the product will look like, or how it will be developed.

Geertz (1973) goes one step further to point toward systems of thought underlying the tension between universalist and relativistic perspectives, arguing that one solution to the tension is deft handling of the socio-political thought, including ethics, by reframing it as a coherent system of cultural symbols—that is, it is not the ethical rules themselves which are so important, but their meanings. For Christakis, this means accepting a certain amount of ambiguity:

We must navigate, in short, between the simplicity of ethical universality and the evasion and complexity of ethical relativism, between intellectual hubris and moral paralysis. We should not ask, 'Is there a single model for research ethics?' but rather, 'Can there be?' We must face and accept the indeterminacy of ethical variability (Christakis, 1992: 1089).

Thus, Geertz seems to reject the possibility of absolutist and relativistic *integration*, as a fixed condition, and underscores the necessity of *processes*. While the specific results of such navigational processes cannot be forecast, certain general guidelines can be set forth.

First, a fair and comprehensive negotiation of ethics presupposes that the voices of all affected actors and agents have been adequately solicited, and fairly considered. This will most certainly include the array of foundational documents that have much of their origin in Western settings, but would also include, just to mention a few, relevant input from the fields of anthropology and transcultural communication, researchers indigenous to developing country settings, cultural critics, research participants, and representative community leadership in developing country or community settings.

A current challenge is to increase the involvement of researchers in developing countries to help frame the relevant ethical questions, and lead discussion about their resolution. Hyder and colleagues (2004), aware of the limited literature on research in developing countries, surveyed developing country researchers with a series of 169 questions that covered IRB reviews, informed consent, relationships with collaborators, availability of interventions, general ethical issues, US and international rules and guidelines, and a variety of other topics. Most of the 203 respondents who completed valid surveys were from Asia (42%), followed by Africa (29%) and South America (23%). Among a variety of findings, the majority (95%) reported that US ethical guidelines ensure high ethical standards in research, and more than two-thirds of developing country researchers reported relying on U.S. ethical regulations for guidance. But developing country researchers also leveled significant criticisms: 83% reported a belief that U.S. IRB regulations are not attuned to cultural norms and traditions that prevail outside of the U.S. and 57% reported that U.S.-based IRBs are more concerned with politics than with the protection of research participants. It is important to note, however, that when asked the same question about IRBs based in developing country IRBs, 63% of the respondents indicated a belief that these IRBs are more focused on politics. Hyder and colleagues (2004: 71) concluded that developing country researchers "have valuable experience with international and local ethical reviews, which contributes to global thinking on these issues"—a conclusion which clearly begs for more research on developing country researcher perspectives. These findings argue powerfully to a more prominent role for developing country researchers in the articulation of ethical questions and answers in their homelands—including the array of ethical and procedural questions that have to do with the relationship between developed-country sponsors of research, and developing-country professionals who help implement the research within their own homelands.

Second, a negotiation of the application of international guidelines to local conditions requires a willingness to be flexible. For Moodley (2002: 197), this means developing strategies to account for cultural differences: "The important concepts of informed consent, risk/benefit ratio and fair treatment of trial participants are interpreted differently in traditional, rural African communities, where a moderate form of communitarianism referred to as 'Ubuntu' or 'communalism' is still prevalent." As a practical strategy—since consent is still vital, but may be mediated through a community context—the provision of waiting periods might address both universalist requirements and local conditions: after explanation of a consent form, a potential research participant might be given a period of time to think it over, and presumably, consult with others, before signing.

Third, researchers must develop appropriate strategies for the resolution of conflicts between "ways of seeing" specific ethical challenges. Creation of and provision of adequate technical support to locally-based ethical review boards, which are composed of a reasonable cross-section of acting/affected voices, can assist in that process. Ensuring appropriate transparency and openness of review board discussions and deliberations can also help "authorize" local boards as mediating bodies in the eyes of local communities. The role of IRBs and local ethics review boards can also be expanded from only a review function to the provision of education. In addition, the use of social marketing processes, rewards for compliance, wider publicity for and marginalization of those doing unethical research, and efforts to more closely align human rights and bioethics programs can effectuate change in perspectives (Clark, 2002).

Fourth, the developmental of navigational procedures, protocols, and strategies should be viewed not simply as requirements attached to specific studies, but long-term initiatives that seek to foster increased bilateral understanding of ethical issues and the means for resolving them in on-the-ground practice. In this vein, Jonathan Mann has pointed to a tension between micro-ethics and macro-ethics in AIDS care. HIV prevention trials in developing countries (and in some cases, in developed countries as well) are limited in terms of time or duration, location, and options; the research interaction between investigators and participants is generally short-term. But HIV is, or can be, a long-term condition that presents itself within social contexts that often structure access to resources in ways that severely restrict health care access for some (if not the majority), and the resource-access decisions of national communities are often profoundly influenced by inter-state relations and the policies of international institutions. Micro-ethics and macro-ethics, Mann (1998) suggested, must converge into a coherent whole that commits the researcher to a broader and more complex set of obligations.

And fifth, researchers should commit to grappling with the realities of one of the most vexing dilemmas of research in international settings: the reality of widely varying standards of care. As a navigational issue, one that confronts the tension between absolutism and relativism, varying standards of care is thorny at best, but cannot be avoided, for it speaks to the perception (and the reality) of inequity at a time when globalization of information and media make such inequity starkly evident to much of the world.

Standard of Care

"Standard of care," as indicated elsewhere, is not simply a matter of equal access to clinically-indicated formularies or procedures, though that in itself is an enormous challenge. In the United States, access to treatment is supported by a vast (if fragile) network of ancillary services for people with HIV/AIDS, including transportation, housing assistance, food and nutritional services, support groups, mental health counseling, substance abuse treatment, and more. Clearly there is a shared understanding in the United States and many other developed countries that in order to maximize the medical outcomes that are *possible* because of antiretroviral therapy, there must be social investments in services and supports that have demonstrated effectiveness in improving adherence, clinic utilization, prevention of opportunistic infections, and prevention of transmission of the virus.

Currently, national, bilateral, international, and industry initiatives, combined, do not generate sufficient resources to cover the cost of only medications—the addition of further costs in order to provide ancillary support services is probably outside of the willingness, at least, of various funders to support. This should not, for a moment, be used as a justification for two standards of care, nor does it absolve researchers and other committed parties of the obligation to advocate for equitable access to the highest possible standard of resources. Indeed, nothing less than an insistence on a single standard of care that is medically-articulated could be viewed as ethical from a global perspective. Thomas (1998) has termed this a "maximalist" perspective.

But it does present a practical dilemma for researchers now carrying out initiatives, in that there are critical research demands and needs, and insufficient funds to meet them according to the highest ethical standards: that is, we will not conduct research in international settings unless and until we can guarantee participants the same high-level, comprehensive care, including ancillary care, that is available in developed countries.

Shapiro and Benatar (2005) offer, as a possible solution, a staged strategy, arguing that "contributing to sustainable improvements in health by progressive ratcheting the standard of care upwards for [HIV/AIDS] research participants and their communities is an ethical obligation of those in resource-rich countries who sponsor and implement research in poorer ones." They highlight a number of challenges, emanating from the principles of beneficence and justice, that are often generated in HIV/AIDS research in developing countries: dual standards of care in local communities where research is taking place; the potential of higher standards of care delivered in research contexts acting as an undue inducement to participate in clinical trials; the inability to sustain higher standards of care once research trials are completed; the possibility that higher salaries paid to research clinicians will drain qualified health care workers from the public sector; sponsorship limitations, especially those that do not permit provision of direct services; treatment challenges; and the possibility that entire communities could be stigmatized due to the presence of research initiatives. After reviewing much of the available sources for guidance—from the National Bioethics Advisory Commission (2001), the

Council of International Organizations for Medical Sciences (2002), the Helsinki Declaration (2000), the UNAIDS *Ethical Considerations in HIV Preventive Vaccine Research* (2000), and the Nuffield Council on Bioethics (2002), Shapiro and Benatar conclude that

as it is unlikely that an overall universal standard of care can be rapidly achieved in research projects in developing countries, the goal should be to implement reasonable standards that are significantly higher than available in the host country, and closer to standards in the sponsoring country . . .in a way that progressively ratchets the standard of care upward, both for subsequent research projects and for the benefit of the local system through genuine partnerships and capacity building, leaving participants and their communities better off after the trial and not merely 'no worse off' (Shapiro & Benatar, 2005: 46).

In some cases, the concept of equipoise may apply. Clinical equipoise indicates that there is an honest and reasoned difference of opinion about preferred treatment for a specific illness or condition. If a particular treatment is not superior to another and the physician or researcher knows as much, then ethical reasoning requires that the superior treatment be given. But if there is genuine uncertainty or no standard of care exists for a certain disease or condition, then clinical equipoise exists (Freedman, 1988). The concept is equally appropriate when considering the provision of ancillary support services as it is for direct medical care.

Recommendations

Beneath the specific challenges of negotiating universality and locality, and of equity in provision of standards of care, we might consider the installation of a number of broad guidelines for the development of an ethics-based framework within which international research can be conducted, and the ethical questions that research generates can be addressed.

Correcting the Imbalance: Adopting a Bilateral Approach

A necessary tonic to the legacy of colonial and imperial relations is the adoption of a bilateral approach to the sharing of tools, knowledge, and findings. At first blush this may appear illogical; in fact, there is much to be said for its elevation as a primary principle. In practice, it means this: researchers and research participants, no matter what their nationality, will approach research tasks as bilateral opportunities for the sharing of knowledge, ideas, and acquired wisdom, and no forms of "knowledge" will be illogically privileged over others. In early discussions of "technology transfer" in relation to HIV/AIDS, there was an implied understanding that "technology"—that is, tools and knowledge—were primarily possessions held by developed-country researchers, and that transfer essentially implied a downloading of those tools and knowledge to developing country researchers and research participants who, for their part, possessed none. This is hardly the case; developed-country researchers have much to learn from the technological innovations, research strategies, and local wisdom of developed country researchers and research participants. But it is only by approaching the research process, from conception to completion, with a deeply-held conviction that bilateral knowledge in fact exists—or more bluntly, that arrogance of knowledge and skill is not logically justified—that such bilateral transfer of skills and knowledge becomes possible.

Long-Term Commitments

Developing-country researchers and research participants have long been vocal critics of developed-country researchers who gain entrance in a local community, solicit participant involvement, produce research findings, and then depart. Research is best conducted when it is carried out in the context of larger and longer-term commitments to mobilize the resources that will put research findings into practice in the participant community, and that will increase indigenous capacity to carry out autonomous research once the initial research project has been completed. This requires that researchers develop partnerships with public institutions, local academia, and NGOs to amass the resources necessary to put local research into local practice, and the developed-country researchers will work collaboratively with local researchers to increase research capacity. A failure to bring about such an approach as a more comprehensive means of conducting research in developing-country settings will only lead to lingering distrust and difficulty in securing participant involvement in research, as well as ongoing developing-country dependence on external parties to gather and analyze necessary data.

The United Nations General Assembly Special Session (UNGASS) declaration is relevant in this regard. The Declaration of Commitment emerging from the UNGASS meeting on HIV/AIDS in 2001 included a number of broad aspirations, and while it did not bind any of the member states to specific goals and objectives, the articles in the Declaration of Commitment are nevertheless significant. Signatories commit to ramp up prevention to reduce HIV incidence in the most affected countries; develop strategies to strengthen health care systems; and expand sustainable care, support, and treatment. What is most notable about the Declaration, however, is the attention it pays to human rights, calling for multisectoral involvement in national AIDS research, care, and prevention efforts (including involvement by marginalized groups and communities), the active participation of people living with HIV/AIDS, and the right to the "highest attainable standard" of physical and mental health (International Council of AIDS Service Organizations, 2005). Clearly implied is that short-term commitments and relationships will be wholly insufficient.

Transfer of Research Findings

Researchers should make a commitment not merely to transfer research tools and findings through traditional venues, such as peer-reviewed journals, but make additional efforts to disclose research findings to research participants them-selves, and to the entire local community or society in which the research is being carried out. This can be effectuated through open community forums in which research findings are reported out to members of the research participant commu-nity, through community flyers and bulletins written in accessible language, and through the development of relationships with local media so that research findings can be reported in print and electronic media.

Because so much research is publicly financed, there is a compelling argument to be made for an implied contractual agreement to report research findings in this manner. But additionally, it can be argued that "subjects" are co-participants in the research process, and have a reasonable and ethical claim to the findings derived therefrom. Research competition—both in terms of protecting data for publication, and in terms of the role that publication plays in an individual's ability to secure future research funding—offers few incentives for rapid reporting of research find-ings to participants and participant communities. Nevertheless, researchers must begin to more proactively consider their ethical obligations in this regard.

Addressing Other Inequities

In a recent article in the *Fletcher Forum of World Affairs*, De Waal (2003) paints a chilling picture of possible connections between the HIV/AIDS, and accelerated famine in Africa:

Ultimately, we can expect widespread entitlement collapse, either gradual or sudden, brought on by an external shock that suddenly lowers returns to labor. In short, famine. Moreover, this will be a qualitatively different kind of famine from those that have been witnessed in peacetime in recent years. Most famines are marked by skillful use of coping strategies by households, which husband their resources and utilize diverse means of acquiring income and food to see them through the period of hardship, prior to recovery. However, successful famine coping strategies require adult labor and considerable skill and experience. With lower LEA [adult life expectancy] and higher dependency ratios, those coping strategies will be much less effective. In essence, the coping strategies have already been used to their limit in adjusting to the protracted lower LEA over preceding years . . .In short, not only can we expect famine, but we can expect a particularly virulent manifestation of famine marked by a much reduced ability to recover. Segments of society will in fact face indefinite destitution (De Waal, 2003: 14).

The fundamental principle of nonmaleficence is associated with the maxim of *primum non nocere* – "first, do no harm." "Harm," in this context, is widely understood to mean not only harm to individuals participating in research, but harm to the local or national community as well. But "harm," in the context of both research and in the larger context of development activities, can also mean political, economic, or cultural myopia. It is now abundantly clear that the epidemic of HIV/AIDS in the developing countries arose, in part, because of the scientific phenomenon of a new virus. However, it achieved widespread diffusion, to no small degree, as a result of the entanglements that the virus enjoyed with a host of other phenomena, of a more social nature that reflected preexisting inequalities: HIV/AIDS has always been about poverty, human rights and personal liberties, injustice, and the enduring, uncomfortable notion that, if policy manifests reality, then the persistent reality is the notion that some lives are worth more than others.

It is also now, as De Waal illustrates with just the example of famine or food scarcity, that HIV/AIDS has an enormous capacity to deepen the preexisting wounds, to widen inequities between those who have, and those who do not. In some discussion, the term "viral apartheid" has been recently employed, to signify not only the distinction between those living with HIV and those free of the virus, but also the great divide between those who have access to HIV/AIDS health-related resources, and those who, in increasing numbers, will never acquire those resources, and whose lives will, as a consequence, be short and marked by illness, pain, and rapidly decreasing capacity.

That the virus was fueled by such conditions, and insofar as the effects of the virus, on peoples and countries, has a tremendous capacity to only make those conditions worse, should give somber pause to researchers and development professionals working in the field or in international settings. The implications are clear: failure to articulate, and forcefully advocate against, underlying and subsequent inequalities attached to HIV/AIDS is to legitimate, by silence, those conditions. It is therefore difficult to imagine the ethical conduct of research in international settings that does not also oblige the researcher to advocate for more just relations and fair distribution of the world's resources.

One way researchers can fulfill such an obligation is by visibly allying with activists and advocacy NGOs in local communities. Such alliance need not require absolute agreement about policy or strategy options; nevertheless, many of the activists and advocacy NGOs in developing countries have articulated cogent analyses of inequalities related to HIV/AIDS, and have implemented skillful strategies for addressing those inequalities by challenging power relations. Since 1998, for example, South Africa's Treatment Action Group (TAC)—seen as a model, around the world, for AIDS activism—has been vigorously advocating for free or very-low-cost drugs that can prevent vertical transmission of HIV, antiretroviral medication, and drugs for the treatment of opportunistic infections associated with HIV. TAC has achieved some notable successes in this regard, as has ACT-UP in the United States. If we—the community of researchers, health authorities, and affected populations—are to reach the goal of universal access, or staged "ramping-up" of standards of care, then changes in pricing, patents, and delivery systems are a prerequisite. To accomplish that goal, the active involvement of activists and advocacy NGOs is absolutely necessary.

Intrinsic to each of the four aforementioned guidelines is the twin ideas of "relations" and "ownership." Some of those relationships—between researchers and research participants, between researchers and the community, for example— are explored in more depth elsewhere in this volume. Here, it is sufficient to underscore that the research process unfolds through a series of relationships between individuals, groups of individuals, and institutions, and each of those relationships serves as potential grounds for inequality that, itself, can undermine the highest standards of ethical behavior. And within each of those relations, there is an exchange of something of value—intellectual property, scientific understanding, local wisdom, and the goods of medicine itself—to which some parties will claim greater ownership than others. It is this question of "who owns what, and why," that must be continually investigated, not simply because the highest ethical standards in HIV/AIDS research demand such inquiry, but because without such questioning, we may, without intending to do so, merely set the stage and reinforce conditions that will permit the next epidemic, perhaps even more consequential and even more vampiristically feeding off inequality, that will kill so many millions more.

References

Anderson, E.S. (1999). What is the point of inequality? *Ethics, 109(2)*, 287–337.

Angell, M. (1997). The ethics of clinical research in the third world. *New England Journal of Medicine 337*, 847–849.

Baker, R. (1998). A theory of international bioethics: The negotiable and the non-negotiable. *Kennedy Institute of Ethics Journal, 8(3)*, 233–273.

Baldwin, P. (2005). *Disease and Democracy: The Industrialized World Faces AIDS.* Berkeley, California: University of California Press.

Barry, M. (1988). Ethical considerations of human investigation in developing countries: The AIDS dilemma. *New England Journal of Medicine, 319(16)*, 1083–1085.

Christakis, N.A. (1992). Ethics are local: Engaging cross-cultural variation in the ethics for clinical research." *Social Science and Medicine, 35(9)*, 1079–1091.

Clark, P.A. (2002). AIDS research in developing countries: Do the ends justify the means? *Medical Science Monitor, 8(9)*, ED5–16.

Council of International Organizations of Medical Sciences. (2002). *International Ethical Guidelines for Biomedical Research Involving Human Subjects.* Geneva, Switzerland: Author.

De Waal, A. (2003). Why the HIV/AIDS pandemic is a structural threat to Africa's governance and economic development." *The Fletcher Forum of World Affairs, 27(2)*, 6–24.

Del Rio, C. (1998). Is ethical research feasible in developed and developing countries?" *Bioethics, 12(4)*, 328–330.

Edejer, T.T. (1999). North-South research partnerships: The ethics of carrying out research in developing countries. *British Medical Journal, 319*, 438–441.

Faden, R. & Kass, N. (1998). Editorial: HIV research, ethics, and the developing world. *American Journal of Public Health, 88(4)*, 548–550.

Fortin, A.J. (1991). Ethics, research and medical power: AIDS research in the Third World. *AIDS and Public Policy Journal, 6(1)*, 15–24.

Freedman, B. (1988). Equipoise and the ethics of clinical research. *New England Journal of Medicine, 317*, 141–145.

French, H. (1997). AIDS research in Africa: juggling risks and hopes. *New York Times,* October 9, A1, A14.

Geertz, C. (1973). Ideology as a cultural system. *The Interpretation of Cultures* (pp. 193–233). New York: Basic Books.

Hane, D. (1997). AIDS researchers resign over editorial. *The Denver Post,* Oct. 16, A-4.

Hellsten, S.K. (2005). Bioethics in Tanzania: Legal and ethical concerns in medical care and research in relation to the HIV/AIDS Epidemic. *Cambridge Quarterly of Healthcare Ethics, 14,* 256–267.

Hyder, A.A., Wali, S.A., Khan, A.N., Teoh, N.B., Kass, N.E., & Dawson, L. (2004). Ethical review of health research: A perspective from developing country researchers. *Journal of Medical Ethics, 30,* 68–72.

International Council of AIDS Service Organizations (ICASO). (2005). *Community Monitoring and Evaluation: Implementation of the UNGASS Declaration of Commitment on HIV/AIDS.* Last accessed January 1, 2007; Available at http://www.icaso.org.

Lurie, P. & Wolfe, S. (1997). Unethical trials of interventions to reduce perinatal transmission of the human immunodeficiency virus in developing countries. *New England Journal of Medicine, 337,* 853–856.

Mabunda, G. (2001). Ethical issues in HIV research in poor countries. *Journal of Nursing Scholarship, 33(2),* 111–114.

Mann, J.M. (1998). HIV/AIDS, micro-ethics, and macro-ethics. *AIDS Care, 10(1),* 5–6.

Moodley, K. (2002). HIV vaccine trial participation in South Africa–An ethical assessment. *Journal of Medicine and Philosophy, 27(2),* 179–195.

National Bioethics Advisory Commission. (2001). *Ethical and Policy Issues in International Research: Clinical Trials in Developing Countries.* Bethesda, Maryland: Author.

Nuffield Council on Bioethics. (2002). *The Ethics of Research Related to Healthcare in Developing Countries.* London: Author.

Orentlicher, D. (2002). Universality and its limits: When research ethics can reflect local circumstances. *Journal of Law, Medicine & Ethics, 30,* 403–410.

Roman, J. (2001). U.S. medical research in the developing world: Ignoring Nuremberg. *Cornell Journal of Law and Public Policy, 11,* 441–460.

Shapiro, K. & Benatar, S.R. (2005). HIV prevention research and global inequality: steps towards improved standards of care. *Journal of Medical Ethics, 31,* 39–47.

Stolberg, S. (1998). Placebo use is suspended in overseas trials. *New York Times,* February 19. Last accessed Jan. 4, 2007; Available at http://query.nytimes.com

Stolberg, S. (1997a). Top U.S. officials defend Third World HIV research. *New York Times,* October 2. Last accessed Jan. 4, 2007; Available at http://www.query.nytimes.com

Stolberg, S. (1997b). U.S. AIDS research abroad sets off outcry over ethics. *New York Times,* September 18. Last accessed Jan. 4, 2007; Available at http://www.nytimes.query.com

Thomas, J. (1998). Ethical challenges of HIV trials in developing countries. *Bioethics, 12(4),* 320–327.

UNAIDS. (2000). *Ethical Considerations in HIV Preventive Vaccine Research: UNAIDS Guidance Document.* Geneva, Switzerland: Author.

Varmus, H. & Satcher, D. (1997). Ethical complexities of conducting research in developing countries. *New England Journal of Medicine, 337,* 1000–1005.

World Medical Association. (2000). Declaration of Helsinki. Ethical Principles for Medical Research Involving Human Subjects. 52nd World Medical Association General Assembly, Edinburgh, Scotland, October.

Case Study Nine
Ethical Challenges in a Multi-Site International HIV Prevention Study with Serodiscordant Couples

Patricia Marshall, Ph.D., Janet W. McGrath, Ph.D., Moses Kamya, MbChb, Andrew Fullem, M.P.H., and David D. Celentano, Sc.D.

Introduction

The statistics on the AIDS epidemic are daunting (UN/World Health Organization, 2006). By the end of 2005, there were more than 38 million people living with HIV/AIDS; more than 24 million of these individuals reside in Sub-Saharan Africa. Since 1981, more than 25 million people have died of AIDS. These figures call attention to the urgent need for HIV prevention research that is effective and ethically sound. International and national policies and recommendations provide guidelines for ethical conduct in biomedical and behavioral research (World Medical Association, 2000; Council for International Organizations of Medical Sciences, 2002; United States Department of Health and Human Services, 1991; Nuffield Council on Bioethics, 2002), including HIV/AIDS research specifically (UNAIDS, 2000). However, investigators working in the field are confronted with the task of balancing accepted ethical standards with the social, economic, and political constraints of local realities. In resource-limited settings, HIV/AIDS investigators must address pragmatic issues associated with weak health infrastructures and inadequate access to treatment and drugs. In some areas, traditional customs surrounding decisional authority may be in conflict with requirements for informed consent. Moreover, cultural and social beliefs about gender roles, sexual behavior, and HIV/AIDS contribute to the moral complexity that surrounds the design and implementation of HIV prevention studies.

Ethical challenges in the treatment of HIV/AIDS and in HIV prevention research have been present throughout the duration of the AIDS pandemic (Schuklenk, 2001; Gostin and Lazzarini, 1997). In clinical settings, ethical issues are associated with a broad spectrum of concerns such as the protection of confidentiality, disclosure of HIV status, access to medical care, and treatment decisions, particularly for non-adherent patients (Anderson and Barrett, 2001; Beggs and Jernigan, 2001; Sollito et al., 2001). In research settings, investigators confront these same ethical challenges and others in the implementation of clinical trials, the development of HIV vaccines, and in HIV prevention studies, particularly among vulnerable populations in culturally diverse resource-limited settings (Dickens, 1997; Mugerwa et al., 2002).

Within the last ten years, reports of ethical misconduct associated with scientific research in resource-limited countries have resulted in ongoing debates among health professionals, people living with HIV and activists over appropriate standards of care, the use of placebos in clinical trials, fair benefits and obligations to study participants and their communities, and informed consent (Angell, 2000; Levine, 1998; Macklin, 2001; Shapiro and Meslin, 2001; Varmus and Satcher, 1997; The Participants in The 2001 Conference on Ethical Aspects of Research in Developing Countries, 2004; Emanuel et al., 2004; Bhutta, 2004). Research on HIV and AIDS has figured prominently in these debates, most notably the controversies surrounding clinical trials of treatment regimens to reduce maternal-fetal transmission of HIV among pregnant women in the developing world (Lurie and Wolfe, 1997; Angell 1997; Killen et al., 2002; DeZuleuta, 2001; Singh and Mills, 2005; Levine, 2000)

In this case analysis, we describe ethical issues that arose in the design and implementation of a multi-site, international study to promote condom use among sero-discordant couples to reduce the risk of HIV transmission. While there are myriad ethical quandaries that have relevance for this study, we focus our attention on issues surrounding informed consent, disclosure of HIV status, and local cultural beliefs about gender roles and the appropriateness of discussing sex or sexual relationships. We conclude with recommendations for best practices in multi-site, international HIV prevention research.

Description of the Case

This collaborative study, "Phase 1 Trial of an Intervention to Increase Condom Use by HIV Serodiscordant Couples," was conducted at sites in three countries: Pune, India; Chiang Mai, Thailand; and Kampala, Uganda. The study was designed by U.S. and local investigators and study staff to evaluate the feasibility of using a group-based couples intervention to promote condom use with HIV serodiscordant couples; this common intervention was used at all three sites (McGrath et al., 2006). The study was funded by National Institute of Allergy and Infectious Diseases, NIH (NIAID), as part of the Family Health International (FHI) HIVNET consortium.. The Uganda project also received support from the Center for AIDS Research at Case Western Reserve University.

The sample included individuals who had been identified as HIV-infected in other studies at each site. For research purposes, these HIV positive men and women were called "index cases." Each participant was asked to bring their primary sexual partner into the clinic for HIV counseling and testing. After determining that the couple was serodiscordant, they were consented individually. If both partners agreed to join the study, they were then enrolled to participate in the group intervention. Due to the requirement that the positive partner initiate participation, couples had discussed the study prior to presenting for the partner's test and consenting. For this reason, no cases arose in which one partner consented and the other did not.

The intervention required participants to attend a total of four group sessions, generally within two weeks: two same sex groups of mixed serostatus and two couple groups. Confidential interviews were conducted with participants at baseline, after the intervention, and at one month and three months after completing the group sessions. The assessments included sexual practices, condom use, HIV/AIDS knowledge and beliefs, and issues related to disclosure, stigma and discrimination.

All three study sites in India, Thailand, and Uganda had established Institutional Review Boards for evaluating scientific protocols, including review of written consent forms and study designs. The study protocol and informed consents were reviewed by IRBs in the U.S. and at participating host institutions. Conflicts did not arise during the protocol review at any site. This was most likely because the study was incorporated into existing research structures that employed staff who had been directly involved in conducting the HIV research from which the index cases were identified initially. These members of the research teams had extensive experience in conducting research with individuals who were HIV positive.

Ethical Challenges

The design and implementation of this study raised a number of ethical concerns for the research teams working at the three sites that varied culturally and epidemiologically. Investigators and research staff confronted challenges associated with informed consent, disclosure of seropositivity to a partner, and beliefs about appropriate gender roles.

Informed Consent

Multiple levels of consent were required by the study protocol. In order to participate, HIV infected participants had to be willing to: 1) disclose their HIV status to their sexual partner; 2) ask their partner to go the clinic for HIV testing and counseling; and 3) participate in same sex group counseling as well as group counseling with their partner in the intervention groups. Conversely, the partners of the HIV infected "index cases" had to agree to undergo HIV testing and counseling, disclose their status to their partners, and agree to participate in individual and couple intervention groups, if found to be HIV negative. If the partner was not HIV negative the couple was not eligible to join the study. The research team made the decision to have each individual provide consent only one time, rather than have a multi-stage consent requiring a second consent to participate in the group intervention after determining their eligibility. This decision was made so as to avoid misunderstandings about the consent process. Multiple consents can lead to confusion if the participant thinks they are only consenting to an HIV test. Both the content and the process of informed consent for HIV prevention studies and other types of research are influenced by a range of factors: US and host government rules and regulations, the cultural environment in which the project is conducted, the project goals and objectives, communication issues that influence comprehension of information, and discrepancies in social and economic power of the participants, communities and researchers (Marshall, 2006; Woodsong and Karim, 2005). Informed consent depends upon effective communication of information, comprehension, and voluntary participation. The growing literature on informed consent for research conducted in resource limited settings with culturally diverse populations calls attention to the complexities surrounding the consent process (Molyneux et al., 2005; Pace et al., 2005; Leach et al., 1999; Karim et al., 1998; Fitzgerald et al., 2002; Pace et al., 2005; Adams et al., 2005). In some cases, language barriers may inhibit effective communication, particularly regarding the translation of difficult scientific concepts. Even when both the researcher and the potential participant speak the same language, an individual's ability to understand the goals and procedures of a study and its associated risks may be compromised by lengthy consent forms that may be hard to read or comprehend (Freeman 1994; Kass, Maman, Atkinson, 2005; Preziosi et al., 1997).

In some situations, the legalistic presentation of consent forms and the need for signatures—or thumbprints or 'marks' for those who are illiterate—may be intimidating for participants, particularly if they are uncomfortable signing a form because of its implications for possible sanctions against them (Beyrer and Kass, 2002). Additionally, the western paradigm for informed consent emphasizes the importance of individual decisionmaking, but in many non-western settings family members or the community play an important role in the decision-making process for participation in research (Marshall 2006; Dawson and Kass, 2005; Tangwa, 2004; Levine 1991). Moreover, some individuals and communities may be vulnerable to coercion or undue influence to participate in research because of their poverty, social status, or gender, or because of their involvement in practices that are viewed as socially unacceptable or illegal (e.g., sex work or drug use). Individuals may also be vulnerable because they have a potentially stigmatizing illness such as HIV/AIDS.

In this HIV prevention study with serodiscordant couples, careful attention was given both to the translation of consent forms into multiple local languages and the actual process of obtaining consent. The research team emphasized the importance of considering the potential risks involved and the ability of participants to withdraw from the study at any time. Beliefs about the risks or harms that may result from participation in a study vary considerably. Information about the nature and possibility of risks provided to individuals asked to join a study reflect the concerns identified by researchers. However, the way in which this information is interpreted and assessed by potential participants may not necessarily correspond with the researchers' views.

In this study, researchers paid close attention to concerns about issues associated with disclosure, right to privacy, and the negative partner's "right to know" that their partner is HIV infected. As noted above, participants had to disclose their HIV status to their sexual partner if they had not already done so. Other concerns about privacy issues were associated with the need for participants to be interviewed individually about their attitudes and sexual practices and that volunteers were required to participate in single sex and mixed sex group activities. Group activities did not require that participants disclose their status to others (with the exception of their partner), except to acknowledge that the groups were for HIV serodiscordant couples. However, group sessions did expect couples to discuss barriers to condom use and issues related to sexual communication and negotiation. These discussions were facilitated using a variety of approaches including role plays, listing exercises, and games in order to increase comfort level and permit individuals the opportunity to participate without revealing personal information. Additionally, participants were also asked to do "homework" in which they practiced some of their skills and reported back to the group about what they did. Importantly, many positive and negative participants did disclosure their serostatus to others in the group exercises.

Respect for Local Values Regarding Appropriate Gender Roles

An important concern for the investigators in developing the protocol focused on the appropriateness of including intervention sessions which required couples to attend multiple group sessions with their partner. A central point addressed by the research team was the possibility that the group sessions might constitute a violation of local norms that prohibit sexual discussion among men and women. After extensive discussion, the investigators concluded that their concerns were balanced by the need to provide a potentially effective intervention for serodiscordant couples, an underserved cohort at identifiable risk of HIV transmission. As the study progressed, groups at all sites openly discussed 'taboo' subjects. This perhaps reflects the expertise and ability of the research staff to facilitate group interactions about sensitive subjects and the level of trust experienced by the participants in the research teams where the study at the three sites was conducted. It may also reflect a significant unmet demand of serodiscordant couples to have a venue to share with others in the same situation experiences and approaches to address condom use within on-going partnerships.

Another issue that had to be addressed by the project team was the potential for harms to participants as a result of disclosure to their partner and efforts to reduce risk. The investigative team and research staff were particularly concerned about the ability and consequences of HIV infected women who disclosed to a sexual partner. In disclosing their HIV seropositivity, the women participating in the study risked identifying themselves as having previous or concurrent sexual partners. This could increase the risk of physical harm to themselves as well as possible emotional, economic, or social harms. Concerns about this issue were balanced against the risks to the individual who was HIV negative of having unprotected sex with a partner not known to be positive.

Similarly, there was apprehension about the ability of HIV uninfected women to successfully bring up the need for condoms and negotiate their use. This issue raises questions about cultural beliefs concerning sex and gender roles, and specifically, the appropriateness of women discussing sex with their partner. In addition, even in cases where a partner is known to be HIV infected, requesting that he use condoms may violate norms regarding women's duty to be subordinate to men.

The possibility of domestic violence —both physical and verbal — against women who either disclose their infection or request condom use posed special challenges. The study investigators, study staff, and sponsors discussed various ways to handle this risk. It was ultimately decided that participants would be ruled ineligible if they reported any history of domestic violence with this current partner. Additionally, participants were urged to report any instance of domestic violence that occurred during the study. No women reported the occurrence of domestic abuse. However, accurate assessment of the occurrence of domestic violence or psychological abuse was dependent upon the willingness of the women to report these events. Investigators and local research assistants at all sites were trained to ask women questions that would address areas of concern regarding physical or psychological abuse. The high level of trust between researchers and participants that was foundational to the success of the project, combined with the ongoing attention to this issue, contributed to a research environment that diminished the possibility of domestic abuse.

Conclusion and Recommendations

The stakeholders involved in this case represent a range of individuals, communities and institutions including, for example, the research participants both as individual volunteers and as a couple, the broader community of persons infected with HIV at all sites, the international team of investigators and staff, the Institutional Review Boards responsible for study oversight and ethical conduct at each site in every country, and the funding institutions sponsoring the study. In the course of designing and implementing this study, a number of social values came into conflict The researchers developed an approach to promote condom use in serodiscordant couples for reducing the risk of HIV infection that potentially required transgressing social norms about gender roles and discussions of sexual behavior. Respect for individual autonomy and protecting the confidentiality of participants were key considerations for all stakeholders. However, study participants, particularly the women who were HIV positive, were at risk for social or physical harms associated with disclosing their HIV status to partners who were HIV negative. This situation highlights the inherent tension between the right to privacy for the infected partner versus "the right to know" for the uninfected partner. The success of the study depended upon effective communication at all levels –between the research team and the IRBs, between the research team and individuals participating in the study, and between the couples involved in the study.

This study demonstrates that it is possible to successfully promote sexual risk reduction for HIV prevention among HIV discordant couples. When the study was initiated, some observers believed that it might not be feasible to implement because of difficulties associated with informed consent and, in particular, because of concerns about disclosure and the protection of confidentiality. There was also skepticism raised at all sites by project staff that women would not participate and that open discussion of sex might be difficult to promote. As the study gained momentum, it became apparent that even in such widely diverse cultural contexts as the three sites in India, Thailand, and Uganda, cultural taboos concerning disclosure, gender roles, discussion of sex, and violence against women can be mitigated through intensive staff training and careful attention to the needs of both men and women confronting AIDS. Specifically, the intervention format which was designed to *decrease* conflicts and *increase* skills to resolve conflicts contributed to a high level of comfort on the part of the participants.

In order to enhance the possibility for successful HIV prevention research outcomes, we recommend the following three guidelines for good practice:

• *Culturally appropriate methods for obtaining informed consent should be implemented. Efforts should be made to insure that participants fully understand the implications and risks of being involved in the research project.*
   HIV prevention researchers working in diverse cultural contexts must develop culturally appropriate methods for obtaining informed consent to insure voluntary participation and adequate comprehension of the goals, procedures, risks, and benefits associated with the study. When developing models for HIV prevention strategies, investigators must be aware of local cultural norms governing gender relations and the negotiation of sexual relationships and contraceptive use. Regardless of whether or not an HIV prevention study is well conceived, well funded, and well staffed, interventions that do not address the local cultural environment are unlikely to succeed. Moreover, conducting successful HIV prevention research in culturally diverse settings requires local knowledge about gender roles and the negotiation of sexual activities, including discussions about sexual issues.

• *The successful implementation of a study depends upon highly trained research staff who are sensitive to local cultural context and the special needs of persons infected with HIV.*
   Members of the research team at all sites must receive intensive training not only on particular research methodologies and interventions but also on ethical issues associated with HIV prevention research. This should include special attention to the process of obtaining informed consent and the protection of study participants from physical or social harms. Adequate training

reinforces the potential for team members to respond to the specific needs of study participants and enhances the possibility of successful results.

• *Collaborative partnerships must be strengthened between researchers in resource- poor and resource-rich settings.*

Finally, every effort should be made to promote a truly collaborative relationship between investigators at all sites. This must include attention to capacity building in resource-poor settings where studies are conducted. Efforts to build a cohesive and highly trained research team are enhanced when local and international investigators are equal collaborators in the design and implementation of studies. Although the study on the promotion of condom use among serodiscordant couples was financially supported by institutions in the U.S., investigators from all sites provided a broad range of knowledge and expertise relevant to study design. Capacity-building and reciprocity are key to successful collaborations. In this case, investigators and members of the research team at sites where the study was being conducted were well positioned to provide important information about local cultural context and traditions that could influence the research design and implementation. Thus, decisions about the specific goals and procedures used in the study were informed by local knowledge about beliefs and social values. Multi-site, international, collaborative research projects require a certain degree of flexibility and creativity in balancing the need for methodological consistency across research sites with the need to consider the unique cultural environments where studies are implemented.

Acknowledgements. This analysis builds upon a case prepared by Janet McGrath, Moses Kamya, David Celentano, and Andrew Fullem, for: Marshall, Patricia, *Ethical Issues in Research Design and Informed Consent to Biomedical and Social Research in Resource Poor Settings.* Geneva: World Health Organization, Special Topics in Social, Economic, and Behavioral Research Series of Programme for Research and Training in Tropical Diseases (TDR), (in press, 2006). The study from which this is derived was supported by Family Health International (FHI) with funds from the National Institutes of Allergy and Infectious Diseases, NIH (NIAID). The views expressed in this paper do not necessarily reflect those of FHI or NIAID. The Uganda site also received support from the Center for AIDS Research at CWRU (NIH Grant no. AI36219, M. Lederman, PI). This case analysis was supported by the Center for AIDS Research at Case Western Reserve University (NIH grant no. AI36219, M. Lederman, PI), the NHI/National Human Genome Research Institute (HG02207), and the Center for Genetic Research Ethics and Law (NIH/NHGRI P50-HG), Case Western Reserve University. We are deeply appreciative of the efforts of the research teams at all sites and the couples who participated in the study.

References

Adams, V., Miller, S., Craig, S. et al. (2005) The challenge of cross-cultural clinical trials research: case report from the Tibetan Autonomous Region, People's Republic of China. *Medical Anthropology Quarterly, 19(3),* 267–289.

Angell, M. (2000). Investigators' responsibilities for human subjects in developing countries. *New England Journal of Medicine, 342,* 967–969.

Angell, M. (1997). The ethics of clinical research in the Third World. *New England Journal of Medicine, 337,* 847–849.

Anderson, J. R., & Barrett, B. (Eds) (2001). *Ethics in HIV-Related Psychotherapy: Clinical Decision-Making in Complex Cases.* Washington DC: American Psychological Association.

Beggs, J. M., & Jernigan, III, I. E. (2001) Mandatory HIV/AIDS testing: an ethical issue *International Journal of Value-Based Management, 14,* 131–146.

Bhutta, Z. (2004). Beyond informed consent. *Bulletin of the World Organization, 82(10),* 771–777.

Beyrer, C., & Kass N. (2002). Human rights, politics, and reviews of research ethics. *Lancet, 360,* 246–51.

Council for International Organizations of Medical Sciences (CIOMS). (2002). *International Ethical Guidelines for Biomedical Research Involving Human Subjects.* Geneva, Switzerland: Council for International Organizations of Medical Sciences (CIOMS).

Dawson, L., & Kass, N. E. (2005). Views of US researchers about informed consent in international collaborative research. *Social Science and Medicine, 61,* 1211–1222.

DeZuleuta P. (2001) Randomized placebo-controlled trials and HIV-infected pregnant women in developing countries. Ethical imperialism or unethical exploitation? *Bioethics, 4,* 289–311.

Dickens, B. (1997). Research ethics and HIV/AIDS. *Medicine and Law,* 16:187.

Emanuel, E., Wendler, D., Killen, J., & Grady, C. (2004). What makes clinical research in developing countries ethical? The benchmarks of ethical research. *Journal of Infectious Diseases, 189,* 932–7.

Fitzgerald, D. W., Marotte, C., Verdier, R. I., Johnson, W. D. Jr., & Pape, J. W. (2002). Comprehension during informed consent in a less-developed country. *Lancet, 360,* 1301–2.

Freeman, W. L. (1994). Making research consent forms informative and understandable: the experience of the Indian health service. *Cambridge Quarterly of Healthcare Ethics, 3,* 510–521.

Gostin, L. O., & Lazzarini, Z. (1997). *Human Rights and Public Health in the AIDS Pandemic.* New York: Oxford University Press.

Karim, Q. A., Karim, S., Coovadia, H. M., & Susser, M. (1998). Informed consent for HIV testing in a South African hospital: is it truly informed and truly voluntary? *American Journal of Public Health, 88,* 637–40.

Kass, N.E., Maman, S., & Atkinson, J. (2005). Motivations, understanding, and voluntariness in international randomized trials. *IRB: Ethics and Human Research, 27(6),* 1–8.

Killen, J. Grady, C. Folkers, G, & Faucie A.(2002). Ethics of clinical research in the developing World. *Nature Reviews, 2,* 210–215.

Leach, A., Hilton, S., Greenwood, B. M., Manneh, E., Dibba, B., Wilkins, A., et al. (1999). An evaluation of the informed consent procedure used during a trial of a Haemophilus influenzae type B conjugate vaccine undertaken in The Gambia, West Africa. *Social Science and Medicine,* 48(2), 139–48.

Levine, R. J. (1991). Informed consent: some challenges to the universal validity of the western model. *Journal of Law, Medicine and Health Care, 19*, 207–13.

Levine, R.J. (1998). The best proven therapeutic method standard in clinical trials in technologically developing countries. *IRB: Ethics and Human Research*, 20, 5–9.

Levine, R.J. (2000). Some recent developments in the International Guidelines on the Ethics of Research Involving Human Subjects. *Ann. N.Y. Acad. Sci*, 918, 170–178.

Lurie, M. P. H., & Wolfe, S. M. (1997). Unethical trials of interventions to reduce perinatal transmission of human immunodeficiency virus in developing countries. *NEJM, 337*, 853–856.

Macklin, R. (2001). After Helsinki: unresolved issues in international research. *Kennedy Institute of Ethics journal, 11*, 17–35.

Marshall, P, Adebamowo, C, Adeyemo, A, et al. (2006). Voluntary participation and informed consent to international genetic research. *American Journal Public Health, 96(11)*, 1989–1995.

McGrath, J. W, Celentano, D. D. Chard, S. E., Fullem, A, Kamya, M, Gangakhedar, R.R., et al. (2006, in press). A group based intervention to increase condom use among HIV serodiscordant couples in India, Thailand and Uganda. *AIDS Care*.

Marshall, P. (2006). Informed consent in international health research. *Journal Empirical Research on Human Research Ethics, 1*, 25–41.

Molyneux, C. S., Wassenaar, D. R., Peshu, N., & Marsh, K. (2005). 'Even if they ask you to stand by a tree all day, you will have to do it (laughter . . .!:) Community voices on the notion and practice of informed consent for biomedical research in developing countries. *Social Science and Medicine, 61*, 443–454.

Mugerwa, R. D., Kaleebu, P., Mugyenyi, P., et al. (2002) First trial of the HIV-1 vaccine in Africa: Ugandan experience. *BMJ, 324*, 226–229.

Nuffield Council on Bioethics. (2002). *The Ethics of Research Related to Healthcare in Developing Countries*. London: Nuffield Council on Bioethics.

Pace, C., Talisuna, A., & Wendler, D. (2005). Quality of parental consent in a Ugandan malaria study. *American Journal of Public Health, 95(7)*, 1184–9.

Pace, C., Emanuel, E., Chuenyam, T., Duncombe, C., Bebhuk, J. D., Wendler, D., et al. (2005). The quality of informed consent for a clinical research study in Thailand. *IRB: Ethics and Human Research, 27(1)*, 9–17.

Preziosi, M. P., Yam, A., Ndiaye, M., Simaga, A., & Simondon, F., & Wassilak, S. (1997). Practical experiences in obtaining informed consent for a vaccine trial in rural Africa. *New England Journal of Medicine, 30,336(5)*, 370–3.

Schuklenk U. (2001). AIDS; *Society, Ethics and Law*. Broofield, VT: Ashgate Publishers.

Shapiro, H., & Meslin, E. (2001). Ethical issues in the design and conduct of clinical trials in developing countries. *New England Journal of Medicine, 345,*139–142.

Singh, J., & Mills, E. J. (2005). The abandoned trials of pre-exposure prophylaxis for HIV: what went wrong? *PLos Medicine 2(9)*, e234:824–827.

Sollito, S., Mehlman, M., Youngner, S., & Lederman, M. (2001). Should physicians withhold highly active antiretroviral therapies from HIV-AIDS patients who are thought to be poorly adherent to treatment? *AIDS, 15(2)*, 153–159.

The Participants in The 2001 Conference on Ethical Aspects of Research in Developing Countries. (2004). Moral standards for research in developing countries: from reasonable availability to fair benefits. *Hastings Center Report, 34*, 17–27.

Tangwa, G. (2004). Between universalism and relativism: a conceptual exploration of problems in formulating and applying international biomedical ethical guidelines. *J Med Ethics*, 30(1), 63–7.

United Nations Programme on HIV/AIDS (UNAIDS). (2000). *Ethical Consideration in HIV Preventive Vaccine Research*, Geneva UN-AIDS. Geneva, Switzerland: Author.

UN/WHO. (2006). Report on the Global AIDS Epidemic. Geneva: UN/WHO.

United States Department of Health and Human Services. (1991) *Protections of human subjects*. Washington, D.C.: Author.

United States Department of Health and Human Services, (45 CFR, 46).

Varmus, H., & Satcher, D. (1997) Ethical complexities of conducting research in developing countries. *New England Journal of Medicine, 337,* 1003–1005.

Woodsong, C., & Karim, Q. A. (2005). A model designed to enhance informed consent: experiences from the HIV Prevention Trials Network. *American Journal of Public Health, 95,* 412–419.

World Medical Association. (2000). Declaration of Helsinki: Ethical Principles for Medical Research Involving Human Subjects. Amended by the WMA 52nd General Assembly, Edinburgh, Scotland.

Chapter 16
Activity-Defined Populations

Research involving activity-defined populations can embrace enormous potential breadth, incorporating populations and communities as diverse as injection drug users, other drug (such as crack cocaine) users, commercial sex workers, individuals engaged in sadomasochistic exchange of blood, transfusion recipients, and people who have received clotting factor for hemophilia. They would, under most conditions, have little to say to each other. But for the most part, as concerns the epidemiology of HIV infection, those activity-defined populations of most significant concern are those that are also most often marginalized and subject to both criminal sanction and human rights abuses—injection drug users, and transactional or commercial sex workers.

This chapter will focus on commercial sex workers, in part because research related to injection drug users has been *comparatively* better-explored in the literature, and in part because more the more recently-published literature has begun to more adequately address the needs, issues, and challenges faced by both commercial sex workers, and the scientists who are concerned about their well-being, in relation to HIV/AIDS. For example, the XVI International AIDS Conference, held in Toronto in August of 2006, yielded over 4,500 abstracts; of those, 170 were directly related to commercial sex work and commercial sex workers, and easily twice that number, or another 340, focused on commercial sex work to some degree—representing, combined, 11.3% of the total abstracts presented. Clearly, there is (appropriately) heightened concern about commercial sex work, and the men, women, girls and boys who engage in it.

Selecting "commercial sex work"—or more traditionally, "prostitution"—is, however, immediately problematic, since feminists and sex workers have offered compelling theoretical evidence that sex work is, in nearly every setting, an exceedingly complex category of labor and social experience, which is marked by tremendous variation in individual significations. "Sex work," then, can include everything from transactional sex in monogamous, temporarily monogamous, or polygamous marriages; occasional sexual intercourse provided along with other productive services; short-term sexual relationships between young women or young men (typically teenagers) and older men in return for money or other commodities ("sugar daddies"); legalized, state-governed prostitution; hotel- and bar-related exchanges; street-based solicitation and sexual transactions; brothel-based transactions; and more (Davis, 2000; Doyle, 1994; Mbilinyi and Kaihula, 2000 Obbo, 1993; Outwater, 1996; Tandia, 1998). Men and women may also be involved in diverse forms of sex work over time, as their needs and realities change (Schoepf, 1997; White, 1990).

A thorough analysis and typology of the range of behaviors that might be included under the heading "sex work" (or "prostitution" or "transactional sex") is far beyond the scope of the present volume, as is (which is only hinted at here) the diversity of meanings, motivations, conditions, behaviors, and significations that shape sex work practices and identities. But a recognition of the aforementioned complexities is the essential first task of those engaged in research with sex workers: whatever one might presume to know about sex workers is probably wrong, or at least only partially accurate some of the time; the duty, then, is to engage in research with a profound sensitivity to the degree to which incomplete understandings can contribute to bias or prejudicial contamination at some point in the research process.

For discussion purposes, most of the following will focus on a broad meaning of "sex work" that includes any negotiated transaction which exchanges sex for money, commodities, or security (such as a safe place to sleep at night). It also focuses, due the lack of available research on male sex workers, on girls and women engaged in sex work.

Risk

The prevalence of HIV infection among commercial sex workers who have been tested varies enormously across geographic domains, from a low of 1.4% among 966 women tested during the late 1990s in China, to a high of 73.7% among 373 women tested in 1998 in Ethiopia (Akilu et al., 2001; Alary et al., 2002; Laurent et al., 2003; Simonsen et al., 1990; van den Hoek et al., 2001). One can surmise, at minimum, that in those cases where high prevalence exists alongside low treatment access, a collective problem exists; generally, those problems are likely to be structural in nature. Data about treatment access for HIV+ commercial sex workers are scant at best, but a 2005 study of HIV-positive female sex workers in Vancouver raises deep concerns about the degree to which they are accessing available medical and support services. Of 159 women studied, only 9% were on HAART (Shannon, Bright, Duddy, and Tyndall, 2005). These findings are supported by others showing low uptake of antiretroviral treatment among injecting drug users who are also engaged in survival sex work (Altice, Motashari, and Friedland, 2001; Palepu et al., 2001; Strathdee, 1998).

But essentializing HIV+ female commercial sex workers as mere vectors of HIV/STI infection—or as morally-tainted, recalcitrant patients—has been a common temptation, and a serious mistake. Issues facing *all* women living with HIV/AIDS are considerable, and include differential diagnoses, disclosure, discrimination and stigma, economic self-sufficiency, women's experience of health services, decisionmaking about pregnancy and child-bearing, access to sexual and reproductive health services, childrearing (for mothers), and power in relation to both local and national systems of law and policy, but power in relation to other women, especially power differences in intimate relations with men, in their personal lives. None of these issues disappear if the women in question are also engaged in commercial sex work; unfortunately, however, the patriarchal gaze can rarely see much of the woman beyond what she does sexually, for whom, and at what price. For women involved in sex work, there are also a host of other issues related to autonomy and human rights, including police activities and violations, the threat and reality of physical violence from a client or pimp, detention based on inadequate evidence, coercion for sex, bribes and extortion, neighborhood displacement, compulsory HIV/STD testing, substandard enforcement of laws in cases involving harm to sex workers, and the daily reality of prejudicial treatment in the media.

For many women involved in commercial sex work, depression is also likely to be a persistent condition. In a Chinese study of 278 rural-county female sex workers, approximately 62% had depressive symptoms; the presence of depressive symptoms was significantly linked to failure to use condoms consistently (19.8% vs. 40.2%, p<.001) (Hong, Lo, Fang, Zhao, and Lin, 2006). Undiagnosed and untreated rates of childhood sexual abuse are also likely to be high: a New York City study of 821 women self-defined as sex workers found that near 2/5 (39.5%) self-reported that they had been sexually abused as children (Weiner and Lorber, 2006). Given the nature of the psychological effects of childhood sexual abuse, such as memory repression and cultivation of shame, the real percentage is likely to be higher.

And the backdrop of poverty, a condition amplified by gender in most countries, remains a powerful influencing factor in the lives of many women (and men) who are or who have been involved in transactional sex. Supporting the proposed link between structural conditions (such as poverty) and sex work, Cohan and colleagues (2005) found that in a population-based, cross-sectional survey of 2,543 low-income women in northern California, 1 out of 11 reported sex work (defined as an exchange of sex for money or drugs), with nearly half engaging in sex work in the previous six months. A South African study, similar in many ways to the California research, found very high levels of transactional sex among female migrant or seasonal workers in South Africa, noting that 52% of female workers reported having sex in exchange for money, gifts, food, accommodation, clothing, lighter workloads, or extension of contracts (Hill-Mlati and Rijks, 2006). And a Canadian study of 1,656 street youth—defined as youth between the ages of 15–24 inclusive, who had spent at least 3 consecutive nights away from home—across seven Canadian cities found that 22.6% reported ever having trading sex for money or goods, with females (29.2%) more likely than males (14.9%). Average age at first exchange of sex for money or goods was 15.8 years (Agbola, Sisushansian, and Wong, 2006). These all point to the critical link between transactional sex and the structural powerlessness and violence in which it is often embedded.

Finally, while researchers may be operating from a paradigm that places HIV and other STD risk as the highest risk concern for commercial sex workers, CSWs themselves may have different priorities. In a Washington, D.C. study of women with strong ties to strip clubs and trick houses, violence and safety concerns were paramount; HIV and health concerns came in second (Hickey and Douglas, 2006). Again, we are reminded of the importance of viewing communities, and in this case, potential research participant communities, from *their* perspective, with all its complicated (and admittedly sometimes contradictory) multidimensionality, rather than merely from the more narrow perspective that perceives only that which relates to, or appears only under the study of, that which is being researched.

Interactions: Health Care and Research

The relationship between sex workers and health care providers, or sex workers and the research community, is fragile at best, and plagued by misunderstanding and, consequently, missed opportunities. Vanwesenbeeck (2001: 242) conducted an overview of the social science research on prostitution published in the decade between 1990 and 2000, finding that "The literature is still much more about sex than it is about work. In addition, although an increasing number of authors have criticized the dominance of a deviance perspective over work perspectives on prostitution, the literature still reveals many of the features of stigmatization." If Vanwesenbeeck's assessment is even reasonably accurate, then we should also reaffirm the recognition that marginalized communities are often acutely aware of dominant-culture, external characterizations of those same marginalized communities, and oftentimes hypervigilant about misperceptions. If this is so, then commercial sex workers, by and large, would not offer a good grade to the accumulated efforts of researchers.

Commercial sex worker experience with health care institutions often does not fare much better. A study of HIV+ commercial workers in India for that an overwhelming percentage (about two-thirds) had a negative experience at hospitals or rehabilitation centers (Deb, 2006). In a surprisingly frank report by an Indian physician working with sex workers in India, delivered at the XVI International AIDS Conference (and a report that, by implication, served as a self-critique of health care provider interactions with commercial sex workers), Mukherjee (2006) noted that doctors must 1) challenge their moral values about sex workers, 2) identify the socioeconomic needs of sex workers, firmly address the structural and livelihood issues associated with HIV risk in sex work, and 3) support capacity building and advocacy skill development among sex workers. These needs, the presentation concluded, can be addressed through formal orientation programs on sex work for health care workers and researchers working in the field.

Recently, perceptions of unethical treatment of commercial sex workers by HIV researchers erupted into public, sustained conflict in a number of countries. In early 2005 randomized trials against placebo of tenofovir (a reverse-transcriptase inhibitor made by Gilead) were cancelled in Nigeria, Cameroon, and Thailand as a result of advocacy from prostitute-rights and HIV/AIDS advocacy groups (such as ACT-UP Paris), which accused the researchers of acting unethically by not supplying treatment after the study, and by choosing to conduct trials in women in parts of the world where a study is cheap. Planned by the US Centers for Disease Control and Prevention and Family Health International, the proposed trials were met by opposition from groups such as the Women's Network for Unity, a grassroots alliance of sex workers based in Phnom Penh, Cambodia. The group wrote to Cambodian Prime Minister Hun Sen, accusing researchers and the NGOs who were recruiting for them out of self-interest, and failing to supply sex worker organizations with information about the proposed trials in advance (Loff, Jenkins, Ditmore, Overs, and Barbero, 2005).

Two important ethical dilemmas have emerged from the controversy, which has yet to be resolved: the responsibility of researchers to involve potential research participants at all stages of the research development process, not just at the point of study recruitment; and the responsibility of researchers to provide a "standard of care"—itself a very elusive concept—for participants in the trial. The latter issue is even more complex, since in the specific case of the tenofovir trials activists were insisting on post-study provision of a standard of care to all study participants. The tension lies in the interpretation of the 2000 Declaration of Helsinki, which requires that study participants be "assured" of access to the same standard of care available in the study. By what does "assure" mean? A list of referrals (one end on a spectrum of possibilities) or lifelong provision of care (the other end)? This question will no doubt find passionate advocates on either side, but one thing is clear: "meeting" with proposed research participants, as the researchers did in Thailand, is not sufficient; planning for research, if it is to be both ethical and successful, must integrate the active involvement of research participants in the research process, most especially in cases such as injection drug users and commercial sex workers; their history with those who conduct research and those who provide medical care has been, often as not, a history of neglect, stigmatization, and sometimes, abuse, and their trust must be regained. As a 2003 editorial in *The Lancet* put it, "In most countries, sex workers are stigmatized, discriminated against, and harassed. They are seen as immoral people or as victims of unscrupulous traffickers who exploit the lack of opportunities of deprivileged inhabitants of mostly poor countries. Unfortunately, public health workers and researchers can share these attitudes" (Wolffers and van Beelen, 2003).

Ethics, Structural Changes, Resolutions

One of the most important unresolved questions that continues to arise at the intersection of research, care and treatment, ethics, and commercial sex work relates to operational paradigms: is sex work to be viewed as a manifestation (or primarily so) of structural inequality, reinforced, in most cases, by gender inequality and violence; or is it to be viewed as a manifestation (or primarily so) of elective individual behavioral responses to social conditions and choices?

(It is a question that contains enormous social and policy significance: if sex work is viewed primarily as a behavioral phenomenon, then society is justified in implementing responses aimed at the individuals who engage in those behaviors—which could mean punitive responses (a criminal approach), or treatment and rehabilitation responses (a neoliberal approach). If, on the other hand, if sex work is viewed primarily as a manifestation of structural and gender inequality, and if society is committed to addressing the harms related to sex work, then a commitment to correcting those inequalities is required. The latter, of course, carries with it a much heftier price tag, and a more fundamental usurpation of prevailing attitudes and beliefs.)

There is no doubt some truth in both perspectives but, increasingly, researchers and analysts, adopting a human rights approach to commercial sex work as it interacts with HIV and STIs, have come to the conclusion that countries and local jurisdictions must, as an essential first step, eliminate repressive or coercive legal restrictions on sex work, essentially adopting a public health model instead of criminalization model (Betteridge, 2005; Brussa, 2006; Johari and Murthy, 2006; Yu, 2006). This a proposal that will invite continuing, and vigorous, debate, but one thing is clear: removal of criminal sanctions against prostitution would facilitate reduced-barrier partnerships between researchers and health care providers on the one hand, and commercial sex workers on the other, since 1) by virtue of legal revision, some of the moral "taint" of prostitution is reduced, and 2) the fear of increased visibility, and therefore possible reprisal by police, will be somewhat attenuated for commercial sex workers interested in receiving care or participating in research.[1]

In the absence of changed laws, however (or perhaps parallel to legal reform), HIV/AIDS research and care providers need to re-center commercial sex workers themselves in the development and implementation of research agendas and care systems: "Helping enhance the ability and willingness of sex workers to organize among themselves should be a major priority . . . The support and assistance of projects, government agencies, and other entities and committed individuals are vital. However, only sex workers themselves are able to fully articulate what they want and need—and forcefully protect the human rights, health, and well-being of themselves and their peers" (Central and Eastern European Harm Reduction Network, 2005).

A number of initiatives have been carried out with such a viewpoint as the driving paradigm. Murugan and associates (2006) worked in partnership with female sex workers in India to develop and utilize spot analysis, contact mapping, and network mapping tools to analyze gaps in prevention outreach and develop possible solutions, concluding that building community capacity in the use of scientific approaches to information gathering can contribute to improved program delivery. The implications for research are evident: sex workers can successfully partner with researchers in the scientific process.

The United Nations Food and Population Fund (UNFPA) Haiti and its partners designed a program to engender self-empowerment among people living with HIV/AIDS, men who have sex with men, and "street girls" (commercial sex workers) in Port-au-Prince that utilized adapted human rights-based education to help participants understand sexual and reproductive health and rights, and the role that poverty plays in relation to those rights, finding that participants were better able to define their rights and take steps to protect them. The initiative also led to implementation of a database to document human rights violations, a referral system for victims of violence, a condom distribution network, and national advocacy efforts (Kunin-Goldsmith and Laurenceau, 2006).

And while addressing IDUs rather than, specifically, commercial sex workers (though there is likely to be significant overlap), a Canadian project aimed at increasing involvement of IDUs in research and program services makes numerous recommendations based on their experience, including explicit recognition by governments of the value of such participation; provision of funding for drug-user organizations; and creating conditions under which drug users can effectively participate in the consultative process (Jurgens et al., 2006). (The case study by Dr. Linda Lloyd that follows this chapter provides one example of such involvement.) These recommendations, it seems, could be equally well applied to collaboration with commercial sex workers on research projects.

A Special Case: Current U.S. Policy

In 2003 the U.S. Congress passed, and President Bush signed into law, legislation requiring foreign organizations receiving U.S. global HIV/AIDS funds to have policies explicitly opposing prostitution, and barring the use of such funds to "promote, support, or advocate the legalization practice of prostitution" (United States Leadership against HIV/AIDS, Tuberculosis, and Malaria Act, 2003; Trafficking Victims Protection Reauthorization Act, 2003). In 2004, the Department of Justice ruled that the restrictions applied to U.S. organizations as well.

The legislation has led to significant protest by HIV/AIDS, human rights, and sex worker organizations both in the U.S. and globally, who have argued that decriminalization or legalization of prostitution may in fact prove to be an effective way to address the health and human needs of sex workers and their customers, and who have contended that forcing organizations to adopt public stands against prostitution will 1) threaten the ability of such organizations to establish trusting, non-judgmental relations with sex workers, and 2) effectively bar sex worker rights organizations, which can have a tremendous capacity to reach peers in local communities, from receiving funds (Roehr, 2005; Schleifer, 2005a, b). In May of 2005, Brazil rejected $40 million in anti-HIV/AIDS funding from the U.S. because of the restrictions; Dr. Pedro Chequer, head of Brazil's national AIDS program, said that the funding requirements could undermine the very programs that have help contribute to Brazil's success in reducing new infections, and providing treatment to all Brazilians living with HIV (Phillips and Moffett, 2005).

The U.S. restrictions on HIV/AIDS funding remain, in the minds of many, the single largest current barrier to effective work with commercial sex workers, and the clearest evidence that U.S. policy is based on a moral position, rather than a human rights-informed public health position. For researchers and those implementing scale-up programs for prevention and treatment, it is not merely a potentially-nagging obstacle—it is an ethical issue as well. How should researchers and health care/behavioral health professionals respond to the restrictions, and work in environments where the restrictions apply?

Ethical research requires respect for the fundamental principles of autonomy, beneficence, and distributive justice, and it is difficult to see how those principles are protected by current U.S. policy. If, as most analysts have noted, the reality of prostitution is structurally-engendered for many of the world's people, than a focus on criminalization of individual behavior misdirects the focus of necessary attention. And forcing aid recipients to adopt policies that will likely drive the intended audience further away, rather than bringing them closer in to systems of care, supports categorical exclusions, and inhibits the rights of individuals to access services that meet basic human needs.

Researchers should be vocal and public in their opposition to the U.S. policy, and consider carefully the price paid for accepting certain U.S. funds under current conditions, if for no other reason than that it will be important to prospective research community participants—in this case, sex workers—to hear researchers take such a stand.[2]

Omissions

The preceding discussion has glaring omissions, which are largely the consequence of omissions in the research itself. Sex work among boys and men, and how that work is contextualized under various social conditions around the globe, remains woefully understudied. Understanding of the interplay between social forces and variable degrees of social equality and inequality remains imprecise, and deserves much greater analysis. Theoretical analysis of sex work from a gender perspective needs more intensive interrogation; it is now possible to argue, with equal force, for and against decriminalization, with each side claiming a feminist legitimacy; the fact of such contradictory duality remains problematic.

Two domains of sex work that are scarcely touched by researchers involve research "subjects" who are not sex workers at all: researchers and customers. Knowledge of, attitudes toward, and interactive behavior with sex workers by members of the research community have not been studied, and deserve to be. After all, at least some researchers are likely to have been involved with sex workers as customers (a probable taboo to raise), and their experience, and subsequent conclusions about that experience, constitute legitimate areas of study if we are to improve research partnerships between the community (however inexact it is) of researchers and the community (however diffuse it may be) of sex workers. And the perspectives, perceptions, motivations, behaviors, and needs of sex work customers (primarily men, but in some contexts, women as well) in general demand much more research; if sex work is appropriately termed a transaction, research continues to focus almost exclusively on one actor in that transaction.

Those motivations—like all motivations related to sexual behavior in gender-unequal societies that commoditize sex, prize youth, and reject homoerotic affection—are likely to be much more complicated than sexual desire alone. A recent Swiss study, for example, found that men who solicit male sex workers do so from a variety of "frames," or situational definitions, including perceptions of romantic content that may or may not have been present, and a frame in which customers are acting out idealized encounters with "types" whose desirability is reinforced by social and media messages. This has significance from an HIV point of view, the researchers concluded, since the frame that customers operated from also influenced the HIV protective behaviors they adopted in the encounter (Gredig, Pfister, Parpan-Blaser, and Nideroest, 2006).

Recommendations and Reconsiderations

Many of the recommendations made elsewhere in this volume apply in the case of research involving commercial sex workers as well: participant representation on research committees; engagement of participant community leadership (especially from sex worker rights organizations); establishment of structures for community-decision about the design, implementation, and dissemination of research; participatory development of a research agenda based on a multiplicity of causes and effects, and on interlocking individual and structural minority community issues; researcher engagement in advocacy designed to improve the lives and working conditions of sex workers (beyond the specific concern of HIV infection); and research community acknowledgement of, and attempt to correct, deep mistrust of the medical establishment that underlies and contaminates so many interactions between sex workers and researchers.

Researchers must also acknowledge and respect the fragility of day-to-day existence for many sex workers, and the resulting generalized "chaos of life," that results from an occupation that is, overwhelmingly, criminally sanctioned and morally condemned. Sex work is risky, with few protections; stepping forward to participate in research, or to seek out medical care, means potential exposure and increased risk. These are not light considerations, and must be addressed as a fundamental part of the research design. Beyond assurances of confidentiality, consent, and autonomy within the research setting itself, researchers must be prepared to provide assurances of safety for participating in the research in the first place. This may require adoption of special provisions in the determination of research locations, times, and records, and vigorous protection of those records against any attempted seizure by police or legal authorities. Above all, it requires a thorough and nuanced knowledge of the ways in which sex work occurs in a local community— its codes, rituals, transactional rules, patterns, and risks—in order to design research methods that fully protect sex worker research participants against further harm.

Activity-Defined Populations: Implications from Sex Workers

This chapter has examined research ethics and commercial sex workers as a representative example of an "activity-defined population," but it takes little effort to recognize that many of the points raised herein will also apply to other activity-defined populations—especially to injection drug and other drug-using populations, whose lives and experience, like sex workers, are shaped as much by social attitudes, criminal codes, and structural conditions as they are by individual behavior. The complexities of research ethics in and for activity-defined population should serve, once again, as a reminder that HIV often involves behavior explicitly condemned or rejected by vast numbers of people, and, simultaneously, behaviors in which millions of people also engage—the condemnation, and activity, sometimes residing in the same individual. This makes the development of an ethics-based, human rights-orientated research agenda exceedingly complex, because it makes open discussion of morally- and legally-sanctioned behaviors impermissible in the first place: it is hard to craft ethics out of an awkward, embarrassed, shame-based silence. Be that as it may, it goes nearly without saying that such silences have served as enormous fuel to the HIV epidemic since its presence was first noted. In the case of activity-defined populations such as sex workers and drug users, there are only our neighbors, ourselves, and the social fictions about "the other" that repress the kind of open conversation that truly needs to take place.

Notes

1. Another possibility is the adoption of specific criminal sanctions focused on the *customers* in sex work. In most jurisdictions solicitation or procurement—or other legal variations—constitute criminal behavior, but in the Merauke regency in the Indonesian province of West Papua officials introduced Bylaw No. 5/2003, which permits men to be fined up to Rp 5 million (approximately $500 U.S.) for refusing to wear a condom with sex workers. The bylaw also permits sex workers to refuse service to clients who will not wear a condom during sex. Officials recognize the complexity of enforcement, but the bylaw is significant because of the legal burden it puts on customers in the obligation to reduce transmission of HIV (Herget, 2005).
2. In 2006 courts in New York and Washington, DC, in response to lawsuits brought by U.S. advocacy groups, ruled the U.S. requirements unconstitutional and issued injunctions against enforcement. New York District Court Judge Victor Marrero, in his ruling, found the government's stance to be "offensive to the First Amendment as it improperly compels speech by affirmatively requiring Plaintiffs to adopt a policy espousing the government's preferred message." The legal status of the requirement, however, remains unclear; it is likely there will be additional legal activity before the matter is resolved (Bristol, 2006).

References

Agbola, O., Siushansian, J., & Wong, T. (2006). Risky business: sex trade as a risk factor for HIV and other sexually transmitted infections (STIs) in Canadian street youth. Presented at the XVI International AIDS Conference, Toronto, Canada, August 13–18 [abstract no. WEPE0851].

Aklilu, M., Tsegaye, A., Tsigireda, B., Miriam, D.H., & van Benthem, B. (2001). Factors associated with HIV-1 infection among sex workers of Addis Ababa, Ethiopia. *AIDS, 15,* 87–96.

Alary, M., Mukenge-Tshibaka, L., Bernier, F., Geraldo, N., Lowndes, C., & Meda, H. (2002). Decline in the prevalence of HIV and sexually transmitted disease among female sex workers in Cotonou, Benin. *AIDS, 16,* 463–470.

Altice, F. L., Mostashari, F., & Friedland, G. H. (2001). Trust and the acceptance of and adherence to antiretroviral therapy. *Journal of Acquired Immune Deficiency Syndromes, 28,* 47–58.

Betteridge, G. (2005). Legal network calls for decriminalization of prostitution in Canada. *HIV/AIDS Policy & Law Review, 10(3),* 11–13.

Bristol, N. (2006). US anti-prostitution pledge decreed 'unconstitutional'. *The Lancet, 368,* 17–18.

Brussa, L. (2006). Prostitution in Europe and need for holistic strategies. Presented at the XVI International AIDS Conference, Toronto, Canada, August 13–18 [abstract no. MOPE0991].

Central and Eastern European Harm Reduction Network. (2005). *Sex Work, HIV/AIDS, and Human Rights in Central and Eastern Europe and Central Asia.* Last accessed January 19, 2007; Available at http://www. soros.org/initiatives/health/focus/sharp/articles_publications/publicatons/sexwork_20051018/sex%20work%20in%20ceeca_report_2005.pdf

Cohan, D. L., Kim, A., Ruiz, J., Morrow, S., Reardon, J., Lynch, M. et al. (2005). Health indicators among low income women who report a history of sex work: The population based North California Young Women's Survey. *Sexually Transmitted Infections, 81,* 428–433.

Davis, P. J. (2000). On the sexuality of 'town women' in Kampala. *Africa Today, 47(3/4),* 28–60.

Deb, S. (2006). Experience of CSWs with HIV/AIDS: An exploratory study in Kolkata, India. Presented at the XVI International AIDS Conference, Toronto, Canada, August 13–18 [abstract no. CDD1300].

Doyal, L. (1994). *AIDS: Setting a Feminist Agenda.* London: Taylor and Francis.

Editorial. (2005). The trials of tenofovir trials. *Lancet, 365,* 1111.

Gredig, D., Pfister, A., Parpan-Blaser, A., & Nideroest, S. (2006). Customers of male sex workers and HIV-protection behavior. Presented at the XVI International AIDS Conference, Toronto, Canada, August 13–18 [abstract no. CDD0564].

Herget, G. (2005). Indonesia: Men refusing to wear a condom with sex workers are liable to fines. *HIV/AIDS Policy and Law Review, 10(3),* 30–31.

Hickey, D. & Douglas, D. (2006). Fluid identities and workplaces: working with women of color in venues where sexual exchange occurs. Presented at the XVI International AIDS Conference, Toronto, Canada, August 13–18 [abstract no. CDD0555].

Hill-Mlati, J. & Rijks, B. (2006). Transactional sex – findings from HIV vulnerability assessment on commercial farms in South Africa. Presented at the XVI International AIDS Conference, Toronto, Canada, August 13–18 [abstract no. CDD0569].

Hong, Y., Li, X., Fang, X., Zhao, R., and Lin, D. (2006). Relationship of depression and condom use with clients among female sex workers in China. Presented at the XVI International AIDS Conference, Toronto, Canada, August 13–18 [abstract no. WEPE0847].

Johari, V. & Murthy, L. (2006). Sex workers and police violence: rights and risk. Presented at the XVI International AIDS Conference, Toronto, Canada, August 13–18 [abstract no. CDE0382].

Jurgens, R., Elliott, R., Csete, J., Palmer, D., Livingston, A., Liang, G., et al. (2006). Nothing about us without us: Greater, meaningful involvement of drug users in the response to HIV/AIDS. Presented at the XVI International AIDS Conference, Toronto, Canada, August 13–18 [abstract no. TUAE0101].

Kunin-Goldsmith, J. & Laurenceau, B. (2006). Meaningful involvement: Enabling the marginalized to defend themselves. Presented at the XVI International AIDS Conference, Toronto, Canada, August 13–18 [abstract no. MOPE0929].

Laurent, C., Seck, K.A., Coumba, N., Kane, T.B., Samb, N.C., Wade, A.A., et al. (2003). Prevalence of HIV and other sexually transmitted infections, and risk behaviours in unregistered sex workers in Dakar, Senegal. *AIDS, 17,* 1811–1816.

Loff, B., Jenkins, C., Ditmore, M., Overs, C., & Barbero, R. (2005). Unethical clinical trials in Thailand: A community response. [letter]. *Lancet, 365,* 1618–1619.

Mbilinyi, M. & Kaihula, N. (2000). Sinners and outsiders: The drama of AIDS in Rungwe. In Baylies, C., Bujura, J. & the Women and AIDS Group. *AIDS, Sexuality, and Gender in Africa: Collective Strategies and Struggles in Tanzania and Zambia.* London: Routledge.

Mukherjee, S. (2006). Gaining acceptance of the target community – a real challenge for a doctor working in sex workers project. Presented at the XVI International AIDS Conference, Toronto, Canada, August 13–18 [abstract no. WEPE0854].

Murugan, S., Bhattacharjee, J. F., Blanchard, J.F., & Anthony, J. Community-based tools for planning and monitoring HIV/AIDS prevention programs in female sex worker (FSW) communities in India. Presented at the XVI International AIDS Conference, Toronto, Canada, August 13–18 [abstract no. MOPE0578].

Obbo, C. (1993). HIV transmission through social and geographic networks in Uganda. *Social Science and Medicine, 36(7),* 949–955.

Outwater, A. (1996). The socioeconomic impact of AIDS on women in Tanzania." In Long, L.D. & Ankrah, E.M. (Eds.). *Women's Experiences with HIV/AIDS: An International Perspective* (pp. 112–122). New York: Columbia University Press.

Palepu, A., Tyndall, M., Yip, B., Li, K., Hogg, R., O'Shaughnessy, M. et al. (2001). Adherence and sustainability of antiretroviral therapy among injection drug users in Vancouver. *Canadian Journal of Infectious Diseases, 12(suppl B),* 32B.

Phillips, M.M., & Moffett, M. (2005). Brazil refuses U.S. AIDS funds, rejects conditions. *Wall Street Journal,* May 2, at A5.

Renaud, M.L. (1997). *Women at the Crossroads: A Prostitute Community's Response to AIDS in Urban Senegal.* Amsterdam: Overseas Publishers Association/Gordon and Breach Publishers.

Roehr, B. (2005). Charity challenges US 'anti-prostitution' restriction. *British Medical Journal, 331,* 420.

Schleifer, R. (2005a). United States: Challenges filed to anti-prostitution pledge requirement. *HIV/AIDS Policy & Law Review,* 10(3), 21–22.

Schleifer, R. (2005b). United States: Funding restrictions threaten sex worker rights." *HIV/AIDS Policy & Law Review,* 10(2), 26–27.

Schoepf, B.G. (1997). AIDS, Gender, and Sexuality during Africa's Economic Crisis. In Mikell, G. (Ed.). *African Feminism: The Politics of Survival in Sub-Saharan Africa* (pp. 310–322). Philadelphia: University of Pennsylvania Press.

Shannon, K., Bright, V., Duddy, J., & Tyndall, N. W. (2005). Access and utilization of HIV treatment and services among women sex workers in Vancouver's downtown eastside. *Journal of Urban Health,* 82(3), 488–497.

Simonsen, J.N., Plummer, F.A., Nguyi, E.N., Black, C., Kreiss, J.K., Gakinya, M.N., et al. (1990). HIV infection among lower socioeconomic strata prostitutes in Nairobi. *AIDS, 4(4),* 139–144.

Strathdee, S.A. (1998). Barriers to use of free antiretroviral therapy in injection drug users. *Journal of the American Medical Association, 280,* 547–549.

Tandia, O. (1998). Prostitution in Senegal. In Kempadoo, K. & Doezema, J. (Eds.). *Global Sex Workers: Rights, Resistance, and Redefinition* (pp. 240–245). New York: Routledge, 1998.

Trafficking Victims Protection Reauthorization Act of 2003, 22 U.S.C. § 7110(g)(2) (2003).

United States Leadership against HIV/AIDS, Tuberculosis, and Malaria Act of 2003, 22 U.S.C. § 7631(f)(2003).

van den Hoek, A. (2001). High prevalence of syphilis and other sexually transmitted diseases among sex workers in China: Potential for fast spread of HIV. *AIDS, 15,* 753–759.

Vanwesenbeeck, I. (2001). Another decade of social scientific work on sex work: A review of research 1990–2000. *Annual Review of Sex Research, 12,* 242–289.

Weiner, A. & Lorber, K. (2006). Childhood trauma and the relationship to drug and condom use, and HIV status among streetwalking sex workers in New York City. Presented at the XVI International AIDS Conference, Toronto, Canada, August 13–18 [abstract no. CDD0533].

White, L. (1990). *The Comforts of Home: Prostitution in Colonial Nairobi.* Chicago: University of Chicago Press.

Wolffers, I. & van Beelen, N. (2003). Public health and the human rights of sex workers. *Lancet, 361,* 1981.

Yu, D. (2006). Decriminalizing sex work: implications for HIV prevention and control in China. Presented at the XVI International AIDS Conference, Toronto, Canada, August 13–18 [abstract no. MOPE0989].

Case Study Ten
Research Relating to Injection Drug Users

Linda S. Lloyd, Dr.P.H.

Background

The Alliance Healthcare Foundation (AHF) conducted a comprehensive needs assessment in order to better understand the needs of injection drug users (IDUs) across the continuum of active abuse of alcohol and other drugs to entry into treatment and recovery. The results were used to help the foundation identify effective program strategies for preventing the transmission of HIV/AIDS and hepatitis among IDUs, their syringe and other drug paraphernalia-sharing partners, and their sexual partners. Several forms of data collection were employed, including in-depth semi-structured interviews with IDUs both in and out of treatment by outreach workers, many of whom were in recovery themselves. Detailed information was collected on drug use history and types of drugs used, sharing of drug paraphernalia, with whom the equipment was shared, previous experiences with treatment programs, knowledge and/or use of the San Diego underground syringe exchange program, and use of condoms. The interviews lasted between 30 and 60 minutes.

Ethical Issues

The ethical issues addressed as part of designing the needs assessment were: (1) hiring interviewers who are in recovery themselves to work on a study that required them to be in contact with active drug users, (2) whether to provide incentives for the in-depth interviews, (3) whether to compensate both the outreach worker and the person being interviewed, and (4) what type of incentive should be provided.

The AHF convened the Injection Drug Use (IDU) Study Group to provide direction to the design and implementation of the needs assessment. This working group was composed of individuals with direct experience working with IDUs, individuals with expertise in HIV/AIDS prevention, care and treatment, and individuals with experience in substance abuse treatment and recovery.

Hiring Individuals in Recovery

The Study Group was asked to address the issues of hiring interviewers who are in recovery and whether to provide incentives for the collection of primary data via interviews. The majority of the individuals to be interviewed were actively injecting drugs, while the balance were IDUs in a treatment setting, generally embarking on their first phase of the recovery process when relapse is of particular concern. The individuals to be recruited to conduct the interviews were outreach workers familiar with the target population, often by way of their own past IDU history, and generally already employed by non-profit agencies serving the target population through a variety of programs.

The group engaged in an open and frank discussion on the marginalization of drug users, in particular IDUs, by mainstream society. The Study Group members felt that individuals who were in recovery could have greater success gaining the trust of active drug users and obtaining accurate information given their own drug-use experience. In light of the very real challenges individuals have finding employment because of a history of drug use, the consensus of the Study Group was that it would be hypocritical to not allow individuals in recovery to freely decide whether they wanted to participate in the needs assessment process. Some members noted that participation in this study also afforded these individuals another opportunity to "give back" to their community, an important aspect of the recovery process. And finally, because the outreach workers were already working with active injection drug users through their employment with local non-profits, Study Group members did not feel that this project *per se* would endanger their recovery more than their current employment. Most of the agencies offered Early Intervention Programs as a benefit and interviewers could be encouraged to take advantage of that should the need arise during the interviewing phase of the study.

Incentives

Study Group members acknowledged that the provision of an incentive was an important way to recognize the key role of both the outreach workers and interviewees in a successful needs assessment process. With respect to the first two ethical issues surrounding the use of incentives, *(1) Should incentives be offered as part of the data collection process* and *(2) Should both outreach workers and interviewees be offered an incentive,* Study Group members agreed that incentives were appropriate and should be offered to both the outreach workers conducting the interviews and to the individuals agreeing to be interviewed.

However, opinions varied within the group regarding the third issue – *What type of incentive should be offered.* The group discussed the merits of providing vouchers for food or other sundry items versus a small monetary incentive, and whether different incentives should be offered to the outreach workers (e.g., a monetary incentive) and the interviewees (e.g., vouchers). Some suggested that only vouchers be given since the interviewees were either actively using drugs or in recovery. These group members felt that providing a financial incentive to substance abusing individuals would serve as an enabler for the continued purchase of drugs or could endanger an individual's recovery process. Other members disagreed, stating that IDUs were entitled to the same level of respect for persons that would be provided for other population groups. This respect included the individual's right to determine how to best use a financial incentive they might receive in exchange for their participation in the interview, rather than attempting to limit use of the incentive. The group also acknowledged that it could not control how people used their incentive regardless of what was given to them since vouchers can also be converted into a cash value. Some felt that the assumption that funds would be used exclusively to purchase drugs was prejudicial and discriminatory.

Before making a final decision on the matter of incentives, the Study Group requested that the two lead consultants on the assessment process seek the input of the San Diego Association of Community Health Outreach Workers (ACHOW). ACHOW was a coalition of staff and volunteers from a variety of County and community agencies conducting field outreach to target populations, including IDUs, to link them to HIV testing and services. The consultants requested time on the monthly ACHOW agenda to present and discuss the project and implementation strategies, and solicit participation of outreach workers already working with the target populations. A written summary description of the proposed assessment was provided in addition to a verbal presentation at the meeting. ACHOW members acknowledged the importance of the project and many expressed interest in recruiting and interviewing individuals for the study. The consultants collected names and contact information of these individuals.

It was at that point, after receiving voluntary commitment to the project from several ACHOW members, that the consultants raised the concept of compensating interviewers and interviewees. While the outreach workers felt that incentives were appropriate for both groups, especially since most of the interviews were to be conducted on their own time, there were some differences of opinion on the type of incentive that should be offered. After a rather short discussion that somewhat mirrored the issues raised by the Study Group, the ACHOW members concluded that a small monetary incentive was appropriate and should be offered to both interviewers and interviewees. Acknowledgement of each individual's contribution and respect for their own ability to decide how best to use a financial incentive were deciding factors in the discussion.

The consultants reported the response from ACHOW to the Study Group, which concurred. The group determined that a $20 incentive would be offered for each interview to the outreach worker, and that each interviewee would receive the same amount. In order to allow as many outreach workers to participate as possible, a limit of between 15 and 20 interviews per person was given. Twenty outreach workers participated, each conducting an average of six to eight interviews.

Conclusion

The Study Group members determined that hiring individuals in recovery as interviewers and compensating both interviewers and interviewees for their time was not only appropriate, but essential to show this highly marginalized group that their experiences and opinions were valued. By involving outreach workers who were known to the target population, the IDU Study Group was able to conduct an outreach process that maintained trust and did not threaten the confidentiality of the interviewees since names, addresses and telephone numbers did not need to be collected. As a result, the subsequent interviews offered candid insights which largely dispelled many stereotypes about IDUs and presented opportunities for successful interventions. None of the outreach workers experienced relapse during the course of the study, while their important contributions on behalf of the study were acknowledged both in the widely distributed final report and in a follow-up community conference at which the findings and recommendations of the report were released.

References

Lloyd, L.S., O'Shea, D.J., & the IDU Study Group. (1994). *Injection Drug Use in San Diego County: A Needs Assessment*. San Diego: Alliance Healthcare Foundation. Available at www.alliancehf.org.

Chapter 17
Training the Next Generation of Researchers

Unfortunately, relatively little literature exists that focuses on the content and efficacy of training programs for HIV researchers, particularly with respect to training in ethical issues or ethical issues arising in the context of that training. As a result, much of this chapter draws on what has been reported regarding training programs for clinicians that may also be relevant to researchers.

The Need for Training

Studies consistently suggest that individuals preparing for research and clinical careers receive inadequate training with respect to HIV/AIDS. Campos and colleagues (1989) surveyed 92 doctoral programs in psychology that were approved by the American Psychological Association (APA), as well as 169 predoctoral internship programs. They found that 75% of the doctoral programs and 40% of the internship programs did not systematically provide information about HIV/AIDS and slightly less than half of the graduate programs even offered training in human sexuality. Those that did provide HIV/AIDS training on a systematic basis often failed to include critical issues within their programs. For instance, 68% offered little training on behavioral medicine, 30% did not include training related to substance abuse, 67% failed to address primary prevention of HIV/AIDS, less than half (41%) provided instruction on urban or minority issues, and more than two-thirds (67%) failed to provide training in community psychology. A survey of training in 115 APA-approved clinical psychology doctoral programs found that AIDS was identified as a specialized area of training in only 7% of the programs (Sayette and Mayne, 1990). Only 15 faculty members were identified as conducting research related to HIV/AIDS.

As recently as 1998, another survey of 585 practicing psychologists found that more than half had not received any formal training or education about the clinical presentation, treatment, or transmission of HIV/AIDS (Schmeller-Berger, Handal, Searight, and Katz, 1998). The majority of respondents received their HIV-related information through radio, television, and newspapers. Although the survey related to practicing psychologists, it is likely that at least some psychologists engaged in research would have under-taken similar programs.

A study of health care providers in Mexico found that 75% had received some training related to HIV/AIDS (Infante et al., 2006). Despite this training, how-ever, almost one-quarter of the respondents indicated that they would not buy food from an HIV-infected individual and 16% thought that HIV-infected per-sons should be banned from public services. More than one-third believed that employers and administrators should have the right to know the HIV status of employees, and many distinguished between "innocent victims" and "guilty" ones, thereby justifying delays and stigmatization in the context of health care. The extent to which the providers also engaged in HIV research was not indicated.

It has been suggested that it may be particularly important to train minority researchers in the field of HIV (Marin and Diaz, 2002). Although access to minority populations for research may be particularly difficult, it may be some-what easier for minority researchers to develop collaborative relationships with these communities. However, minority researchers in this field are relatively scarce (Marin and Diaz, 2002).

The training of researchers is critical not only to facilitate needed research, but also to develop a cadre of individuals sufficiently familiar with principles of research to review research protocols, both for potential funding and in the context of ethics review committees. For instance, although many ethics review committees in parts of Africa will likely be required to review protocols for HIV vaccine trials, most have no or only moderate capacity to do so, both with respect to the ethical issues and the scientific aspects involved (Milford, Wassenaar, and Slack, 2006).

Developing Program Content

Researchers, like clinicians, must have basic knowledge about HIV, its epidemiol-ogy, its prevention, and its impact on individuals, families, and communities. Table 17.1 presents the various domains that a comprehensive HIV/AIDS training program would include, together with a listing of some elements encompassed within each such domain.

Mechanisms for Training

Research training grants may be available to fund training programs for HIV researchers in a variety of disciplines. The Fogarty International Center, part of the National Institutes of Health, has partnered with several institutions in the United States to fund training programs for HIV researchers based in a number of countries (United States Department of Health and Human Services, 2004). These and other NIH-funded programs have trained individuals at the master's

TABLE 17.1 Recommended content for comprehensive HIV/AIDS training programs
for researchers

Domain	Elements
Medical Overview	HIV transmission, epidemiology, testing, and prevention, disease course, treatment, and progression; symptoms
Cultural and Historical Context	History of disease "discovery" and identification; disease impact on subgroups; changing demographics of epidemic; population and governmental response to HIV/AIDS in various countries and localities; discrimination and scapegoating;
Psychological and Psychosocial Aspects	Responses to HIV testing; psychiatric and other medical comorbidities; human sexuality; coping strategies; chronic and terminal illness; death and dying; impact on caregivers; family composition and dynamics; impact of HIV treatment; substance use; identity issues; suicide risk; stigma and stigma management;
Community Aspects	Local community response to HIV/AIDS; impact on local community; type and extent of community services available to HIV-infected and -affected persons; local economic impact of disease;
Legal and Ethical Issues	*Participant-focused issues*: Evolution of legal and ethical issues over time; confidentiality and privacy; informed consent; mandatory reporting (infectious disease; unprotected sexual relations, etc.); contact tracing; duty to warn; protection of vulnerable research participants; risks and benefits of study participation; obligations to community and research participants; legal vulnerabilities of/risks to subgroups (e.g., prisoners, sex workers, undocumented immigrants) *Profession-focused issues*: Data ownership, sharing, storage, and retention; professional boundaries; obligations to colleagues;
Human Rights	International protections for HIV-infected persons; stigmatization and discrimination
Research and Methodology	Study design; quantitative data collection methods; sampling; statistical analysis; qualitative data analysis; preparation of manuscripts, presentations, and grant proposals;

and doctoral levels (National Institutes of Health, 2006). Depending upon the specific program, training has focused one or more of the following areas:

• animal models
• applied mental health research
• behavioral sciences
• bioethics
• clinical trials
• communicable disease control
• cultural and social factors
• epidemiology
• family structure and dynamics
• HIV vaccinology
• human rights
• immunology
• qualitative research methods

• social support and networks
• substance use
• tuberculosis
• vaccine preparedness
• virology

Marin and Diaz (2002) described a program developed through the University of California San Francisco Center for AIDS Prevention Studies (CAPS) to train scientists of color, with an emphasis on preparation for the successful submission of grant applications This program included the following elements:

• Summer 1, at CAPS: emphasis on development of the research question, literature review, and conceptualization of the problem
• Academic Year 1, at home institution: conduct of preliminary studies
• Summer 2, at CAPS: analysis and synthesis of preliminary data, draft of research proposal
• Academic Year 2, at home institution: revision and submission of research proposal to potential funding source
• Summer 3, at CAPS: revision of proposal in response to reviewer comments, or preparation to begin conducting research

Other program components included training on research with human subjects; seminars related to qualitative research, intervention planning, recruitment and retention, and grants management; internal peer review; and individualized mentoring (Marin and Diaz, 2002).

Long-term collaborations between countries and institutions may be critical to the professional development of HIV researchers. One such example is provided by the Ethio-Netherlands AIDS Research Project (ENARP), which was developed in order to strengthen both international HIV/AIDS research and the research capacity of Ethiopia (Sanders et al., 2000). Components of this collaborative effort included the addition and renovation of research facilities; the extension of electrical resources; the provision of equipment and materials; degree programs at the master's and doctoral levels; technical programs for laboratory technicians, maintenance engineers, and computer staff; and short-term fellowships abroad for Ethiopian scientists. Ph.D. programs were established in the fields of epidemiology, virology, immunology, and parasitology. Research resulting from this program has included the identification of HIV subtypes in Ethiopia, a study on HIV infection progression, the evaluation of laboratory markers to aid providers in initiating antiretroviral therapy, and a study of the relationship between HIV and intestinal parasitic infections (Sanders et al., 2000).

Other models of HIV training that have been utilized with health care professionals may also be useful in training researchers. These include: short-term training sessions over several days, in-service training for a period of a few months, didactic sessions on a regular basis, mentoring visits by experienced professionals, refresher courses, off-site clerkships, consultation systems, case conferences on a regular basis, and institutional exchanges of information and resources (McCarthy, O'Brien, and Rodriguez, 2006).

Training is also available via the internet (McCarthy, O'Brien, and Rodriguez, 2006). For instance, PLoS Medicine provides free access to all readers (Eisen, Bowen, and Varmus, 2004). The endeavor is supported by charging the authors of manuscripts a publication fee to cover the costs of peer review, editorial oversight, and publication. In addition to providing access to researchers and researchers-in-training, this mechanism also allows access to providers, patients, and research participants because it is "open access." Additionally, courses can be developed that are web-based and permit learners to progress at their own pace (McCarthy, O'Brien, and Rodriguez, 2006).

Evaluations of specific training programs for health professionals have found them to be efficacious. Various studies have found that training may resulted in an increase in providers' knowledge (Wertz, Sorenson, Liebling, Kessler, and Heeren, 1987), a reduction in their level of anxiety about having contact with HIV-infected patients (Sherr and McCreaner, 1989), an increase in their comfort level in dealing with substance users (Mejita, Denton, Krems, and Hiatt, 1988), and an increase in their level of professional responsibility towards HIV-infected individuals (Ezedinachi et al., 2002). However, it is unclear to what extent such findings would be equally applicable to HIV researchers, in contrast to individuals in clinical practice.

The Training Process

Often, training is reduced to a focus on the substantive content, without regard to the process of training or the process by which researchers are socialized into the field of research. It is critical to the development of professional researchers, however, that attention also be paid to these issues.

Training, for instance, can be accomplished both on an informal basis, through direct supervision of training activities, through formal meetings to review progress and accomplishments, and by role modeling. The mentor should be prepared to assist the trainee not only in the development of his or her research skills, but also in the development of scientific integrity, the formation of professional networks and relationships, the choice of a career path, and the development of a "thick skin" that allows the trainee to persevere in the face of professional disappointments. Trainees will also need guidance on such critical issues as time management, the development of skills for grant writing and manuscript preparation, and the preparation of a curriculum vitae.

Ethical Issues

Numerous ethical and legal issues may arise in the context of developing and implementing training programs. Because many of these issues are addressed in the following case study, they will not be discussed in detail here. These issues include access to and ownership of data; confidentiality and privacy concerns related to research participants and to the trainees themselves; ethical obligations owed by trainees to their colleagues; responsibilities of the training mentors to the trainees; and ethical obligations of the mentors/administrators of the training program to the program, their colleagues, and their institutions.

Acknowledgements. This work was supported in part by the Fogarty International Center R25 TW001603, International Research Ethics Training Program.

References

Campos, P.E., Brasfield, T.L., & Kelly, J.A. (1989). Psychology training related to AIDS: Survey of doctoral graduate programs and predoctoral internship programs. *Professional Psychology: Research and Practice, 20,* 214–220.

Eisen, M.B., Brown, P.O., & Varmus, H.E. (2004). PLoS Medicine—A medical journal for the internet age. *PLoS Medicine, 1,* e31. Last accessed September 2006; Available at http://www.plosmedicine.org

Ezedinachi, E.N.U. et al. (2002). The impact of an intervention to change health workers' HIV/AIDS attitudes and knowledge in Nigeria: A controlled trial. *Public Health, 116,* 106–112,

Infante, C. et al. (2006). El stigma asociado al VIH/SIDA: El caso de los prestadores de sevricios de salud en Mexico [The HIV/AIDS-associated stigma: The case of health care providers in Mexico]. *Salud Publica Mex* [Public Health Mexico], *48 (2),* 141–150.

Marin, B.V. & Diaz, R.M. (2002). Collaborative HIV Research in Minority Communities Program: A model for developing investigators of color. *Public Health Reports, 117(3),* 218–230.

McCarthy, E.A., O'Brien, M.E., & Rodriguez, W.R. (2006). Training and HIV-treatment scale-up: Establishing an implementation research agenda. *PLoS Medicine, 7,* e304. Last accessed November 2006; Available at http://plosjournals.org

Mejita, C.L., Denton, E., Krems, M.E., & Hiatt, R.A. (1988). Acquired immuno-deficiency syndrome (AIDS): A survey of substance abuse clinic directors' and counselors' perceived knowledge, attitudes, and reactions. *Journal of Drug Issues, 18,* 403–419.

Milford, C., Wassenaar, D., & Slack, C. (2006). Resources and needs of research ethics committees in Africa: Preparations for HIV vaccine trials. *IRB: Ethics & Human Research, 28(2),* 1–9.

National Institutes of Health. (2006). Last accessed September 27, 2006; Available at http://crip.cit.gov/crisp/

Sanders, E. et al. (2000). Development of research capability in Ethiopia: The Ethio-Netherlands HIV/AIDS Research Project (ENARP), 1994–2002. *Northeast African Studies, 7(2),* 101–118.

Sayette, M.A. & Nayne, T.J. (1990). Survey of current clinical and research trends in clinical psychology. *American Psychologist, 45,* 1263–1266.

Schmeller-Berger, L., Handal, P. J., Searight, H. R., & Katz, B. (1998). A survey of psychologists' education, knowledge, and experience treating clients with HIV/AIDS, *Professional Psychology: Research and Practice, 29(2),* 160–162.

Sherr, L. & McCreaner, A. (1989). Summary evaluation of the national AIDS counseling training unit in the U.K. Special issue: AIDS counseling: Theory, research, and practice. *Counselling Psychology Quarterly, 2(1),* 21–32.

United States Department of Health and Human Services. (2004). Fogarty International Center announces research training grants to tackle AIDS and tuberculosis. NIH News, November 30. Last accessed September 27, 2006; Available at http://www.fic.nih.gov/

Wertz, D.C., Sorenson, J.R., Liebling, L., Kessler, L., & Hereen, T. (1987). Knowledge and attitudes of AIDS health care providers before and after education programs. *Public Health Reports, 102,* 248–254.

Case Study Eleven
Ethical Issues in Training the Next Generation of Researchers

Oscar Grusky, Ph.D.

Introduction

Faculty and staff have a responsibility not only to prepare students for the technical aspects of their work but also for providing them with ethical guidance and monitoring the learning process. This essay builds on experiences with the author's National Institute of Mental Health (NIMH) supported multidisciplinary research training program on service systems for persons living with HIV/AIDS (PLA) (T32 MH19127, 1989-present). Hence, this is a case study of some ethical issues that arose in the context of a training program (an organization) rather than the case of a single individual.

Although the program is based primarily at UCLA, it also draws upon the Drew University of Science and Medicine and the RAND Corporation for faculty resources and works cooperatively with the Los Angeles Health Department and community-based HIV organizations throughout Los Angeles County to attain shared objectives.

The basic objectives of the program are to instruct trainees about social science behavioral and psychiatric and public health theories and methods in areas such as organizational analysis, health services research, health psychology, epidemiology, and biostatistics, so that trainees will be equipped to undertake research careers in HIV-related research. The training is designed to give participants the conceptual and methodological tools needed to conduct rigorous research.

The program focuses on psychosocial issues and health/mental health service systems for Persons Living with AIDS (PLA). It is led by scholars from medicine, psychiatry, psychology, sociology, health services, epidemiology, social policy, social welfare, nursing, and community health sciences. Each year 6 to 10 students are trained. Four to six pursue the Ph.D. in the social sciences, public health, or psychology and undertake a three-year sequence of courses and supervised research on mental health/HIV/AIDS. Two to four are postdoctoral trainees who already hold a Ph.D. degree and undertake a two-year sequence of courses and advanced research on mental health/HIV/AIDS. All trainees participate in a core seminar on HIV/AIDS that is offered the year round, mentored HIV research, and a carefully designed set of courses and field experiences.

To train persons in services research for persons living with HIV/AIDS it is essential that they be taught fundamental knowledge of the organization of community service systems and the methods and models of HIV service systems research. Trainees must work closely with faculty engaged in ongoing research in this area so that they can master the fundamentals of theory and research design, identify and comprehend potential ethical issues and issues of research responsibility, and learn to deal with the administrative realities of behavioral health sciences field research. There are 24 faculty-HIV scientific mentors associated with the training program.

Since the program's inception in 1989 a total of 75 individuals, 38 predoctoral and 37 postdoctoral fellows have been trained. Over two-thirds are currently engaged in HIV research and over half have submitted an HIV-related grant application. Almost 4 out of 10 predoctoral and two-thirds of postdoctoral trainees in the 2000–2005 funding cycle have published jointly with HIV faculty mentors. The program is closely connected with the NIMH-supported Center for HIV Identification, Prevention, and Treatment Services (CHIPTS) and has links with numerous other UCLA units engaged in HIV behavioral health research.

All identifying individual case information presented in this essay is fictional. Pseudonyms and other devices are used in the cases presented in order to prevent identification of any persons who may be involved. Although the situations are real, any similarity to actual persons is coincidental and is not intentional.

Ethics Training

Formal ethics training came about in the program because it was mandated by the NIMH. The primary manner in which trainees are formally instructed about research ethics and the responsible conduct of scientific research is by means of a special required course on this topic, Sociology 284, which is specifically designed for the program and is offered by the Program Director in alternate years. This seminar is designed to foster sensitivity and understanding of biosocial ethical issues in mental health/HIV/AIDS and health services research. The course provides researchers with the kinds of analytic tools needed to anticipate, understand, and hopefully resolve appropriately ethical conflicts and issues of scientific responsibility and integrity that emerge in social research. Course participants are required to complete the on-line program in protecting human research subjects of the UCLA Office for the Protection of Research Subjects. Students learn about the consent process and many other issues that relate to obtaining institutional review board approval for their research projects. Of course, a number of the issues raised in this course may also arise and are discussed in other courses taken by trainees such as in the required course on health services evaluation taught in the School of Public Health and the program's core seminar. Faculty mentors also provide significant supervised albeit less formal training in this area by serving as role models for trainees.

Ethical Aspects of Social Influence in Organizations

The AIDS research training program has goals, a core technology, and boundaries, and, therefore, by definition is an organization (Aldrich, 1999). A major feature of organizations is hierarchy. A hierarchy refers to a pyramid of influence whereby those at the top of the power structure (superordinates) have the opportunity and rights and privileges to influence those at lower levels in the organization (subordinates). Training program faculty are superordinates and students or trainees are subordinates. For example, the faculty and staff develop rules and requirements that trainees are required to fulfill. The organization's norms and values serve as limits on the process by which superordinates can influence subordinates and subordinates can influence superordinates. The AIDS research training program exists within a larger academic organization, UCLA, and consists of faculty, staff, postdoctoral students, and graduate students. The faculty may influence students directly and indirectly by transmitting information to them. The relationship between faculty and students is asymmetric as faculty generally have greater direct influence over students than students have over faculty. Nevertheless, it is a two-way influence process.

The underlying theme of this essay is the relationship between organizational power imbalance, social influence processes, and ethical issues. Many years ago, Lipset, Coleman, and Trow (1956) argued that despite a deep commitment to democracy in the U.S. most of the nation's complex organizations are oligarchies. This is because incumbent elites have a disproportionate amount of power, are able to stay in power for a long time, and ordinary members of the organization tend to avoid participating in its internal political processes. The UCLA training program to some extent fits this model. The faculty has overwhelming power compared to students, maintain their power over time, and lower level members such as students often avoid direct internal political participation. Lipset and his colleagues argue that this power imbalance contributes to the development of flawed social influence processes including misunderstanding and mis-communication that inhibit the effective transmission of information. Consequently, less than ideal ethical decisions may result in these systems.

This essay is concerned with selected ethical aspects of the influence process that characterize training in the program. Kelman (2001) has identified four moral principles in evaluating organizational social influence processes:

1. Autonomy. Is the influence attempt respectful of the values and concerns of the person being influenced?
2. Nonmaleficence. Is any harm involved in the influence attempt?
3. Beneficence. Is the welfare of the person being influenced promoted?
4. Justice/fairness. Are the rights of the person being influenced respected?

As Kelman (2001) points out, there is a continuum varying from persuasion to coercion along which social influence processes may be evaluated. Overall, persuasive methods such as conversation, dialogue, and discussion, superficially, at least, may often represent approaches to social influence that are respectful of the subordinate, non-harmful, considerate of the person's welfare, and are respectful of the person's rights. In contrast, coercive social influence tends to be none of these and is associated with perceived disrespect, is harmful to the target person's self-esteem, and involves very little dialogue. However, appearances may be deceptive. A finer and more nuanced analysis requires careful examination of the specific social and organizational context within which each aspect of persuasion and coercion occurs. There may be situations such as enforcement of racial desegregation decisions in the U.S. where coercion is both legitimate and justified. Likewise there may be situations where persuasion, or what appears to be persuasion, is quite similar to manipulation and coercion. Only an intensive examination of the context within which the influence process takes place and the changes that occur in the social situation and in the characteristics of the participants can provide the information needed to assess accurately where it should be situated on the persuasion-coercion dimension.

The Ethical Dilemmas

Letters of Evaluation and CVs As Influence Devices: A Need for Evaluative Neutrality?

The Ethical Issue

The first issue to be examined is faculty and students' ethical responsibility for describing themselves and others accurately in documents such as resumes, curriculum vitae (CV), and letters of evaluation. Two ethical questions are considered: To what extent are faculty and students obligated to be accurate in framing a personal document such as a CV or letter of evaluation? Should faculty and students strive for evaluative neutrality?

The Thompson case is one where a trainee candidate seriously misrepresented the status of her publications and apparently incorrectly described her teaching skills in her CV. In addition, faculty letters on Thompson's behalf stated only positive aspects of the student and omitted serious weaknesses in her performance despite the likelihood that the evaluators were aware of the candidate's deficiencies in teaching, publication, and other areas. Faculty may feel obligated to help their student gain employment by accenting their positive attributes in letters and omitting negative ones. Framing a letter in a generally positive rather than a negative manner is not uncommon. Although it may be understandable and admirable for a faculty member to try to help a student, the question is how far should a faculty member go in positively framing an evaluation of the candidate's characteristics? To what extent should a candidate's glaring weaknesses in fundamental areas such as teaching, research, and collegial relationships be downplayed or overlooked?

The Factual Situation

Thompson was accepted into the training program by the program's review committee to a considerable extent because of her extensive publication record. This included a book listed as in press with a major publisher, six peer-reviewed journal articles, and uniformly strong faculty letters of evaluation. The mentors asserted that she was an outstanding researcher, an excellent teacher (she had been an assistant professor for several years), and a very helpful colleague. For example, the faculty letters stated:

"Thompson is among our very best Ph.D. recipients in the last decade. She has a good balance of theoretical, substantive, and methodological strengths that enable her to engage in a broad range of teaching and research activities. . . . I feel that I know her very well . . . She had considerable teaching and research experience during her years at the university and since joining the faculty of R University as an assistant professor, where she has added many new courses to her portfolio. She has reported to me many times the very positive teaching evaluations she has received from students."

"I've known Thompson since 1988 when she came here . . . One of her strengths is her level of maturity. During her first couple of years at the university, she seemed to be devoting most of her time to teaching, which she enjoys and apparently does well. She has changed now how she uses her time, so that she has several papers and a book published in the last few years."

"Of the faculty in our department, Thompson has consistently had the highest student evaluations . . . I am quite familiar with her research publications since we share an interest in her discipline . . . I highly recommend Dr. Thompson to you. She is highly regarded by her students and . . . has a promising research program."

Unfortunately, Thompson's experiences as a trainee were largely unsuccessful. She produced no research publications. Her PowerPoint conference presentations were poorly organized and sometimes were almost incoherent. Unlike other trainees, she refused to practice her presentation in a training session prior to her first yearly student conference on the grounds that she was a highly experienced teacher and did not need the practice. Unfortunately, her presentation at the annual conference was dismal. It was by far the worst of over 20 student oral presentations made at a conference in front of an audience of about 80 persons including graduate and postdoctoral students, staff, faculty and representatives of community-based HIV organizations. The low point in her presentation occurred when she presented a slide with a table. The slide was so poorly constructed that it was indecipherable even to those in the first row. Worse yet, she seemed at the time completely unaware that the audience could not comprehend the table and could not understand anything that was said about it. When the Director discussed her presentation with her after the conference was over, she was contrite. She said she appreciated the honest feedback about the weaknesses of her presentation and she acceded to the request that prior to her next conference presentation she should this time give a practice presentation so that she would be better prepared. One year later she did present at a practice session. As was standard for all trainees, the Director worked with her before and after the practice presentation to improve her slides and her presentation style. The result was a conference presentation that was somewhat better than her previous effort and was judged to be adequate.

Thompson also delayed for several months selecting an HIV scientific mentor until pressured to do so by the Director. She finally selected a mentor. However, she had a very unsuccessful relationship with the mentor, a very prominent HIV behavioral scientist. The problem as described by the mentor was that Thompson kept reporting by email on her progress, but avoided face-to-face meetings and never produced promised drafts of a joint manuscript. The mentor sent the Director an email: "I have been meaning to write you for some time. I am concerned because I haven't seen Thompson in several months. I have emailed her about her progress and she keeps saying she is busy with presentations and writing. To date, I haven't seen a draft of her paper based on the data I provided her. Please advise."

Each year trainees are required to prepare a plan for the forthcoming year of research training. At the end of the year they are asked to evaluate their accomplishments in light of that plan. Thompson summarized her final year in the program by asserting that her goals for publication and training were "exceeded." However, when she first applied for her postdoctoral position she claimed that she had a book with a respectable publisher dated 2000 that was "in press." Two years later, at the completion of her postdoctoral training, the same book was still listed as "in press," but the date was changed. When the Director asked Thompson how she could claim that her expectations had been exceeded when she completed the year with no manuscripts and no publications, she was unable or chose not to respond. Thompson spent two years in the program and did manifest some modest improvement in her ability to communicate with others. However, overall she was unable to work effectively with her fellow students, with the program director, and with her HIV scientific mentors.

Analysis

The Thompson case reveals the possible harm that can result from the gross misrepresentation of a candidate's qualifications. One recent report cited the reprehensible behavior of a famous scientist who routinely wrote very negative letters of evaluation for his best postdocs so that they would remain in his laboratory for a long time (De Vries, Anderson, & Martinson, 2006). Although Thompson succeeded in obtaining a postdoctoral fellowship, her flawed performance in the program undoubtedly contributed to her inability to obtain a position once she completed the program. Thompson's CV and supporting faculty letters, viewed as influence attempts, violated the moral principles of autonomy, nonmaleficence, beneficence, and justice/fairness. The documents were disrespectful of the values and concerns of the target. Because they were misleading they were harmful to the training program staff and students. Indeed, the program faculty and students were publicly embarrassed by Thompson's poor teaching skills. Hence, the welfare of the program's members was reduced. Still another injustice or unfairness occurred by appointing Thompson since her appointment meant that a more highly qualified applicant did not receive an appointment.

Some constructive changes did emerge from the Thompson case. One of the positive developments of the Thompson experience was the institution by the program's leadership of improved trainee selection and evaluation procedures. Selection procedures were improved by the addition of telephone calls to faculty evaluators of the top candidates. The program leadership also decided to institute a Quarterly Evaluation of Trainee Performance Form which consists of an evaluation of each trainee by their HIV scientific mentor(s) every three months. This procedure has helped provide the program leadership with up-to-date information on the process and outcomes of the relationship between the HIV scientific mentor and the trainee.

Evaluative neutrality in CVs and letters of evaluation is fundamentally an ideal rather than a practical standard to be implemented. Injustice and unfairness are often difficult to discern because one's values impact on perceptions of injustice and unfairness. Also, often there are structural issues associated with the social influence process such as elite control over the agenda, the availability of information, and consequently the manner with which controversial issues are framed. Those with the most power may also be better able than others less well-positioned to give the impression of fairness even when undeserved. Hence, an individual or organization may appear to be neutral, but actually have unnoticed preferences which are demonstrated in subtle and non-obvious ways.

Space limitations preclude a fuller and more comprehensive discussion of letters of reference. For example, it would be useful to examine the potential influence on such letters of lawsuits by terminated employees (such as postdoctoral fellows) against their former employers for failure to provide positive references to subsequent prospective employers (Salter, 2002; Sayko, 2004). Concern over possible defamation lawsuits has apparently led to a practice by some employers of providing only dates of employment, salary, and title (McCord, 1999). The possibility of these lawsuits illustrates the potential tension that may exist between legal realities, on the one hand, and ethical obligations, on the other.

Data Ownership, Acknowledgement of Assistance, and Authorship: A Need for Guidance?

Background

When research is conducted with the financial support of the federal government, the university holds legal title to the data collected and the principal investigator is generally delegated to be the person with ownership rights to their use (Fishbein, 1991). The National Institutes of Health, the largest U.S. supporter of research, and other federal agencies now request that grant applicants propose a plan for data-sharing including a timetable, mode of data-sharing, whether a data-sharing agreement is required, and other issues. Although it appears that the legal issue of data ownership is relatively clear, the case of Mark demonstrates that a number of students and perhaps faculty as well may be unaware that principal investigators own the data from their projects. In addition, there remain many ethical issues associated with the complex problem of data access, data use, and authorship.

Students in training programs are frequently deeply interested in developing their research careers. The prime method of doing this is through publication, preferably in top peer-reviewed journals. Access to datasets and authorship are obviously important elements in the publication process. Postdoctoral and advanced Ph.D. students also view their position as an opportunity to strengthen their attractiveness in the academic and non-academic market by generating publications. At the same time, a primary index of a training program's success is effective mentorship. A key index of program research effectiveness is the number of student publication authorships or co-authorships produced under the aegis of the program. Since authorship is so vital to the careers of students and faculty alike, it is not surprising that this ethical area is potentially volatile. We turn next to Mark, a case where data ownership was a significant concern.

The Factual Situation and Ethical Dilemma

Mark was a graduate student research assistant hired to work as a participant observer on a multi-year National Institute of Health (NIH) research project. During the orientation session for the ethnographers and interviewers, the Principal Investigator (PI), who was an HIV scientific mentor in the training program, and the Project Manager explained in detail the nature of the project to the group of graduate research assistants with whom they met weekly over a period of many months to train them in the use of the project protocols and to supervise data collection. Copies of the narrative of the NIH research application were also provided to the students. The issue of data ownership was addressed and it was stressed that all project data, including the observational and interview data, were the property and responsibility of the project and the PI. At the same time, joint authorship and possibly first authorship, was described as a possibility depending on who actually takes the main leadership role with regard to manuscript conceptualization, writing, and other duties. Later, one of the research assistants (not Mark) was granted first authorship on a published project paper. All of the project research assistants including Mark were given co-authorship on a publication that involved, in part, a comparative analysis of the numerous sites that were studied. The NIH study compiled both qualitative and quantitative datasets. To date there have been a dozen peer-reviewed publications produced by the project with most involving current or former graduate students as co-authors. Two students are currently using the project's quantitative dataset for a Ph.D. thesis and three others are working on manuscripts based on that dataset.

Meanwhile, Mark was taken ill and was unable to continue working on the project. After about two years had passed, April, a postdoctoral trainee working under the PI's supervision, was asked to prepare an abstract and paper with others based in large part on the observational and interview data collected mostly by

Mark, but processed and analyzed by several others. April planned to submit the abstract and present it at a conference. The PI suggested and April agreed to send the paper abstract to Mark to see if he would be interested in participating in the writing of the manuscript and being a co-author with April, other graduate assistants, the PI, and the Project Director. Mark replied that he would be uncomfortable having his name on the paper because it was in an area in which he had already presented a sole-authored paper at a national conference and he was concerned that there may be a lot of overlap. Mark also insisted that he had obtained the PI's prior approval to work solely in the conceptual area of the paper and hence to write and present this paper.

The PI was sent a copy of the email. The PI promptly informed Mark that his understanding was incorrect. He reminded Mark (and the Project Director confirmed) that Mark had been informed at the outset of his work as had the other graduate research assistants that the data were the sole property of the NIH project and the PI. Moreover, the PI stated that no agreement was ever made with Mark or with any other research staff member assigning sole rights to any conceptual area. Also, the PI expressed concern that he was not previously informed of Mark's manuscript or of Mark's paper presentation at a national meeting based on data from the NIH project. He noted that the NIH project was funded for this work and that all resources, IRB approvals, and permissions to undertake the ethnographies were obtained by the PI. He also requested a copy of the paper. Mark replied by sending a copy of the paper. He apologized for not informing the PI of the paper that was presented one year previously. However, he continued to insist that he received oral consent to produce a single-authored paper and that he had discussed the topic with the PI in advance.

When the PI read Mark's paper he discovered that the material in it was directly related to the NIH project's specific aims. Also, he learned that Mark had even failed to acknowledge that the data were collected with the support of the project. The manuscript never mentioned the NIH study, the PI, the Project Director, or the other research staff who contributed to the research. Since Mark had been ill for some time and was in poor health, no other actions were taken other than informing him of the situation as described above,

Analysis

This case illustrates a major communication breakdown regarding data ownership and acknowledgment responsibilities. It suggests that other graduate students may have serious misunderstandings about data ownership. It is likely that a number of students and faculty would benefit by PIs presenting them with written statements that discuss clearly who owns the data, authorship issues, responsibilities with regard to acknowledgement of assistance (including sample wording of acknowledgements), and the need to keep the PI informed of working manuscripts. Such a document could be used as the basis for a comprehensive group discussion of these and other ethical issues and issues of scientific responsibility.

Acknowledgement. The author gratefully acknowledges the very helpful advice of Sana Loue on this chapter and the support of the National Institute of Mental Health by T32-MH19127, R01-MH62709, and P30 MH58107 (Rotheram-Borus, PI).

References

De Vries, R., Anderson, M.S., & Martinson, B.C. (2006), Normal misbehavior: scientists talk about the ethics of research. *Journal of Empirical Research on Human Research Ethics*, 43–50.

Fishbein, E.A. (1991). Ownership of research data. *Academic Medicine, 66*(3),129–133.

Kelman, H. C. (2001). Ethical limits on the use of influence in hierarchical relationships. In John M. Darley, David M. Messick, and Tom R. Tyler (Eds.), *Social Influences on Ethical Behavior in Organizations* (pp. 11–20). Mahwah, N.J.: Erlbaum Associates.

Lipset, S. M., Trow, M. A., & Coleman, J. S. (1956). *Union Democracy.* Glencoe, IL: Free Press.

McCord, L. B. (1999, Spring). Defamation vs. negligent referral: A policy of giving only basic references may lead to liability. *The Graziadio Business Report.* Retrieved July 28, 2006, from http://gbr.pepperdine.edu/992/referral.html.

Salter, L. (2002, February 17). The rules about references. *California Job Journal.* Retrieved July 28, 2006, from http://www.jobjournal.com/article_printer. asp?artid = 429.

Sayko, J. D. (2004). When employers get "something for nothing": The need to impose a limited obligation to disclose in employment reference situations. *Suffolk University Law Review, XXXVIII*(1): 123–145. Retrieved July 28, 2006, from http://www.law.suffolk.edu/highlights/stuorgs/lawreview/docs/Sayko.pdf.

Chapter 18
Training Community

Training

The literature on HIV/AIDS is fairly dense with the language of personal and community "empowerment." Common concepts and keywords include "consultation," "input," "community planning," "mobilization," "participation," "inclusion," and many others. These ideas or aspirations are operationalized through a series of continually evolving but definable structures: community advisory boards (CABs), research review committees, and advisory councils.

There is nothing intrinsically wrong with any of the concepts or their manifesting agents. Quite the opposite: they are important examples of the degree to which the desire to involve participant communities in HIV prevention, care and treatment, and research has permeated the field, especially in recent years. CABs and various other advisory bodies have played an important role in the conduct of research and implementation of programs; their significance cannot be minimized.

But it is worth asking whether what has been accomplished is good enough, and whether the field has permitted the surrogate markers of inclusion and partnership—say, for example, the presence of, and the conduct of regular meetings by, a local CAB—to become an end in itself.

Recently Emily Bass raised tough questions about lingering dichotomies in the establishment of clinical trials: participants versus researchers, partnership versus paternalism, and "talking *at* versus talking *with* trial volunteers" (Bass, 2006: 12). Bass quotes Morenike Ukpong, of the Nigerian HIV Vaccine and Microbicide Advocacy group, who questions whether there is an authentic partnership between researchers and communities, asserting bluntly: "The scientific community consults with itself to come up with expensive answers" (Upkong, 2006).

These questions, in turn, have raised once again the question of what constitutes a community. On the one hand, the global nature of the HIV/AIDS epidemic suggests a kind of universality. On the other hand, the specific epidemiologic patterns of HIV illness suggest both an overwhelming number of communities characterized by marginalization, and unusual "viral alliances": the "HIV community" in the United States, for example, includes or has included gay and bisexual men, women of color, African Americans and Latinos, people with hemophilia, injection drug users, and transfusion recipients—an array of "communities" marked by diverse histories, experiences, and agendas, even HIV-related agendas. Even the utilization of geographic community, for the sake of research initiatives, is problematic, because individuals may occupy the same territorial domain (i.e., a Nairobi neighborhood), but also occupy vastly different conceptual or existential domains (men who have sex with men, female sex workers, military recruits) that may well interact, but very rarely truly converse. The notion of "the HIV/AIDS community," therefore, has always and everywhere been ambiguous at best. The tension between universality and a fluid locality has been manifested in the contradictions of prevention campaigns, which have on the one hand declared that everyone is at risk of HIV infection and, on the other hand, targeted programs toward very specific populations (such as men who have sex with men).

How, then, can the highest ideal of community partnership in HIV/AIDS research be realized, based on a meaningful definition of community? How do we "train community"?

That such partnership has become an integral value is nearly unquestionable. It is one of the essential guidelines in the 2006 Consolidated Version of the Joint UNHCHR/UNAIDS *International Guidelines on HIV/AIDS and Human Rights*:

"GUIDELINE 2: *States should ensure, through political and financial support, that community consultation occurs in all phases of HIV policy design, programme implementation and evaluation and that community organizations are enabled to carry out their activities, including in the fields of ethics, law and human rights, effectively*" (United Nations High Commissioner for Human Rights and the Joint United Nations Programme on HIV/AIDS, 2006: 24).[1] (Italics in original.)

Notable in the stated Guidelines is the obligation of states to provide "political and financial support" that will facilitate community consultation and that the fields of "ethics, law, and human rights" are included under the activities that community organizations are enabled to carry out.

In the Commentary on Guideline 2, the *Guidelines* make an assertion with far-reaching significance: "Community partners have knowledge and experience States need in order to fashion effective State responses." (UNHCR/UNAIDS, 2006: 25). The same point could be made about community partners and research institutions, and therein lies a foundational principle for maximizing the potential of community partnerships and training community: knowledge and experience are bilateral commodities. "Training community," therefore, goes in both directions, from researchers to the community and vice versa.

The nature of the "content" in which a community will train researchers will be extraordinarily varied, but Stewart and Rappaport (2005) underscore the importance of community narratives as embodiments of knowledge, beliefs, and values in a community. To Stewart and Rappaport (2005: 76), "community narrative" certainly includes traditional texts or lessons, such as ethnographic analysis, but it also includes "the arts, media and discourse analysis, literature and literary criticism, theology, gender and queer studies, African American studies, history and oral history, linguistics, marketing, communications, political science, and party planning (yes, really)." And, one might add—for the omission is glaring—personal narratives and genealogies, oral and written accounts of community lives, both formal and informal, and how those accounts are connected through time.

One might suggest, then, that the first step the researcher takes in the process of training the community is to ask that community, "what do we need to know, about you as a community, and how can we come to know it?" The community (not the researchers) should define the syllabus, and the form of the learning experience. This may well prove to be an awkward question and a difficult process: communities are not accustomed to such a question, especially when posed (as it would almost invariably be) *by* representatives of relative status and privilege (researchers, scientists) *of* the marginalized communities that typify HIV/AIDS and HIV/AIDS research communities. Nevertheless, it is important if for no other reason that it symbolizes a *partnership position* of equality of knowledge, and can serve as a remedy against the encroachment of a paternalistic tone of which so many "researched" communities are weary and wary.

There is little research in this area; it deserves substantially more attention. Needed are models of how communities can develop and implement training for researchers, on what "things" the community deems essential to know, and how that training can be used to inform the research process. There is considerably more (but still insufficient) literature on training participant communities on research. (For one example, see the case study by O'Shea that follows this chapter.)

Among the most significant community training "scale-up" efforts now taking place are community preparation for vaccine trials, and for microbicide trials. Some advocates for vaccine and microbicide research and deployment are addressing the complexities of preparation, often with considerable attention to the points raised through this volume—this, for example, from a report on an international consultation on microbicides:

[T]he general trend has been to rely far more directly upon consultation with the communities and individuals who are directly affected. Typically, clinical researchers now turn to social scientists, nongovernmental organizations, and community-based organizations. They solicit input from both individuals and communities to help prioritize benefits that contribute more broadly to justice. Inevitably, new questions arise on the ethics of the process: Who represents and speaks for the community? What precisely qualifies as 'input'? What is meant by 'involvement'? (Global Campaign for Microbicides, 2005: 15).

The quote demonstrates the degree to which some planning efforts—and in this case, specifically for microbicide research—are grappling with difficult questions. Setting aside for a moment, then, "Who represents and speaks for community," the core content of training for the community should be outlined.

A Lexicon

Inevitably the process of training communities will involve the use of terms and concepts that may be difficult to understand, or may be unstable in their meaning.

This would include both terms and language used within the community, and terms and language used by researchers. One suggestion might be to develop a shared and evolving lexicon of terms and meanings, written (and illustrated) to be as accessible as possible.

Science, as it were, is a kind of language in and of itself, and carries considerable power that must be negotiated in the process of shared community usage: "The language of science often obliterates the language and voice of peoples' lived experiences . . .", according to Stewart and Rappaport (2005: 75), and a community-researcher lexicon can serve as the site in which the obliterating potential of scientific language can be attenuated, and meanings negotiated to reach shared understanding.

Research Principles

As earlier chapters have demonstrated, it is too often the case that standard scientific principles (consent, randomization, and so on) the researcher believes are understood are not understood at all. To correct this confusion HIV/AIDS advocates have developed a number of especially useful tools and resources that can be locally redesigned. Again, current efforts in vaccine and microbicide research preparation are relevant. The International Council of AIDS Service Organizations has created a largely-accessible (readability needs more scrutiny) document for potential trial participants, *Finding Your Way: A Guide to Understanding Ethical Issues Related to Participation in Clinical Trials for Preventive HIV Vaccines* (Lee, 2005), that helps explain the nature of vaccines, how research should be designed, and the core principles that will govern the development and implementation of trials. (See also IVAC, 2002). The previously-mentioned Global Campaign for Microbicides has also done an excellent job in developing individual- and community-oriented informational materials on all aspects of the microbicide research process, including explication of core research principles. The challenge will be to continue making these and similar materials as accessible as possible.

Ethics

Communities have the right to learn not only about traditional understanding of bioethics and its foundational concepts, but about metabioethical debate about the tension between universalist and relativist perspectives as well. According to Baker (1998: 254), "fundamentalism has been bankrupted by the multicultural and post-modern critiques," and "international bioethics has always been a matter of negotiated conventions" (Baker, 1998: 255). This suggests that mere enunciation of the principles of autonomy, beneficence, and distributive justice (among others) alone will not be sufficient; communities can learn about research ethics *as researchers learn about them*, which is as a set of foundational ideas, unfolding in historical contexts, that are sharpened, revised, localized, critiqued, expanded, reduced, and re-envisioned in practice using a variety of investigational and critical tools.

In ethics, to a degree greater than perhaps any other area of research, questions and resolutions are crafted through dialogue about accrued values and lived experience, so "teaching ethics" is inherently a dialogic process. Hellsten (2005: 265) suggests that there are three critical ethical dilemmas to be resolved about the conduct of HIV in the developing world, and these should constitute a topic of discussion in community training: (1) the distribution of all resources and the problem of global and local inequity; (2) conflicting rights and duties, in relation to (a) individual rights versus (collective) public health, (b) individual rights versus the social responsibilities of individuals, and (c) individual rights versus collective group rights; and (3) universality of codes of medical ethics and medical research in relation to cultural values and practices, and in relation to available resources.

While a handful of groups have attempted to develop translational documents that will "teach" ethics to local communities, the current focus seems to be on doctrinal instruction rather than teaching communities to "do ethics" through the various prisms of culture, context, and community. This will no doubt be a complex undertaking, but a necessary one. The alternative is to further solidify the evolutionary results of Western bioethics thinking as universally canonic, and to maintain developing-world dependency on Western tools, discourses, and strategies in the process of "doing" science.

It should be noted that with the engagement in ethics that constitutes part of the community training process, disclosure and transparency must constitute part of the dialogue. The roles, investments, and anticipated benefits of sponsors and funders, researchers and their supporting academic bodies, government agencies (both national and international), and other constituents ought to be subject to full disclosure, and analyzed as a part of the process of ethical review. No one should consider, from the outset, that such discussions are impossible in resource-poor and low-literacy settings. It should be assumed, instead, that it can be accomplished, but that the commitment will have to come in the form of necessary allocations of time, commitments to developing appropriate educational materials, and dedication to utilizing culturally appropriate strategies for education.

Research Management and Monitoring

The process of research—from acceptable approaches to framing the research question, through IRB approval, the participant recruitment enrollment, and so on—is so standardized, in its essential scaffolding, which experienced researchers may not even think about the architecture much; they just carry it out according to internalized rules and protocols.

But to the inexperienced, nothing is a given, and may all seem, therefore, mystical, imbued with an abstract and unattainable power. In truth, elucidating the research process, its standard trajectory and protocols, is a fairly straightforward exercise; once again, only time and a commitment to community-centered education are required.

Community representatives should also be taught to independently carry out monitoring of the research study, and relay those results back to the community and to researchers. A number of groups have developed effective community monitoring tools, which can be flexibly adapted to a variety of monitoring subjects. ICASO, for example, has published a manual for community groups to monitor adherence to the UNGASS (United Nations General Assembly Special Session) Declaration of Commitment on HIV/AIDS, forged in 2001 (ICASO, 2006b).

The Specific Study(ies)

The community deserves a full explanation of the specific planned trial or other study, what research question it is intended to answer, what protocol it will follow, how participants will be recruited, what services they will receive during and after the investigation, and the full range of possible risks and benefits. To be sure, much of this content falls under what is traditionally viewed as the informed consent process, but it can be more. Ideally, the community can come to understand how the specific study fits into the larger HIV/AIDS research agenda and the development of HIV/AIDS science, how a range of study results (positive, negative, unanticipated) could affect the future of disease management and prevention efforts, and how science interprets—and can mistakenly misinterpret—research findings.

Rights and Obligations

In addition to a full recitation of the specific rights and responsibilities of each constituent in a specific research study, community training should also address the more general rights—the human rights—of community participants. Admittedly these are ethical questions as well, but also involve, as the chapter on researcher-community relations indicated, researcher obligations throughout the study (and afterwards); standard of care decisionmaking; responsibilities in relation to advocacy (especially advocacy about structural conditions that give rise to disproportionately distributed diminished health in communities); the right to protest, including protest of the study or its sponsor; and the rights of not only the participant research community as a whole, but subsets of that community (i.e., commercial sex workers) as well.

Integrating Community Knowledge and Research Knowledge

From the outset we understand that if we are viewing community training as a bilateral process, then we involve two distinct sets of knowledge (from the community's perspective, and from the researcher's) that may or may not overlap. In the process of training, then, those sets of knowledge need to "speak" directly to each other, as discourses or world-views, in order to maximize the potential richness of such cross-training. This can be accomplished through direct, facilitated dialogue, structured co-presentations, or other formats.

Community

To return to an earlier, unresolved question, the question of what boundaries, however permeable, frame a "community," and who resides within those boundaries, must be resolved as a primary question before community training can be conceptualized and initiated. To quote, once again, Stewart and Rappaport,

> But community—as concept and as experience—also presents a number of challenges. Not the least of these is the lack of a stable definition and the fluidity of changing experience. Many of us are aware of the complexities presented by the experience of inhabiting different communities, particularly when those community memberships and the identities implied by them are in one or more sense incompatible. Perhaps the most obvious example in relation to HIV/AIDS is gay or bisexual men of color, and the liminality they may experience both in gay communities and in their 'cultural communities.' (Stewart and Rappaport, 2005: 70).

One way the author has heard this expressed in Cleveland, on numerous occasions, is: "Here, you can be Black, and you can be Gay, but you can't be both at the same time at the same place"—which raises a profoundly troubling question of whether community is organized around the presence of identity, the excision of identity, or both.

Community, too—at least in many industrialized countries—is also highly plastic, and fueled by technological innovation. The rapid rise of MySpace and FaceBook has expanded (or corrupted) the meaning of "community" and "friend'; it is now possible to construct virtual boundaries around an idea, an interest, a proclivity, an identity, or a thousand other existential features and achieve something like a community.

In many contexts, it is also important to recognize that "communities" associated with HIV were formed and continue to evolve as oppositional structures or "spaces," and may operate within or across the boundaries of other, non-marginalized communities:

> The meanings of sexuality and drug use are engendered within networks of face-to-face communications and within cultural productions (counter-cultural practices, the media, art, rituals of partnering, styles of dress) which cut across the 'communities' articulated for the purpose of engaging in the political languages of civil rights and claims for the apportionment of social resources (Patton, 1990: 8).

In other words, communities are woven into each other, and redefine their boundaries based on the needs of the moment. And the "needs of the moment" may be externally-driven, and change rapidly. For sex workers and injection drug users, whose behavior forces them to fashion community networks under the watchful gaze of state police power, the ability to navigate such change may be an act not merely of socialization, but of survival as well.

None of this brings us any closer to a definition of community that is tidy and easily adopted for the present purpose. It only, and rather, confounds a facile definition—which is as it should be. Research is usually focused on a notion of community that is often geographic, and that may be entirely too limiting. Stewart and Rappaport, in their contention that community is shared narratives, suggest that a description of community comes through an understanding of language practices, patterns of collaboration, contextualizing narratives, and "exemplars"—concrete models of valued practices. It is an elusive set of analytical principles, but may yield definitional results that are far more useful than mere geography.

That said, the significance of geography, for many of the world's people— who may be born, live their lives, and die entirely inside a neighborhood with a very small radius, or a relatively remote village—cannot be understated; in such circumstances, community as a defined space can be a totalizing reality. What is important, here, is that researchers be as clear as possible about what they mean by the "community" they are partnering with for research, and that that community shares with the researchers the same core definitional values.

Models

Much of what has been said throughout this volume argues for an approach to research partnerships that embraces an expanded view of ethics and human rights rather than one that is narrowly prescriptive; for an open-ended structuring of the research chronology rather than one limited to the term of a particular study; for a broadly contextualizing analysis of the research problem rather than an individual, solely medicalizing investigation; for a stance of full disclosure and transparency rather than revealing only what "needs" to be known; and for the assumption of a bilaterality of knowledge between researchers and participants or participant communities. In other words, it is, fundamentally, a rights-based approach predicated on an equality of actors or constituencies.

There are precious few models that can carry the burden of all these requirements. One, discussed earlier, is community health partnerships (CHPs), which are defined as "voluntary collaborations of diverse community organizations, which have joined forces to pursue a shared interest in improving community health" (Mitchell and Shortell, 2000: 242). CHPs have not been without failures, as noted earlier, but it is precisely because of their multi-dimensionality that they may serve as a useful model. First, they are founded on a general principle of community organization, which has been defined as a process of enabling communities to engage in planned collective action to address shared problems within a shared democratic framework of values (Kramer and Specht, 1969). This provides a foundation much more in line with the aspirations argued for above than the kind of "project management" approach that can typify research initiatives. But CHPs also diverge from the traditional community organization model in their "greater emphasis on *cross-sectoral*, public-private participation and collaboration." (Mitchell and Shortell, 2000: 246. Emphasis in the original.) They are charged (or enjoin themselves) with multiple tasks, including problem analysis, the development of a strategic agenda, resource procurement, ensuring representative community involvement over time, partnership management, mission alignment, community accountability, and infrastructure development and maintenance (Mitchell and Shortell, 2000.) All these tasks require ongoing education of CHP partners, or continuous training.

The main advantage of a CHP, or a multi-sectoral community organization model, is that it offers more durability, greater capacity to carry the weight of the community training or capacity-building required—training on research principles, ethics, research management and monitoring, rights and obligations, and so on—than CABs or advisory committees *on their own*. A CHP offers an architecture that would likely include those CABs and advisory committees, but would include them as one element in a larger and more comprehensive community plan to change knowledge and conditions, and mediate diverse values and world-views. And a CHP inherently attempts to universalize what is more narrowly hoped for in a specific research study: community growth and development (in this case, in relation to health and infectious disease) that will measurably, and continually, improve the lives of citizens.

There are other models as well, such as the Hartford Model in Connecticut in the United States (Singer and Weeks, 2005). The model has its roots in Action and Advocacy Anthropology (Schensul, 1974; Singer, 1990). The Hartford Model is characterized by

1) long-term, community-based partnerships between activist public-health researchers and research-informed communities; 2) highly collaborative team efforts guided by a participatory action orientation to research and the transfer of research skills; 3) closely linked community-based research with research-informed intervention; and 4) an interdisciplinary or blended methodological approach to formative research, needs assessment, and program process and outcomes evaluation. (Singer and Weeks, 2005: 155).

Its salient characteristics include a strong theoretical orientation, and intervention strategy (which, applied here, would translate into a research strategy), an organizational plan, and a methodological approach.

Critical to the Hartford Model is the proposition that social change or community development is driven by awareness-building processes which use "science to produce knowledge and action that is directly useful to the community," and the use of science "to enhance the locus of effectiveness in the community of concern by helping participants identify and build their own knowledge, apply that knowledge, and assess the impact of their efforts (Singer and Weeks, 2005: 166). These goals are accomplished through training workshops and seminars, community forums and discussions, the development of accessible educational materials, and community-researcher co-participation in different phases of research, such as community ethnography. Clearly, the Hartford Model is not concerned merely with specific research studies or the transfer to others of the particular knowledge gained in those studies. Rather, the model is embedded in an analysis about "the role of structural factors, including structural violence, in the generation of risk, the spread of disease, and the production of social misery" (Singer and Weeks, 2005: 171).

Not discussed here is the question of evaluation: if communities and researchers form partnerships intended to promote bilateral diffusion of knowledge across boundaries, then the partnership must define its goals in terms of knowledge acquisition, and develop a strategy to assess whether that learning has occurred. It is fair warning to suggest that evaluation for such a complex process is likely to be complex itself. To take one example: post-learning assessment of changes in knowledge about biomedical ethics could simply tabulate correct answers to questions about the definition of terms, or seek student recitation of relevant bioethical codicils. But that is not what is being suggested, or should be assessed. We are proposing that researchers and members of the community, as a result of the training, will be better equipped to "do ethics" with due attention to universal principles and local mediating factors. In other words, evaluation will need to assess acquisition of skills, and their use in community life, as much as it assesses knowledge.

Notes

1. The 2006 *Guidelines* consolidate the Guidelines adopted at the Second International Consultation on HIV/AIDS and Human Rights, held in 1996, and revised Guideline 6 (access to prevention, treatment, care and support) adopted at the Third International Consultation on HIV/AIDS and Human Rights held in 2002. As a result, the *2006 Consolidated Version* stands as one of the best current comprehensive taxonomies of HIV/AIDS ethics and human rights.

References

AIDS Vaccine Advocacy Coalition. (2002). *Developing Vaccines for HIV and AIDS: An Introduction for Community Groups*. Available at http://www.avac.org

Baker, R. (1998). A theory of international bioethics: The negotiable and the non-negotiable. *Kennedy Institute of Ethics Journal, 8(3)*, 233–273.

Bass, E. (2006). Communities and clinical trials: Adding nuance and changing definitions. *The Microbicide Quarterly, 4(2)*, 12–19.

Global Campaign for Microbicides. (2005). *Rethinking the Ethical Roadmap for Clinical Testing of Microbicides: Report on an International Consultation*. Last accessed January 17, 2007; Available at http://www.global-campaign.org/researchethics.htm

Hellsten, S.K. (2005). Bioethics in Tanzania: Legal and ethical concerns in medical care and research in relation to the HIV/AIDS epidemic. *Cambridge Quarterly of Healthcare Ethics, 14*, 256–267.

International Council of AIDS Service Organizations. (2006a). *Community Involvement in HIV Vaccine Research: Making it Work*. Last accessed January 17, 2007; Available at www.icaso.org

International Council of AIDS Service Organizations. (2006b). *Community Participation in the Monitoring and Evaluation of the Implementation of the UNGASS Declaration of Commitment on HIV/AIDS*. Last accessed January 17, 2007; Available at www.icaso.org

International Council of AIDS Organizations. (2002). *Developing Vaccines for HIV and AIDS: An Introduction for Community Groups*. Last accessed January 17, 2007; Available at http://www.icaso.org/CommunityPrep_VaccinePrimer_WebVersion_EN.pdf

Kramer, R. & Specht, H. (1969). *Readings in Community Organization Practice*. Englewood Cliffs, New Jersey: Prentice-Hall.

Lee, E. (2005). *Finding Your Way: A Guide to Understanding Ethical Issues Related to Participation in Clinical Trials for Preventive HIV Vaccines*. Last accessed January 17, 2007; Available at http://www.icaso.org/FindingYourWay_web.pdf

Mitchell, S.M & Shortell, S.M. (2000). The governance and management of effective community health partnerships: A typology for research, policy, and practice. *The Milbank Quarterly, 78(2)*, 241–289.

Patton, C. (1990). *Inventing AIDS*. New York: Routledge.

Schensul, S. (1974). Skills needed in action research: Lessons from El Centro de La Causa. *Human Organization, 33(2)*, 203–208.

Singer, M. (1990). Another perspective on anthropology. *Current Anthropology, 31(5)*, 548–550.

Singer, M. & Weeks, M. (2005). The Hartford Model of AIDS practice/research collaboration. In Trickett, E. J. and Pequegnat, W. (Eds.). *Community Interventions and AIDS* (pp. 153–175). New York: Oxford University Press.

Stewart, E. & Rappaport, J. (2005). Narrative insurrections: HIV, circulating knowledge, and local resistances. In Trickett, E. J. and Pequegnat, W. (Eds.). *Community Interventions and AIDS* (pp. 56–87). New York: Oxford University Press.

Ukpong, M. (2006). Role of community advocates in defining the microbicides research agenda. Presented at Microbicides 2006 Conference, Cape Town, South Africa, April 23–26. Quoted in Bass, E. (2006). Communities and clinical trials: Adding nuance and changing definitions. *The Microbicide Quarterly, 4(2)*, 12–19, at 12.

United Nations High Commissioner for Human Rights and the Joint United Nations Programme on HIV/AIDS. (2006). *International Guidelines on HIV/AIDS and Human Rights* [UN Publication No. E.06.XIV.4]. Geneva, Switzerland: Author.

Case Study Twelve
San Diego HIV/AIDS Needs Assessment

Daniel J. O'Shea

Background

San Diego County's HIV, STD and Hepatitis Branch (formerly Office of AIDS Coordination [OAC]) conducted seven HIV/AIDS needs assessments between 1996 and 2006. The purpose of these assessments was to gather information from people living with HIV/AIDS (PLWH/A or "consumers") on their care, treatment, housing, support and HIV prevention needs. Consumer input was considered essential for both service evaluation and planning for subsequent years. Community and consumer involvement were key to the success of these needs assessments. A working group, convened for each assessment to oversee all aspects of development and implementation, included members and staff of the planning bodies, consumer groups, service providers and community gatekeepers. Significant outreach was made to include as many community gatekeepers to as many groups as possible in planning and/or implementation.

The following ethical issues related to involvement of human subjects were considered in the HIV/AIDS needs assessments conducted in San Diego County between 1996 and 2006:

- Involvement of human participants
- Characterization of participant populations
- Sources of needs assessment data
- Recruitment of subjects
- Process of data collection
- Consent procedures
- Potential risks to participants and procedures to minimize risks
- Anticipated benefits to participants
- Plan for resolution of problems/issues of protection

Conducting the Needs Assessment

Recruitment

The needs assessment involved the collection of information from between 1,096 and 1,431 PLWH/A for each HIV/AIDS Needs Assessment. Participants were persons infected with HIV disease and/or diagnosed with AIDS and living in San Diego County, representing a variety of demographic characteristics, including gender, age, race/ethnicity and HIV risk behaviors, and socio-economic backgrounds. California law prohibits disclosure of someone else's HIV status without their consent.

Data were collected through written questionnaires targeted to all PLWH/A; focus groups with members of demographic groups affected by HIV/AIDS; and key informant interviews with members of demographic groups affected by HIV/AIDS, or individuals who work closely with and are very knowledgeable of these groups.

For the survey, the working group developed a list of population groups or sub-groups within the larger HIV/AIDS population for whom it would be beneficial to analyze needs separately from the larger group, and to ensure that the survey response was representative of the known demographics of the epidemic. The percentage representation of each subgroup to the total HIV/AIDS population, if available, was used as a guide to determine the goal for number of subjects recruited from that subgroup. An attempt was made to recruit a minimum of 50 subjects from each subgroup.

Participants were recruited by convenience sampling in a variety of ways. These included:

- Face-to-face recruitment in HIV/AIDS service agencies by trained volunteers or staff;
- Face-to-face or telephone recruitment of known personal contacts in various subgroups by trained volunteers or staff;
- Face-to-face or telephone recruitment of homebound individuals by service providers and volunteers who knew and provided services to them;
- Face-to-face recruitment at needs assessment focus group meetings, at which survey forms and return envelopes were available;
- Written recruitment of participants by including survey forms and return envelopes with other materials received (home-delivered meals, food bags, newsletters, etc.), or by mailing the forms directly to clients from the provider; and
- Promotion of the internet website from which the survey could be downloaded, or on which the survey could also be completed and sent in anonymously on-line (2006 only).

For focus groups and key informant interviews, consumers and other volunteers or HIV/AIDS service agency personnel recruited, by face-to-face or telephone contact, known personal contacts who were members or key informants from targeted subgroups. Whenever possible, existing support and social group meetings of individuals belonging to community subgroups targeted for the focus group(s) were utilized with permission of the facilitator(s) and participants.

Staff Training

Prior to distributing survey forms or scheduling focus groups, key contacts for each subgroup, survey interviewers and distributors, focus group facilitators and recorders and key informant interviewers (all comprised of community/consumer volunteers, HIV/AIDS agency personnel or County staff) received at least one and one-half hours of training. This included rigorous instruction on methods to maintain confidentiality and voluntary consent, to keep data anonymous and secure, and to prevent embarrassing participants in public. Emphasis was also placed on referral of individuals to service providers or counseling services they might need as a result of issues raised during their participation in the survey, focus groups or interviews. Focus group facilitators and recorders and key informant interviewers received special additional training on how to conduct meetings.

Data Collection

Survey forms were widely available in English and Spanish languages, distributed through HIV/AIDS service agencies, medical clinics, street outreach, food distribution centers, newspapers, newsletters, etc. Consumers could pick up surveys at these multiple provider sites, and needs assessment volunteers or agency staff sometimes set up tables for distribution, pick-up and completion of surveys at larger volume HIV-specific service sites. Surveys were returned to closed drop boxes with slits in them for inserting completed surveys, or stored in locked file cabinets or other secured containers or spaces until picked up by OAC staff, or returned by mail. Consumers were always given the option of taking the survey form with them to complete at home and return in postage-paid envelopes, also available at all distribution points. For some members of smaller, more vulnerable community subgroups, surveys were conducted as feasible by face-to-face or telephone interviews (e.g., for incarcerated individuals, active injection drug users, caretakers of children, homebound people, etc.) at provider sites or in homes. Assistance was also available from OAC staff over the phone, including a toll-free line, to help, explain or complete the form in English or Spanish during normal working hours. Finally, the survey was available for download from the internet, to complete and mail in; an option to complete and send in the survey online was offered for the most recent survey (2006). With the likelihood of receiving a survey more than once if accessing multiple services in the community, consumers were advised to complete and return the survey just once, and to pass extra surveys confidentially to others who may not have received it.

Completed surveys were collected by or returned to OAC. Staff entered and analyzed data by aggregate number or subgroup within the two two-month timeframe following the cutoff date for surveys. Subgroup analyses were not conducted if the sample size for the subgroup was so small that identification of individuals would be possible from their responses. Data were compared to results from previous years.

Focus groups and key informant interviews. At least one focus group was held for each identified subgroup; in a few instances where a focus group was not feasible, key informant interviews were conducted. Ground rules posted at each meeting and read verbally advised that participants respect each other's opinions, ideas and confidentiality. Information was captured and posted on flip charts for review by participants during the meeting, and later transcribed directly from these into the meeting notes.

Ethical Issues

Consent Procedures

For the survey, any participants completing the survey forms were assumed to have given implied consent by action of completing the form. Written consent was not obtained. For each survey conducted between 1996 and 2006, the instructions noted the purpose and use of the information gathered, included contacts in English or Spanish for assistance or to respond to questions, and highlighted the anonymous and voluntary nature of the survey:

- "This survey is completely anonymous. *Don't tell us your name.* Just tell us what you need;" "Your name will not be connected to the results of this survey."
- "This survey is completely voluntary;" "Your input is *very important.* Please fill out as much of the survey as you can. You don't need to answer any questions that you don't know how to answer. Leave blank any questions you don't want to answer;" "Your HIV/AIDS care and access to services will not be affected if you decide not to answer any or all of the questions."

For focus groups and key informant interviews, participants were assumed to have given implied consent by their action of participation. Written consent was not obtained. A project explanation was given to all focus group participants prior to the session. It was also read aloud at the beginning of each session. In addition to the purpose and use of focus group information, the following statement was included:

"This voluntary session will gather information about the needs of people infected with HIV/AIDS in San Diego County. Your name will not be disclosed in any records or summary of this session. Please answer the questions as completely as you can and feel free to discuss the information with other participants. The results of the focus group sessions will be used to help plan future HIV/AIDS services. You need not answer any questions that you do not know how to answer or any that make you feel uncomfortable, especially questions that you feel are very personal. Your HIV/AIDS care will not be affected if you decide not to answer any or all of the questions. **All of the information collected in this focus group will remain anonymous and confidential.**"

Potential Risks to Participants and Procedures to Minimize Risks

Each training session for staff or volunteers assisting with surveys, focus groups or interviews included a discussion of the following areas of risk, followed by brainstorming of solutions to mitigate the risk:

- *Risk:* Answering or considering a question or idea raised while filling out the survey or participating in a focus group or key informant interview may cause anxiety related to the future course of the individual's disease or other issues.
  - *Solutions:* Interviewers and focus group facilitators and recorders were trained and encouraged to watch for and discuss this possibility with participants and provide appropriate information and referrals, including the availability of counseling services, and to note referral information to HIV services and crisis intervention on the last page of the survey form and the focus group/key informant interview handout.
- *Risk:* HIV/AIDS status may be disclosed inadvertently to family, friends, neighbors, etc.
  - *Solutions:* Interviewers and people giving out surveys were trained to clearly stress the confidentiality of the survey or focus group/interview to all participants, that names or other identifying information were not to be recorded on the survey form or other materials. Volunteers and staff using phone contacts were trained not to refer to the purpose of the call or the acronyms "HIV" or "AIDS" on any telephone answering machines or on written notes recruiting participants. Surveys were not to be left in mailboxes, behind screen doors, or in any other personal space where they might be seen by anyone other than the participant. Return mailing envelopes for the survey did not have the acronyms "HIV" or "AIDS" anywhere on the outside. Survey instruction specified that names of participants were not to be written on survey forms. Participant names were not recorded in focus group sessions. Survey data for subgroups with less than ten members were not analyzed separately from the total sample. Interviewer, focus group leader, volunteer, or participant concerns related to this issue were to be discussed immediately with the OAC staff person coordinating the needs assessment.
- *Risk:* Interviewers share information from a survey with an affected service provider.
  - *Solution:* Interviewers were trained to not disclose *any* information from an individual survey with service providers or anyone else.
- *Risk:* An individual feels pressured to fill out a survey, or to respond in a certain way to particular questions.
  - *Solutions:* Survey interviewers and distributors were specifically trained not to coerce or attempt to coerce individuals to fill out a survey if they chose not to, fill out more than one survey, or to influence how the survey is filled out. Individuals who felt that they had been pressured in this way were encouraged to report this as soon as possible to the Needs Assessment Project Coordinator.

Trainings for volunteers and staff stressed that consumers "fill out and return this survey only one time."

- *Risk:* Interviews may not be conducted in a confidential place.
  - *Solutions:* Interviewers were trained to conduct all interviews in a quiet, confidential place out of the hearing range of other people. Focus group sessions were conducted in a closed room. Interviewers needed to make every effort to assure that the interview area was "safe" for all participants. As much as possible, existing support groups and known meeting or service delivery sites, where a "safety zone" has already been established, were used.

In addition to these items, training participants were asked to identify any other potential risks and how the risk or potential harm could be reduced or avoided. Volunteers and staff were encouraged to continue this discussion throughout implementation of the needs assessment.

Anticipated Benefits

We noted anticipated benefits in all materials and trainings to moderate concerns about risk. These included:

- Access to information increased potential to raise awareness of available HIV/AIDS services. Interviewers and facilitators were trained to provide appropriate information and referrals, including the availability of counseling services. Referral information to HIV services and crisis intervention was provided on the last page of the survey form and focus group/ interview materials, and resource guidebooks in English and Spanish were available for pick-up at survey distribution and focus group/interview sites.
- Focus group members benefited by having an opportunity to discuss common issues with other subgroup members, thereby creating an informal support group.
- Potential was enhanced for increasing direct consumer involvement in HIV/AIDS service planning; consumers volunteering as interviewers or distributors became role models to others for increased involvement.
- Data from surveys and focus group/interview sessions were used to help direct HIV/AIDS service planning and financial support, which directly benefited many participants.

Plan for Resolution of Problems/Issues of Protection

Surveys and focus group/interview materials included written information on how to contact OAC for further information or to discuss issues related to the needs assessment process. Oral information regarding OAC contacts to address protection issues was given to focus group subjects and provided at all trainings of survey interviewers/distributors, focus group facilitators/recorders and key informant interviewers. The Needs Assessment Working Group and the OAC maintained an ad hoc committee for protection issues during the data collection period.

Conclusion

In the eleven years of conducting HIV/AIDS needs assessments, there have been no reported breaches of confidentiality, violation of consent procedures or other ethical issues raised related to the HIV/AIDS needs assessments. This is believed to be due to several key factors:

- Widespread community involvement in planning and implementing the assessments, including extensive involvement of PLWH/A;
- Establishment of protocols and procedures for conducting surveys, focus groups and key informant interviews;
- Proactive and rigorous training of survey distributors and interviewers, focus group facilitators and recorders and key informant interviewers on respective protocols and procedures, with discussion of potential ethical risks and anticipated solutions; and
- Focus throughout on these key points: anonymity and confidentiality; voluntary consent; access to information; and anticipated benefits to participants.

Index